Sustainable Solar Housing

Volume 1 – Strategies and Solutions

Edited by S. Robert Hastings and Maria Wall

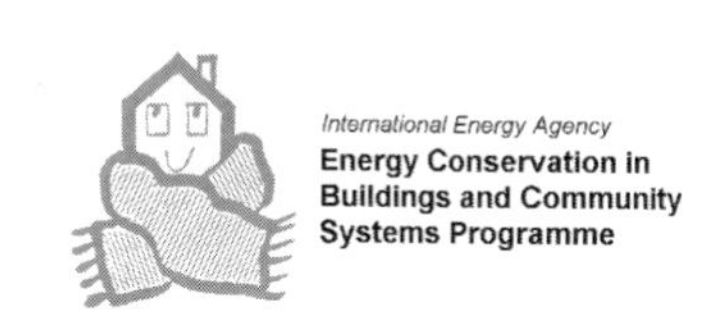

London • Sterling, VA

First published by Earthscan in the UK and USA in 2007

Volume 1: ISBN-13: 978-1-84407-325-2
Volume 2: ISBN-13: 978-1-84407-326-9

Typeset by MapSet Ltd, Gateshead, UK
Printed and bound in the UK by Cromwell Press, Trowbridge
Cover design by Susanne Harris

Published by Earthscan on behalf of the International Energy Agency (IEA), Solar Heating & Cooling Programme (SHC) and Energy Conservation in Buildings and Community Systems Programme (ECBCS).

Experts from the following countries contributed to the writing of this book: Austria, Belgium, Canada, Germany, Italy, the Netherlands, Norway, Sweden and Switzerland.

For a full list of Earthscan publications please contact:

Earthscan
8–12 Camden High Street
London, NW1 0JH, UK
Tel: +44 (0)20 7387 8558
Fax: +44 (0)20 7387 8998
Email: earthinfo@earthscan.co.uk
Web: **www.earthscan.co.uk**

22883 Quicksilver Drive, Sterling, VA 20166-2012, USA

Earthscan is an imprint of James and James (Science Publishers) Ltd and publishes in association with the International Institute for Environment and Development

A catalogue record for this book is available from the British Library

Library of Congress Cataloging-in-Publication Data has been applied for

The paper used for this book is FSC-certified and totally chlorine-free. FSC (the Forest Stewardship Council) is an international network to promote responsible management of the world's forests.

Contents

Foreword *v*
List of Contributors *vii*
List of Figures and Tables *ix*
List of Acronyms and Abbreviations *xxi*

INTRODUCTION

I.1 Evolution of high-performance housing 1
I.2 Scope of this book 4
I.3 Targets 4

Part I STRATEGIES

1 Introduction **9**

2 Energy **11**
2.1 Introduction 11
2.2 Conserving energy 12
2.3 Passive solar contribution in high-performance housing 14
2.4 Using daylight 20
2.5 Using active solar energy 28
2.6 Producing remaining energy efficiently 32

3 Ecology **37**
3.1 Introduction 37
3.2 Cumulative energy demand (CED) 39
3.3 Life-cycle analysis (LCA) 42
3.4 Architecture towards sustainability (ATS) 46

4 Economics of High-Performance Houses **51**
4.1 Introduction 51
4.2 Cost assessment of high-performance components 52
4.3 Additional expenses 59
4.4 Summary and outlook 61

5 Multi-Criteria Decisions **63**
5.1 Introduction 63
5.2 Multi-criteria decision-making (MCDM) methods 63
5.3 Total quality assessment (TQA) 70

6 Marketing Sustainable Housing **77**
6.1 Sustainable housing: The next growth business 77
6.2 Tools 79
6.3 A case study: Marketing new passive houses in Konstanz, Rothenburg, Switzerland 81
6.4 Lessons learned from marketing stories 89

Part II SOLUTIONS

7 Solution Examples **95**
7.1 Introduction 95
7.2 Reference buildings based on national building codes, 2001 96
7.3 Targets for space heating demand 98
7.4 Target for non-renewable primary energy demand 99

8 Cold Climates **103**
8.1 Cold climate design 103
8.2 Single family house in the Cold Climate Conservation Strategy 114
8.3 Single family house in the Cold Climate Renewable Energy Strategy 124
8.4 Row house in the Cold Climate Conservation Strategy 133
8.5 Row house in the Cold Climate Renewable Energy Strategy 142
8.6 Apartment building in the Cold Climate Conservation Strategy 150
8.7 Apartment building in the Cold Climate Renewable Energy Strategy 156
8.8 Apartment buildings in cold climates: Sunspaces 171

9 Temperate Climates **179**
9.1 Temperate climate design 179
9.2 Single family house in the Temperate Climate Conservation Strategy 186
9.3 Single family house in the Temperate Climate Renewable Energy Strategy 196
9.4 Row house in the Temperate Climate Conservation Strategy 202
9.5 Row house in the Temperate Climate Renewable Energy Strategy 211
9.6 Life-cycle analysis for row houses in a temperate climate 221
9.7 Apartment building in the Temperate Climate Conservation Strategy 226
9.8 Apartment building in the Temperate Climate Renewable Energy Strategy 232

10 Mild Climates **237**
10.1 Mild climate design 237
10.2 Single family house in the Mild Climate Conservation Strategy 242
10.3 Single family house in the Mild Climate Renewable Energy Strategy 248
10.4 Row house in the Mild Climate Conservation Strategy 254
10.5 Row house in the Mild Climate Renewable Energy Strategy 260

Appendix 1 Reference Buildings: Constructions and Assumptions 265
Appendix 2 Primary Energy and CO_2 Conversion Factors 279
Appendix 3 Definition of Solar Fraction 283
Appendix 4 The International Energy Agency 285

Foreword

The past decade has seen the evolution of a new generation of buildings that need as little as one tenth of the energy required by standard buildings, while providing better comfort. The basic principle is to effectively isolate the building from the environment during adverse conditions and to open it to benign conditions. Such buildings are highly insulated and air tight. Fresh air is mechanically supplied and tempered by heat recovered from exhaust air. Solar resources are also used for heat, light and power. This is possible as a result of the development of high efficiency heating plants, control systems, lighting systems, solar thermal systems and photovoltaic systems. Enormous improvements in glazing systems make it possible to open buildings to sun, light and views. Finally, through favourable ambient conditions, the envelope can be physically opened and all systems shut down – the most energy efficient operating mode a building can have.

Such buildings are a challenge to design. Buildings of the mid 20th century followed the whims of fashion. Upon completion of the design, the architect turned the plans over to the mechanical engineers to make the building habitable. Resulting energy demands of over 700 kWh/m^2a were not uncommon, compared to carefully crafted low energy buildings of today requiring only 10 to 15 kWh/m^2a!

Achieving such efficiency requires skill, but, like the design of an aircraft, cannot rely on intuition. Two interdependent goals must be pursued: minimizing energy losses and maximizing renewable energy use. This begins with developing a solid concept and ends in the selection and dimensioning of appropriate systems. It is the goal of this book to serve as a reference, offering the experience of the 30 experts from the 15 countries who participated in a 5-year project within the framework of 2 programmes of the International Energy Agency (IEA). The authors of the individual chapters include consulting engineers, building physicists, architects, ecologists, marketing specialists and even a banker. We hope that it helps planners in their efforts to develop innovative housing solutions for the new energy era.

Robert Hastings

S. Robert Hastings
AEU Architecture, Energy and Environment Ltd
Wallisellen, Switzerland

Maria Wall

Maria Wall
Energy and Building Design
Lund University
Lund, Sweden

List of Contributors

Inger Andresen
Architecture and Building Technology
SINTEF Technology and Society
Trondheim, Norway

Tobias Boström
Solid State Physics
Uppsala University, Sweden
Tobias.Bostrom@angstrom.uu.se

Manfred Bruck
Kanzlei Dr Bruck
A-1040 Wien, Austria
bruck@ztbruck.at

Tor Helge Dokka
Architecture and Building Technology
SINTEF Technology and Society,
Trondheim, Norway
Tor.H.Dokka@sintef.no

Annick Lalive d'Epinay
Fachstelle Nachhaltigkeit
Amt für Hochbauten
Postfach, CH-8021 Zurich
Switzerland
abbick.lalive@zuerich.ch
www.stadt-zuerich.ch/nachhaltiges-bauen

Helena Gajbert
Energy and Building Design
Lund University
PO Box 118
SE-221 00
Lund, Sweden
Helena.Gajbert@ebd.lth.se

Susanne Geissler
Arsenal Research
Geschäftsfeld: Nachhaltige Energiesysteme
A-1210 Wien, Austria
susanne.geissler@arsenal.ac.at

Udo Gieseler
Contact: Professor Frank Heidt
Division of Building Physics and Solar Energy
University of Siegen, Germany

Trond Haavik
Synnøve Aabrekk
Segel AS
N-6771 Nordfjordeid
Norway
trond@segel.no

S. Robert Hastings
AEU Architecture, Energy
and Environment Ltd
Wallisellen, Switzerland
robert.hastings@aeu.ch

Anne Grete Hestnes
Faculty of Architecture
Norwegian University of Science
and Technology
Trondheim, Norway

Lars Junghans
Passivhaus Institut
D- 64283
Darmstadt, Germany
www.passiv.de

Berthold Kaufmann
Passivhaus Institut
D- 64283
Darmstadt, Germany
www.passiv.de

Sture Larsen
Architekturbüro Larsen
A-6912 Hörbranz, Austria
www.solarsen.com

Joachim Morhenne
Ingenieurbuero Morhenne GbR,
Wuppertal, Germany
info@morhenne.com

Kristel de Myttenaere
Architecture et Climat
Université Catholique de Louvain
B-1348 Louvain-la-Neuve, Belgium
www.climat.arch.ucl.ac.be

Carsten Petersdorff
Energy in the Built Environment
Ecofys GmbH
D-50933 Köln, Germany
c.petersdorff@ecofys.de
www.ecofys.de

Luca Pietro Gattoni
Building Environment Science
and Technology
Politecnico di Milano, Italy
luca.gattoni @polimi.it

Edward Prendergast
moBius Consult
NL 3971 Driebergen-Rijsenburg,
The Netherlands
Edward@moBiusconsult.nl

Alex Primas
Basler and Hofmann
CH 8029 Zurich, Switzerland
alex.primas@bhz.ch
www.bhz.ch

Martin Reichenbach
Reinertsen Engineering AS
Avdeling for Arkitektur
N 0216 Oslo, Norway

Johan Smeds
Energy and Building Design
Lund University
Lund, Sweden
Johan.Smeds@ebd.lth.se

Maria Wall
Energy and Building Design
Lund University
Lund, Sweden
maria.wall@ebd.lth.se

List of Figures and Tables

Figures

I.1 House interior by George Fredrick Keck 2
I.2 Test house facility 3
I.3 The 'Passivhaus' row houses 4

2.2.1 Energy losses of a row house (reference building in temperate climate) 12
2.3.1 A prototype direct gain house by Louis I. Kahn 15
2.3.2 Window heat balance 15
2.3.3 Vertical south solar radiation on a sunny (300 W) and overcast (75 W) day 16
2.3.4 One-hour internal gains from a light bulb (75 Wh) 17
2.3.5 Heating demands and solar (south) per m^2 heated floor area 17
2.3.6 Reduction of heating demand as a function of window/façade proportions and glass quality for a top-middle and middle-middle apartment 19
2.3.7 Heating peak load versus ambient temperature for the apartment block living and working areas, Freiburg i.B 19
2.4.1 Computer-generated image of the reference room 20
2.4.2 Daylight versus window percentage of façade 22
2.4.3 Reference case 23
2.4.4 Glass door 23
2.4.5 Horizontal window 23
2.4.6 High and low windows (same total area) 23
2.4.7 Corner window 23
2.4.8 Windows on two sides 23
2.4.9 Window flared into the room 24
2.4.10 Effect of room surface absorptances on illumination 25
2.4.11 A tubular skylight in Geneva 26
2.5.1 A solar combi-system with a joint storage tank for the domestic hot water (DHW) and space heating systems 29
2.5.2 Seasonal variations in solar gains and space heating demand in standard housing versus high-performance housing 29
2.5.3 A solar combi-system with the possibility of delivering solar heat directly to the heating system without passing through the tank first 30
2.5.4 Effect of collector tilt and area on solar fraction 30
2.5.5 Suitable collector areas at different tilt angles for a collector dimensioned to cover 95% of the summer demand; solar fractions for the year and for the summer are also shown 30
2.6.1 A high-efficiency woodstove 34
2.6.2 A wood pellet central heating system 35
2.6.3 A compact heat pump-combined heating water and ventilation system 35

3.1.1 Aspects covered by the different methodologies 38
3.2.1 Example of a process chain 40
3.3.1 Structure of a life-cycle assessment 43
3.4.1 Case study of the Hirschenfeld housing development on Brunnerstrasse in Vienna 48
3.4.2 Photo impressions of the milieu of Brunnerstrasse in Vienna 48

4.2.1 Total costs for a compound thermal insulation layer in the wall (area of €/m^2 relates to the wall area) 54
4.2.2 Total costs for a thermal insulation layer between roof rafters 54
4.2.3 Total costs for high-performance windows (€/m^2 window area) 55
4.2.4 Optimized ground plan of an apartment in a social house project in Kassel, Germany; and a view of the air ducts 59

5.2.1 The cyclical process of decision-making in design 64
5.2.2 The hierarchical structure of design criteria 65
5.2.3 Graphical presentation of the weights 68
5.2.4 A star diagram showing the scores for each criterion 69
5.2.5 A bar diagram showing the total weighted scores for each alternative design scheme 69
5.3.1 Concept of total quality assessment and certification 71
5.3.2 Wienerberg city apartment building 73

6.2.1 The six-step process 79
6.2.2 Political, economical, social and technological (PEST) factors that influence the competitive arena 80
6.3.1 Relative energy costs for housing: A marketing argument 81
6.3.2 Product life cycle 83
6.3.3 Defining a potential market niche 83
6.3.4 A passive house block as it was constructed 86
6.3.5 Value chain of Anliker AG's passive house development and marketing 87

7.2.1 Space heating demand for the regional reference buildings with standards based on building codes for the year 2001; the reference climates have been used 97
7.2.2 Non-renewable primary energy demand for the regional reference buildings 98
7.3.1 Factor 4 space heating target for the regional high-performance buildings conservation strategy 99
7.3.2 Factor 3 space heating target for the regional high-performance buildings renewable energy strategy 99
7.4.1 Approximate non-renewable primary energy demand for the regional reference buildings with 50% solar DHW, one quarter of the reference space heating demand and 5 kWh/m^{2}a electricity demand for fans and pumps (multiplied by 2.35) 100

8.1.1 Degree days (20/12) in cold, temperate and mild climate cities 104
8.1.2 Monthly average outdoor temperature and solar radiation (global horizontal) for Stockholm 104
8.1.3 Monthly space heating demand during one year for the reference single family house; the total annual demand is approximately 10400 kWh/a 105
8.1.4 Space heating load and balance point temperature of a single family house, a high-performance case (20 kWh/m^{2}a) and a reference case (69 kWh/m^{2}a) 106
8.1.5 Overview of the total energy use, the delivered energy and the non-renewable primary energy demand for the single family houses; the reference building has electricity resistance heating 108

8.1.6 Overview of the CO_2 equivalent emissions for the single family houses; the reference building has electric resistance heating 108
8.1.7 Overview of the total energy use, the delivered energy and the use of non-renewable primary energy for the row houses; the reference house is connected to district heating 110
8.1.8 Overview of the CO_2 equivalent emissions for row houses; the reference house is connected to district heating 110
8.1.9 Monthly space heating demand during one year; the annual demand is 70,000 kWh 112
8.1.10 Overview of the total energy demand, the delivered energy and the non-renewable primary energy demand for the apartment buildings; the reference building is connected to district heating 113
8.1.11 Overview of the CO_2 emissions for the apartment buildings; the reference building is connected to district heating 113
8.2.1 Space heating demand for the high-performance solution (annual total 1700 kWh/a) and the reference house (annual total 10,400 kWh/a) 116
8.2.2 Space heating peak load for the high-performance solution and the reference building 116
8.2.3 Space heating demand for the high-performance solution (annual total 2950 kWh/a) and the reference building (annual total 10,400 kWh/a) 118
8.2.4 Space heating peak load for the high-performance solution and the reference building 118
8.2.5 Indoor and outdoor temperatures for ventilation strategy 1 120
8.2.6 Indoor and outdoor temperatures for ventilation strategy 2 120
8.2.7 Hours above a certain indoor temperature 120
8.2.8 Effect of window size and orientation on space heating demand; the star shows the actual design of the building according to solution 1b 121
8.2.9 Air tightness of the building envelope; the star shows the actual design for solution 1b 122
8.3.1 Space heating demand (annual total 3701 kWh/a) 125
8.3.2 Space heating peak load 125
8.3.3 The auxiliary demand's dependence upon collector area during the summer months and the total auxiliary annual demand in kWh/m^2 (living area) 128
8.3.4 Remaining annual auxiliary demand as a function of tank size 129
8.3.5 Solar collector tilt effect on the remaining auxiliary annual demand for flat-plate systems 130
8.4.1 Simulation results for the energy balance of the row houses (six units) according to Table 8.4.3 136
8.4.2 Simulation results for the hourly heat load without direct solar radiation; the maximum heat load for an end unit is 1730 W 136
8.4.3 Monthly heating demand for a row house unit (average over four mid and two end units) 137
8.4.4 Number of hours with average indoor temperature exceeding certain limits for an end unit; the simulation period is 1 May to 30 September 138
8.4.5 Number of hours with average indoor temperature exceeding certain limits for a mid unit; the simulation period is 1 May to 30 September 138
8.4.6 Space heating demand for the south window variation in the row house (average unit); U-values shown are for the glazing 139
8.4.7 Space heating demand for the north window variation in the row house; U-values shown are for the glazing 139
8.4.8 Space heating demand for the row house (all six units) for different shading coefficients 140
8.5.1 Scheme of the solar assisted heating system, central and individual solutions: 143
8.5.2 Space heating demand 144
8.5.3 Energy balance of the reference and solar base case 144

8.5.4	Number of hours of the indoor temperature distribution	146
8.5.5	Scheme of the central system	146
8.5.6	Influence of the collector size on usable solar gains	147
8.5.7	Influence of the collector size on usable solar gains for a reduction in heating demand depending upon the supply temperature of the heating system	148
8.5.8	Primary energy demand depending upon collector area	148
8.6.1	Space heating demand of solution 1a	151
8.6.2	Space heating demand	153
8.7.1	The design of the suggested solar combi-system with a pellet boiler and an electrical heater as auxiliary heat sources and two external heat exchangers, one for DHW and one for the solar circuit; the latter is attached to a stratifying device in the tank	157
8.7.2	Monthly values of the space heating demand during one year; the annual total space heating demand is 30,400 and 70,000 kWh/a, respectively, for the high-performance building and the reference building	159
8.7.3	An overview of the net energy, the total energy use and the delivered energy for the high-performance building and the reference building	160
8.7.4	The solar fraction of the system for the whole year and for the summer months	161
8.7.5	Monthly values of auxiliary energy demand for solar systems of different dimensions	161
8.7.6	The energy savings per margin collector area (i.e. how much additional energy is saved if 10 m^2 is added to the collector area, read from left to right)	162
8.7.7	The resulting non-renewable primary energy demand and CO_2 equivalent emissions for different collector areas	162
8.7.8	Annual auxiliary energy demand per living area for different system dimensions based on Polysun simulations	163
8.7.9	The energy demands and the solar gains of the high-performance building and the reference building; the solar gains are shown for different collector areas	163
8.7.10	The influence of tank volume and tank insulation level; the auxiliary energy demand per living area for different tank volumes is shown	164
8.7.11	Annual auxiliary energy per living area for different system dimensions based on Polysun simulations	164
8.7.12	The collector area required for differently tilted collectors in order to obtain a solar fraction of 95% during summer	165
8.7.13	Results from simulations of systems with differently tilted collectors	165
8.7.14	The auxiliary energy demand and solar fractions of systems with different azimuth angles	166
8.7.15	Results from simulations of systems with different collector types	166
8.7.16	The auxiliary energy demand, with varied flow rate and type of heat exchanger in the solar circuit	167
8.7.17	The use of non-renewable primary energy for three different energy system designs: the combi-system in question, with a pellet boiler and an electrical heater; a solar DHW system combined with district heating; and a solar DHW system combined with electrical heating	168
8.7.18	The emissions of CO_2 equivalents for three different energy system designs: the combi-system in question, with a pellet boiler and an electrical heater; a solar DHW system combined with district heating; and a solar DHW system combined with electrical heating	168
8.8.1	Apartment units selected for simulation in the study	171
8.8.2	Sunspace types	172
8.8.3	Space heating demand in relation to area to volume (A/V) ratio	174

8.8.4 Space heating demand and sunspace minimum temperature in relation to U-value of the common wall glazing 174
8.8.5 Sunspace versus mean ambient temperatures (sunspace unheated) 176
8.8.6 Temperature frequency in the sunspace 176

9.1.1 Degree days (20/12) in cold, temperate and mild climate cities 179
9.1.2 Monthly average outdoor temperature and solar radiation (global horizontal) for Zurich 180
9.1.3 Overview of the total energy use, the delivered energy and the non-renewable primary energy demand for the single family houses; the reference building has a condensing gas boiler for heating 181
9.1.4 Overview of the CO_2 equivalent emissions for the single family houses; the reference building has a condensing gas boiler for heating 182
9.1.5 Overview of the total energy use, the delivered energy and the use of non-renewable primary energy for the row houses; the reference house has a condensing gas boiler 183
9.1.6 Overview of the CO_2 equivalent emissions for the row houses; the reference house has a condensing gas boiler 183
9.1.7 Overview of the total energy demand, the delivered energy and the non-renewable primary energy demand for the apartment buildings; the reference building uses a condensing gas boiler 185
9.1.8 Overview of the CO_2 emissions for the apartment buildings; the reference building uses a condensing gas boiler 185
9.2.1 Monthly space heating demand for the proposed solution (19.8 kWh/m^{2}a) and the reference building (70.4 kWh/m^{2}a) 187
9.2.2 The annual temperature duration with only the ventilation strategy 189
9.2.3 The annual temperature frequency with only the ventilation strategy 190
9.2.4 The annual temperature duration with both the solar shading and the ventilation strategy 190
9.2.5 The annual temperature frequency with both the solar shading and the ventilation strategy 191
9.2.6 The influence of the mean U-value of the opaque building envelope on the space heating demand 192
9.2.7 Monthly space heating demand (annual total 965 kWh/a) for the super conservation solution 194
9.3.1 Monthly space heating demand for the proposed solution (25 kWh/m^{2}a) and the reference building (70.4 kWh/m^{2}a) 197
9.4.1 Simulation results for the energy balance of the row house (six units) according to Table 9.4.3 204
9.4.2 Simulation results for the hourly heat load without direct solar radiation; the maximum heat load for an end unit is 1900 W 205
9.4.3 Monthly space heating demand for a row house unit (average over four mid and two end units) 205
9.4.4 Number of hours with average indoor temperature exceeding certain limits; the corresponding simulation period is 1 May to 30 September 207
9.4.5 Number of hours with average indoor temperature exceeding certain limits; the corresponding simulation period is 1 May to 30 September 208
9.4.6 Space heating demand for the south window variation in the row house; U-values are shown for the glazing 208
9.4.7 Space heating demand for the north window variation in the row house; U-values are shown for the glazing 209
9.4.8 Space heating demand for the row house for different shading coefficients 209

9.5.1 Scheme of the solar-assisted heating system with individual and central solutions 213
9.5.2 Space heating demand (base case) 214
9.5.3 Energy balance of the reference and solar base case (columns 1 and 3 are gains, columns 2 and 4 are losses) 214
9.5.4 Indoor temperature of the end house (independent of cases except lightweight construction) 215
9.5.5 Schemes of the two systems: a typical individual solar combi-system; and the building mass as a heat storage 216
9.5.6 Influence of the collector size on useable solar gains 217
9.5.7 Savings in delivered energy (gas) due to collector gains 218
9.5.8 Primary energy demand depending upon collector area 218
9.5.9 Influence of the heating system's design temperature on solar gains and the surface area of radiators 219
9.6.1 Construction types 222
9.6.2 Life-cycle phases, Eco-indicator 99 H/A 223
9.6.3 Building components, Eco-indicator 99 H/A 223
9.6.4 Building components, cumulative energy demand (non-renewable) 224
9.6.5 Influence of the heating system, Eco-indicator 99 H/A 224
9.6.6 Influence of the collector area, Eco-indicator 99 H/A 225
9.7.1 Monthly space heating demand 227
9.7.2 Space heating peak load; results from simulations without direct solar radiation 228
9.7.3 Space heating demand with different glazing areas (double glazing, one low-e coating and argon) 230
9.7.4 Space heating demand for different insulation levels with and without ventilation heat recovery 230
9.8.1 Monthly space heating demand 233
9.8.2 Space heating peak load; results from simulations without direct solar radiation 233
9.8.3 Influence of the collector area on the primary energy demand and CO_2 emissions; solution with biomass boiler and solar combi-system 235
9.8.4 Influence of the choice of energy source on primary energy demand and CO_2 emissions; a comparison between biomass fuel and gas 235

10.1.1 Degree days (20/12) in cold, temperate and mild climate cities 237
10.1.2 Monthly average outdoor temperature and solar radiation (global horizontal) for Milan 238
10.1.3 Overview of the total energy use, the delivered energy and the non-renewable primary energy demand for the single family houses; the reference house has a condensing gas boiler 239
10.1.4 Overview of the CO_2 equivalent emissions for the single family houses; the reference building has a condensing gas boiler 240
10.1.5 Overview of the total energy use, the delivered energy and the use of non-renewable primary energy for the row houses; the reference house has a condensing gas boiler 241
10.1.6 Overview of the CO_2 equivalent emissions for the row houses; the reference house has a condensing gas boiler 241
10.2.1 Simulation results for the energy balance of the single family house according to Table 10.2.3 244
10.2.2 Monthly space heating demand for the single family house 244
10.2.3 Simulation results for the hourly peak load without direct solar radiation 245
10.2.4 Number of hours with a certain indoor temperature: The simulation period is 1 May to 30 September; night ventilation and shading devices during daytime are used 246

10.2.5 Space heating demand as a function of the glazing-to-floor area ratio, in combination with variation of percentage of window area on the south façade (dots on descending lines) and on the north façade (dots on rising lines); the frame area is always 30% of the window area 246
10.2.6 Space heating demand as a function of glazing-to-floor area ratio: All numbers are kept constant as indicated in Table 10.2.2 apart from the south glazing area that varies in order to reach the total glazing area/floor area shown on the x-axis; the different lines represent different wall insulation levels 247
10.2.7 Space heating demand as a function of glazing-to-floor area ratio: All numbers are kept constant as indicated in Table 10.2.2 apart from the south glazing area that varies in order to reach the total glazing area/floor area shown on the x-axis; the different lines represent different types of glazing 248
10.3.1 Simulation results for the energy balance of the single family house according to Table 10.3.3 251
10.3.2 Monthly space heating demand for the single family house 251
10.3.3 Simulation results for the hourly peak load without direct solar radiation 251
10.3.4 Number of hours with a certain indoor temperature: The simulation period is 1 May to 30 September; night ventilation and shading devices during daytime are used 252
10.3.5 Sensitivity analysis for different supply systems in terms of non-renewable primary energy 253
10.4.1 Simulation results for the energy balance of the row house according to Table 10.4.3: Average for the row with two end units and four mid units 256
10.4.2 Monthly space heating demand for the row house: Average for the row with two end units and four mid units 256
10.4.3 Simulation results for the hourly peak load for an end unit without direct solar radiation 257
10.4.4 Number of hours with a certain indoor temperature: The simulation period is 1 May to 30 September; night ventilation and shading devices during daytime are used 258
10.4.5 Space heating demand as a function of the glazing-to-floor area ratio, in combination with variation of percentage of window area on the south façade and on the north façade; the frame area is always 30% of the window area 258
10.4.6 Space heating demand as a function of glazing-to-floor area ratio: All numbers are kept constant as shown in Table 10.4.2 apart from the south glazing area that varies in order to reach the total glazing area/floor area shown on the x-axis; the different lines represent different wall insulation levels 259
10.4.7 Space heating demand as a function of glazing-to-floor area ratio: All numbers are kept constant as shown in Table 10.4.2 apart from the south glazing area that varies in order to reach the total glazing area/floor area shown on the x-axis; the different lines represent different types of glazing 259
10.5.1 Simulation results for the energy balance of the row house according to Table 10.5.3: Average for the row with two end units and four mid units 262
10.5.2 Monthly space heating demand for the row house: Average for the row with two end units and four mid units 262
10.5.3 Simulation results for the hourly peak load for an end unit without direct solar radiation 263
10.5.4 Number of hours with a certain indoor temperature: The simulation period is 1 May to 30 September; night ventilation and shading devices during daytime are used 264

10.5.5 Sensitivity analysis for different supply systems in terms of non-renewable primary energy 264

A1.1 Geometry of the apartment building 266
A1.2 Section of the apartment building 267
A1.3 Geometry of a row house unit 268
A1.4 Geometry of the single family house 269
A1.5 Heat gains and losses: Detached house 276
A1.6 Heat gains and losses: Apartment building 276
A1.7 Heat gains and losses: Row house mid unit 277
A1.8 Heat gains and losses: Row house end unit 277
A1.9 Heat gains and losses divided by degree days: Detached house 277
A1.10 Heat gains and losses divided by degree days: Apartment building 278
A1.11 Heat gains and losses divided by degree days: Row house mid unit 278
A1.12 Heat gains and losses divided by degree days: Row house end unit 278

A2.1 National primary energy factors for electricity; the line represents the EU-17 mix that is used in this book 282
A2.2 National CO_2 equivalent conversion factors for electricity; the line represents the EU-17 mix that is used in this book 283

A4.1 A very low energy house in Bruttisholz, CH by architect Norbert Aregger 291

Tables

2.6.1 Example costs for heating a high-performance house with gas 33

3.1.1 Characteristics of the different methods presented 38

4.1.1 General data for cost calculation 52
4.2.1 Total costs (investment plus energy losses) of insulation added to the exterior wall 53
4.2.2 Total costs for a thermal insulation layer in the roof; all area-specific numbers (€/m²) here are related to the roof area 54
4.2.3 Total extra costs for high-performance windows 56
4.2.4 General data for cost calculation with respect to electric heating and heat pump systems 57
4.2.5 Investment and running costs for a combined system with a heat pump compared to a system with direct electric heating 58
4.3.1 Basic and added construction costs for high thermal performance components 60
4.3.2 Annual rate of total costs 61

5.2.1 Example of main design criteria and sub-criteria 65
5.2.2 The common measurement scale 66
5.2.3 Example of measurement scales for a qualitative criterion (flexibility) and a quantitative criterion (energy use) 67
5.2.4 The weighting scale 68
5.3.1 Weighting factors for energy-use criteria under the category of resource consumption 72
5.3.2 Performance of selected indicators of total quality assessment categories 75

6.3.1 Anliker's strengths, weaknesses, opportunities and threats (SWOT) analysis of the passive house market 84

7.2.1 Mean regional U-values of the building envelope based on national building codes for the year 2001 97

8.1.1 Building component U-values for the single family house 105
8.1.2 Total energy demand, non-renewable primary energy demand and CO_2 equivalent emissions for the reference single family house 105
8.1.3 Building component U-values for the row house 109
8.1.4 Total energy demand, non-renewable primary energy and CO_2 equivalent emissions for the reference row house 109
8.1.5 Building component U-values for the apartment building 111
8.1.6 Total energy demand, non-renewable primary energy demand and CO_2 equivalent emissions for the reference apartment building 112
8.2.1 Targets for the single family house in the Cold Climate Conservation Strategy 114
8.2.2 Building component U-values for solution 1a with supply air heating 115
8.2.3 Total energy use for solution 1a 116
8.2.4 Total energy demand, non-renewable primary energy demand and CO_2 equivalent emissions for the solution with supply air heating and solar DHW heating 116
8.2.5 Building component U-values for solution 1b 117
8.2.6 Total energy use for solution 1b 119
8.2.7 Total energy demand, non-renewable primary energy demand and CO_2 equivalent emissions for the solution with outdoor air to water heat pump 119
8.2.8 Solution 1a: Conservation with electric resistance heating and solar DHW – building envelope construction 122
8.2.9 Solution 1b: Conservation with outdoor air to water heat pump – building envelope construction 123
8.3.1 Targets for the single family house in the Cold Climate Renewable Energy Strategy 124
8.3.2 Building component U-values for solution 2 124
8.3.3 Total energy use for solution 2 126
8.3.4 Primary energy demand and CO_2 emissions for solar combi-system with biomass boiler 127
8.3.5 Primary energy demand and CO_2 emissions for solar combi-system with condensing gas boiler 127
8.3.6 Collector area effect on various system parameters 129
8.3.7 Pros and cons with a roof- or wall-mounted collector for cold climates 130
8.3.8 Solution 2: Renewable energy with the solar combi-system and biomass or condensing gas boiler 132
8.4.1 Targets for row houses in the Cold Climate Conservation Strategy 133
8.4.2 Comparison of key numbers for the construction and energy performance of the row house (areas are per unit) 135
8.4.3 Simulation results for the energy balance during the heating period 135
8.4.4 Total energy demand, non-renewable primary energy demand and CO_2 equivalent emissions for the solution with district heating; all numbers are related to the heated floor area (120 m^2) 137
8.4.5 Details of the construction of row houses in the Cold Climate Conservation Strategy (layers are listed from inside to outside) 141
8.5.1 Row house targets in the Cold Climate Renewable Energy Strategy 142
8.5.2 Building envelope U-values 142
8.5.3 Performance of the building, including the system 144
8.5.4 Total energy use, non-renewable primary energy demand and CO_2 emissions 145
8.5.5 Important system parameters 147
8.5.6 Construction according to the space heating target of 20 kWh/m^2a 149

8.6.1 Targets for apartment building in the Cold Climate Conservation Strategy 150
8.6.2 The building components 150
8.6.3 Total energy demand, non-renewable primary energy demand and CO_2 equivalent emissions for the apartment building with electric resistance space heating and solar DHW system with electrical backup 152
8.6.4 U-values of the building components 152
8.6.5 Total energy demand, non-renewable primary energy demand and CO_2 equivalent emissions for the apartment building with district heating 154
8.6.6 Solution 1a: Conservation with electric resistance heating and solar DHW – building envelope construction 154
8.6.7 Solution 1b: Energy conservation with district heating – building envelope construction 156
8.6.8 Design parameters of the solar DHW system in solution 1a 156
8.7.1 Targets for apartment building in the Cold Climate Renewable Energy Strategy 156
8.7.2 U-values of the building components 157
8.7.3 Total energy demand, non-renewable primary energy demand and CO_2 equivalent emissions for the apartment building 159
8.7.4 A comparison between the high-performance house and the reference house regarding energy use and CO_2 equivalent emissions 160
8.7.5 Collector parameters and corresponding solar fraction and auxiliary energy 166
8.7.6 The auxiliary energy demand for an evacuated tube collector with and without reflectors 167
8.7.7 General assumptions for simulations of the building in DEROB-LTH – construction according to solution 2, space heating target 20 kWh/m^2a 169
8.7.8 Design parameters of the solar combi-system used in the Polysun simulations 170
8.8.1 Space heating demand: Reference case without sunspace 173
8.8.2 Simulation results for the studied sunspace types (unit A) 173

9.2.1 Targets for single family house in the Temperate Climate Conservation Strategy 186
9.2.2 Building envelope components 186
9.2.3 Total energy use 188
9.2.4 Total energy use, non-renewable primary energy demand and CO_2 emissions for the solar domestic hot water (DHW) system with condensing gas boiler 188
9.2.5 Total energy use, non-renewable primary energy demand and CO_2 emissions for the solar DHW system and wood pellet stove 188
9.2.6 Mean U-value for different standards of the building shell 191
9.2.7 Calculated space heating demand for different window constructions; the triple glazing with one low-e coating and krypton is used in the solution 192
9.2.8 Calculated space heating demand for different window distributions 193
9.2.9 Calculated space heating demand for different air tightness standards 193
9.2.10 Calculated space heating demand for different heat exchangers 193
9.2.11 Description of the super conservation level 194
9.2.12 Primary energy demand and CO_2 emissions for the super conservation solution combined with a DHW solar system 194
9.2.13 Constructions according to the space heating target of 20 kWh/m^2a 195
9.3.1 Targets for single family house in the Temperate Climate Renewable Energy Strategy 196
9.3.2 Building envelope components 196
9.3.3 Energy use, non-renewable primary energy demand and CO_2 emissions for the DHW solar system and biomass boiler 198
9.3.4 Energy use, non-renewable primary energy demand and CO_2 emissions for a solar combi-system with a condensing gas boiler 199

9.3.5 Energy use, non-renewable primary energy demand and CO_2 emissions for a solar combi-system with district heating 199
9.3.6 Energy use, non-renewable primary energy demand and CO_2 emissions for a combined earth tube and heat pump system 200
9.3.7 Construction according to the space heating target of 25 kWh/m^2a 201
9.4.1 Targets for row house in the Temperate Climate Conservation Strategy 202
9.4.2 Comparison of key numbers for the construction and energy performance of the row house (areas are per unit) 203
9.4.3 Simulation results for the energy balance in the heating period 204
9.4.4 Analysed example solutions 205
9.4.5 Total energy demand, non-renewable primary energy demand and CO_2 emissions for the solution 1a with oil burner and solar DHW 206
9.4.6 Total energy demand, non-renewable primary energy demand and CO_2 emissions for the solution 1b with a heat pump 207
9.4.7 Details of the construction of the row house in the Temperate Climate Conservation Strategy (layers are listed from inside to outside) 210
9.5.1 Targets for the row house in the Temperate Climate Renewable Energy Strategy 211
9.5.2 Building envelope U-values 212
9.5.3 Total energy use for space heating 213
9.5.4 Maximum peak load at $T_{ambient}$ = –12.2°C 214
9.5.5 Delivered energy for DHW and solar contribution per unit (mean) 214
9.5.6 Delivered and non-renewable primary energy demand and CO_2 emissions 215
9.5.7 Important parameters for the two systems 217
9.5.8 Construction of different cases (building envelope target) 220
9.6.1 Basic parameters of the investigated heating systems; the collector area is per row house unit 224
9.7.1 Targets for apartment building in the Temperate Climate Conservation Strategy 226
9.7.2 U-values of the building components 227
9.7.3 Assumptions for the simulations 228
9.7.4 Total energy use, non-renewable primary energy demand and CO_2 equivalent emissions for the apartment building with condensing gas boiler and solar DHW 229
9.7.5 Variations A, B and C of the glazing area 229
9.7.6 Building envelope constructions 231
9.8.1 Targets for apartment building in the Temperate Climate Renewable Energy Strategy 232
9.8.2 U-values of the building components 232
9.8.3 Assumptions for the simulations 233
9.8.4 Total energy use, non-renewable primary energy demand and CO_2 equivalent emissions for the apartment building with biomass boiler and solar DHW and space heating 234
9.8.5 Building envelope constructions 236

10.2.1 Targets for single family house in the Mild Climate Conservation Strategy 242
10.2.2 Comparison of key numbers for the construction and energy performance of the single family house; for the proposed solution the percentage of the south window frame is 25% instead of 30% 243
10.2.3 Simulation results for the energy balance in the heating period (1 October–30 April) 244
10.2.4 Total energy demand, primary energy demand and CO_2 equivalent emissions 245
10.3.1 Targets for single family house in the Mild Climate Renewable Energy Strategy 248
10.3.2 Comparison of key numbers for the construction and the energy performance of the single family house; for the proposed solution the percentage of the south window frame is 25% instead of 30% 250

10.3.3 Simulation results for the energy balance in the heating period (1 October–30 April) 250
10.3.4 Total energy demand, primary energy demand and CO_2 equivalent emissions 252
10.4.1 Targets for row house in the Mild Climate Conservation Strategy 254
10.4.2 Comparison of key numbers for the construction and energy performance of the row house; for the proposed solution the percentage of the south window frame is 25% instead of 30% 255
10.4.3 Simulation results for the energy balance in the heating period (1 October–30 April) 256
10.4.4 Total energy demand, primary energy demand and CO_2 equivalent emissions 257
10.5.1 Targets for row house in the Mild Climate Renewable Energy Strategy 260
10.5.2 Comparison of key numbers for the construction and energy performance of the single family house; for the proposed solution the percentage of the south window frame is 25% instead of 30% 261
10.5.3 Simulation results for the energy balance in the heating period (1 October–30 April) 262
10.5.4 Total energy demand, primary energy demand and CO_2 equivalent emissions 263

A1.1 General information on the apartment building 267
A1.2 General information on the row house mid unit 268
A1.3 General information on the row house end unit 268
A1.4 General information on the detached house 269
A1.5 U-values of the reference buildings 270
A1.6 U-values of the regional apartment buildings 272
A1.7 Resistance of the regional apartment buildings 272
A1.8 U-values of the regional row house mid units 273
A1.9 U-values of the regional row house end units 274
A1.10 Resistance of the row house 274
A1.11 U-values of the regional detached houses 275
A1.12 Resistance of the regional detached house 275

A2.1 Primary energy factor (PEF) and CO_2 conversion factors 282
A2.2 Primary energy factors for electricity (non-renewable) 283

List of Acronyms and Abbreviations

ach	air changes per hour
ATS	architecture towards sustainability
A/V	area to volume ratio
BI	business intelligence
C	Celsius
CED	cumulative energy demand
CERT	Committee on Energy Research and Technology
CHP	combined heat and power
CI	competitive intelligence
cm	centimetre
CO_2	carbon dioxide
CO_2eq	carbon dioxide equivalent
COP	coefficient of performance
DHW	domestic hot water
ECBCS	Energy Conservation in Buildings and Community Systems
EPS	expanded polystyrene insulation
ERDA	US Energy and Research Administration
EU	European Union
GW	gigawatt
h	hour
HVAC	heating, ventilating and air conditioning
IEA	International Energy Agency
ISO	International Organization for Standardization
K	kelvin
kg	kilogram
kW	kilowatt
l	litre
LCA	life-cycle analysis
LCI	life-cycle inventory
LCIA	life-cycle impact assessment
LHV	lower heating value
m	metre
MCDM	multi-criteria decision-making
MW	megawatt
NO_x	nitrogen oxide
OECD	Organisation for Economic Co-operation and Development
Pa	Pascal
PEF	primary energy factor
PEST	political, economical, social and technological
PV	photovoltaic(s)
SF	solar fraction
SHC	Solar Heating and Cooling Programme

SO_2	sulphur dioxide
SPF	seasonal performance factor
SWOT	strengths, weaknesses, opportunities and threats
TIM	transparent insulation materials
TQA	total quality assessment
UCTE	Union for the Coordination of Transmission of Electricity
UK	United Kingdom
US	United States
VOC	volatile organic compound
W	watt

INTRODUCTION

S. Robert Hastings

I.1 Evolution of high-performance housing

Designing houses to need very little energy was important during the beginning of the 20th century, became irrelevant as oil and gas became plentiful and inexpensive in mid century, but today again has a high priority. It is instructive to briefly review this cyclical development over the last hundred years so that we make no illusions. Houses must serve over decades, some over centuries.

The beginning of the 20th century

At the beginning of the 20th century, houses were typically not heated: individual rooms were heated. The most common heat source in cities was an oil or kerosene stove. Some urban houses had the luxury of coal-fired central heating, though here, too, for reasons of economy, not all rooms were necessarily heated. At this time, however, much of the population lived in rural areas (agrarian society) and wood was the most common heating source. Hot water was heated on the stove top, or in a compartment in the stove, and carried to a big tin basin set in the kitchen each Saturday night (whether one needed a bath already or not).

Relative to salaries fuel was expensive and heating laborious. Fuel had to be carried to the stove. The coal furnace had to be stoked each morning and ash removed. Firewood had to be harvested, split, dried, the stove fed and ash removed. Given the cost and effort of heating, it is surprising that houses were so badly constructed. They had minimal or no insulation and were draughty. To minimize losses from leaky single-glazed windows a 'snake' pillow was laid on the window sill or 'storm windows' were hung over the primary window each autumn and removed each spring. It was a laborious attempt to slow the loss of precious heat out of the house.

These were ideal circumstances for the introduction of a means to produce hot water which required no fuel, needed no cleaning and operated with no maintenance – a solar system. American entrepreneurs took European know-how and developed the first commercial roof solar water systems. Clarence M. Kemp from Baltimore brought his Climax Solar Heater onto the market. Frank Walter improved the concept and marketed a roof-integrated system. A solar water heating industry boomed, particularly in California. Then, in the 1930s, enormous natural gas reserves were discovered, crippling the young, active solar industry (Butti and Perlin, 1980).

Passive solar energy use became a popular topic when Libbey Owen-Ford introduced insulating glass in 1935. It became possible for windows to become net energy producers in cold climates. Architects such as George Fredrick Keck from Illinois built houses with large south-facing windows and high thermal mass interiors. Measurements of the Duncan House showed that by ambient temperatures of –20°C no heating was required between 08:30 and 18:30. This was a sensation for the press.

During World War II house building went through a dormant phase. After the war energy prices fell to record low prices. Central air conditioning led to a decoupling of architecture from climate. Low energy buildings were no longer a topic.

Source: Pilkington North America, Inc

Figure I.1 *House interior by George Fredrick Keck*

The 1972 oil crisis renewed interest in renewable energy as a means to reduce oil dependency. The US Energy and Research Development Agency initiated a massive research and demonstration programme. Passive and active solar housing was instantly a national priority! National competitions were held, test houses and test cells were built to validate computer models, and handbooks were written. This solar movement quickly crossed the Atlantic to Europe.

At the same time, numerous pilot projects demonstrated that even zero-energy housing was possible. One famous example is the Nul-Energihus built in 1974 in Lyngby, Denmark, by Vagn Korsgaard. It combined a large active solar system with a highly insulated building envelope. At this time, windows were still a weakness compared to the thick insulation walls. The solution was movable exterior window insulating panels. During this era the solar collector industry boomed again, thanks to numerous and generous subsidy programmes.

By the 1990s, Europe had become the leader in advancing the state of low energy housing design. The topic again lost priority in the US, and as subsidies were cut off, the solar collector industry nearly disappeared while countries such as Austria achieved world records for the collector production per capita. Fascination with zero-energy houses continued. The Solar House Freiburg, built in 1992, achieved total energy autonomy through its highly insulated transparent insulation envelope, extensive area of active solar thermal and photovoltaic (PV) collectors and production of hydrogen for energy storage (City of Freiburg, 2000). This house, like all zero-energy houses of the past, was a pioneering success but not intended to be affordable in the near future. A more plausible approach was conceived by a German physicist (Wolfgang Feist) and Swedish engineer (Bo Adamson).

Source: S. Robert Hastings, NIST (1978)

Figure I.2 *Test house facility*

Their 'Passivhaus' prototype row houses were extremely well insulated, tightly constructed and heat was efficiently recovered from mechanical ventilation. During much of the year, these houses were self-heating. It is this very simple but effective concept which is the basis for the approach presented in this reference book.

Now, in the beginning of the 21st century, there is a growing recognition that using a non-renewable energy source will result in its depletion. In the meantime, there are now over 4000 Passivhaus projects built across middle Europe and as far north as Gothenburg, Sweden. High-performance components, formerly custom made, are now readily available on the market, including super windows, high-efficiency ventilation heat exchangers, package do-everything mechanical systems and optimized solar thermal systems. Subsidies for photovoltaic systems have resulted in their explosive growth and they are now commonplace as an architectural element.

In the near future, the most noteworthy development is likely not to be a technical breakthrough, but a market breakthrough. Some currently prototype technologies may become standard, such as vacuum insulation. Home automation systems will allow homeowners the same degree of programming control taken for granted in automobiles today. However, the biggest breakthrough is likely to be in the massive penetration in the housing market of this new generation of high-performance housing. Several influences will have to be accommodated in this process – for example, the special requirements of an aging but still active senior population. Comfort expectations will increase (particularly cooling), along with sensitivity to the energy costs of a house.

It is now a good time to plan low energy houses. The topic of sustainability is part of the public consciousness. Substituting renewable energy for expensive fossil fuel-produced energy will help to sell houses as energy prices continue to rise.

Source: W. Feist, PHI, Darmstadt

Figure I.3 *The 'Passivhaus' row houses*

I.2 Scope of this book

In planning very low energy housing it is useful to profit from the experience of already built projects across Europe. This book presents insights from architects, energy consultants and building physicists, as well as marketing specialists and even a banker.

Three housing types are addressed: apartment buildings, row houses and single family detached houses. Solutions for the housing types were optimized for three climates: cold (Stockholm), temperate (Zurich) and mild (Milan). Two different approaches to achieving very low auxiliary non-renewable energy demand were examined: minimizing losses (conservation) and maximizing renewable energy use.

Some housing types in some climates are better suited for one or the other solution, so not all 18 variations ($3 \times 3 \times 2$) were investigated. While it seems obvious that the best solution would be to apply both strategies, some aspects conflict. For example, maximizing passive solar gains requires large window areas. This contradicts with minimizing losses since even the best windows will likely have five to eight times the heat loss rate of the highly insulated opaque envelope. Decisions must be made on where to set priorities.

I.3 Targets

Today, it is 'easily' possible to build an energy autonymous house. The problem is not technical in nature; rather, it is a question of economics. The question, therefore, is how low to set the energy standard relative to the added costs that the market will tolerate. At the beginning of this international research and demonstration project, some experts argued that the energy target should be relatively easy in order to facilitate a strong market penetration. Others argued that this was not ambitious enough to justify the research effort. A factor four improvement for space heating demand was set for the conservation strategy and a factor three for the renewable energy strategy. A factor two was set for the total primary energy needed for space and water heating, as well as electrical equipment to operate these systems. Electricity for fans, pumps and controls was multiplied by a factor of 2.35. This was set considering the primary non-renewable energy needed to produce electricity, given the European mix of types of power generation.

The decision to set a tougher target proved good because, in the meantime, over 4000 housing projects have been built to this standard. Manufacturers of the needed high-performance components have responded to the growing demand, with the result that the cost of such housing is continuously decreasing.

To give the target absolute values, the energy demand of housing built to current building codes in the year 2000 was calculated for countries in each climate region. It was then not difficult to set a target for the climate region. Interestingly, conventional housing in mild climates tends to consume more heating energy than housing in cold climates. This is explained by the tighter building codes of colder climate regions.

Factors considered

Energy is, of course, only one factor to be considered if the goal is to build sustainable housing. What is 'sustainability'? The Brundtland Commission[1] on Environment and Development (WCED, 1987) gave the following definition: 'Sustainable development is development that meets the needs of the present without compromising the ability of future generations to meet their own needs.' Therefore, in building a house to improve one's personal quality of life, three domains must be considered over the long term: society, environment and economy.

While this book focuses on energy, Part I on strategies also includes the topics of ecology and economics. To help in the planning process, Chapter 5 on multi-criteria decision-making is also included. Finally, even the best house solutions will remain one of a kind unless marketing aspects are considered.

The authors hope that their chapters are helpful to others in accepting the challenge to build sustainable low energy housing.

Note

1 The Brundtland Commission was chaired by Norwegian Prime Minister Gro Harlem Brundtland, and its report, *Our Common Future*, published in 1987, was widely known as the Brundtland Report. This landmark report helped to trigger a wide range of actions, including the UN Earth Summits in 1992 and 2002, the International Climate Change Convention and worldwide Agenda 21 programmes. It was the Brundtland Report which inspired towns and cities in Northern Europe to initiate the Brundtland City Energy Network in 1990. The network has taken energy use as a starting point for action.

References

Butti, K. and Perlin, J. (1980) *A Golden Thread*, Cheshire Books, Palo Alto, CA
City of Freiburg (2000) *Freiburg Solar Energy Guide*, City of Freiburg, Germany
Simon, M. J. (1947) *Your Solar House*, Simon and Schuster, New York
WCED (World Commission on Environment and Development) (1987) *Our Common Future*, Oxford University Press, New York

Part I

STRATEGIES

1

Introduction

This section examines a selection of strategies for designing very low energy housing towards the goal of sustainability.

The challenge is to balance strategies, each of a very different nature, to fit the personal values and priorities of the home-owner or investor. Clearly a high priority is given to drastically reducing the amount of non-renewable energy a building needs, while at the same time improving comfort. If a building is to be promoted as ecological, then obviously there are other factors in addition to energy to address and the time horizon has to be extended to the lifetime of the construction.

To systematically weigh plusses and minuses, methodologies have been developed. Such multi-criteria decision tools can also be applied to housing design. It is important to assure quality control from design through construction in the ongoing decision process. Compromises can be cumulative so that good intentions at the beginning of a project are not fulfilled when the building is completed.

Decisions in the planning process should also be made with an awareness of the housing market being targeted. Achieving a market breakthrough for a new product takes specialized skills – and very low energy, ecological and sustainable housing is a new product,. This know-how is not part of the normal formal education of architects, engineers or building physicists.

Many other topics exist regarding sustainable housing design. The topics and strategies presented here represent the work done by the experts within the time and budget of this international project.

2

Energy

2.1 Introduction

Joachim Morhenne

This section examines a selection of strategies for designing very low energy housing towards the goal of achieving sustainability.

The first set of strategies address energy. The objective of the first energy strategy is simple: need as little as possible to provide comfort. Energy not needed is the most ecological energy, so first priority goes to conservation. The energy still needed, after building a highly insulated tight envelope and recovering heat otherwise lost, should ideally be provided from renewable energy sources. Accordingly, the next strategies are targeted at maximizing the use of 'free' energy: passive solar gains, daylight and active solar thermal systems. These sources can cover all remaining demand but it is more economical to cover the last small fraction by conventional means as efficiently as possible.

Strategies aimed at low ecological impact have to address the energy and materials flow throughout the lifetime of the housing. Two approaches to quantifying this impact are the cumulative energy demand analysis and life-cycle analysis. 'Sustainability' is a broader topic that has to be considered, and which encompasses social, economic and energy impacts. A chapter is also included which examines 'architecture' in this broader context.

Low energy, ecological elements must be affordable and from the experience of the projects analysed during the production of this book, we see that these qualities cost more than design where decisions were made based on a short-term perspective. Here we examine which aspects of high-performance design added most to the additional costs and offer an outlook on cost developments based on observed trends.

It is quickly apparent that there are many strategies to choose from. Strategies may require actions which contradict other strategies and, in any case, the budget can hardly finance the application of all strategies. Decisions must be made as to which strategies will be given priority. To help in this decision process two approaches are reviewed: multi-criteria decision-making and total quality assessment.

Finally, this 'wonder housing' must be marketable. Here, experts offer insights from their experience building and marketing sustainable housing, which approaches are most effective and which arguments have little influence on buyers. Marketing is both a science and an art!

Following these strategies selectively, given the constraints and opportunities of a specific project, increases the likelihood of succeeding in building and selling low energy, ecological, affordable housing – the goal of this book.

2.2 Conserving energy

	DESIGN ADVICE
Minimal insulation values:	U-envelope 0.15 W/m^{2}K (walls and roof) (insulation between 25–40 cm thick) U-windows 0.8 W/m^{2}K (average frame + glass) g-value > 0.50 (triple glazing with two low-e coatings and noble gas)
Thermal bridges:	$\psi \leq 0.01$ W/mK
Air tightness:	< 0.6 air changes per hour by 50 Pa
Ventilation air:	30 m^3 / person
Heat exchanger efficiency:	> 0.75
Electric/m3 ventilation air:	≤ 0.4 W/m^3 air

Source: Feist et al (2005)

The goal of consuming very little energy to provide superior comfort can be achieved by two basic approaches:

1 conservation; and
2 use of low or non-emissions resources.

An analysis of built projects demonstrates that both strategies can achieve these goals, but each has limitations. This section will examine the potential and limitations of the conservation path.

Conservation strategies must reduce the energy needed to offset transmission and infiltration losses, supply and temper ventilation air, produce hot water and run technical systems (fans, pumps and controllers). Because the planner has little control over the occupants' selections of household appliances over the building lifetime, this end use is not addressed here.

The proportion of these four principle end uses of energy for conventional housing per building codes is shown in Figure 2.2.1

Opportunities to conserve energy lost along these paths include:

- reducing the demand;
- increasing the efficiency of devices; and
- recovering otherwise lost heat.

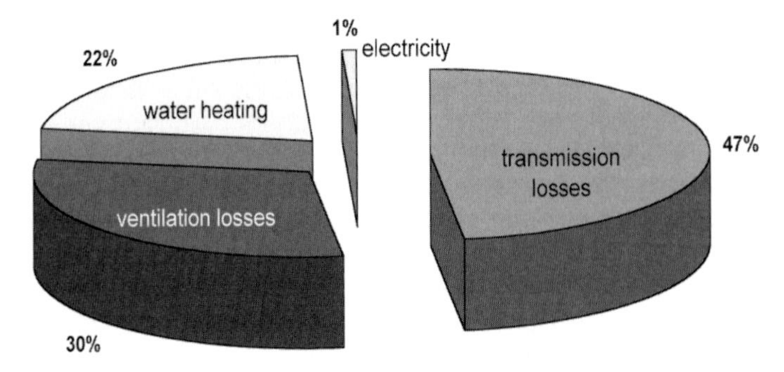

Source: Joachim Morhenne

Figure 2.2.1 *Energy losses of a row house (reference building in temperate climate)*

Which opportunity makes the most sense for which end use varies by end use. Reducing demand is appropriate for reducing transmission losses, but not the first priority for reducing ventilation losses or producing hot water. The minimum air change rate in dwellings is dictated by human and hygienic requirements and cannot be dramatically reduced. Nor can the planner dictate a reduction in hot water use by occupants. For these end uses, recovering heat has a high priority. In the case of electrical consumption for technical systems, the main conservation approach is to specify more efficient devices and reduce the work load imposed on the devices.

2.2.1 Reducing transmission losses

Transmission losses can be drastically reduced by:

- improving the building's insulation;
- using active insulation (transparent insulation materials, or TIM) to compensate envelope heat losses by passive solar gains;
- interrupting thermal bridges across constructions; and
- making the building form more compact to reduce the amount of envelope heat losses for the enclosed heated volume (area-to-volume ratio, or A/V).

The effectiveness of TIM, for example, is affected by building orientation, shading and internal gains. Thermal bridges become more pronounced as the overall insulation level is increased. An interesting fact is that transmission losses can be substantially reduced but not eliminated, even if the insulation is increased beyond all comprehensible thickness. What remains as a means to further conserve energy is to decrease the envelope area for the given enclosed volume (compactness).

A highly insulated envelope has the added benefit of providing better indoor comfort because room surface temperatures are warmer. Specifying a highly insulated envelope is important because the envelope construction should have a long life span. This means that the opportunity to increase the insulation will not come again for many years. By comparison, mechanical systems have a shorter lifetime, providing the opportunity to install a more efficient component when a replacement is needed. For example, in the future a small fuel cell might provide heat and electricity as a package unit for houses.

2.2.2 Reducing ventilation losses

As mentioned, minimum ventilation rates are a given. Typical minimum values are 30 m^3/h per occupant, which should not be reduced further and an increase of the air change rates should be anticipated over the building's lifetime. To reduce the amount of energy consumed for ventilation, the first step is to ensure that no spaces are excessively ventilated. The next step is to reduce the fan power needed to supply this required air volume. Duct lengths and layout should be optimized to reduce hydraulic pressure drops (short is beautiful). Finally, some ventilation systems (fans and heat exchangers) are more efficient than others regarding both heat exchange efficiency and electrical power. The latter is a very important factor given the primary energy conversion factor for electricity. The heat recovery can be an efficient air-to-air heat exchanger or a heat pump to extract still more heat out of exhaust air before it leaves the house.

An air-tight building is essential in high-performance houses, and it is a requirement in many voluntary standards such as the Passivhaus (Feist et al, 2005) or MINERGIE-P (MINERGIE, 2005) standards. These standards require pressurization tests to verify that a high degree of air tightness has been achieved (0.6 air changes by 50 Pa under and over pressure generated by a 'blower door'). A critical issue is that the air tightness achieved to pass the certification testing be long lasting. Taped joints must, for example, be taped with adhesives that will maintain their bond over decades.

2.2.3 Reducing energy needed to heat domestic hot water

The amount of hot water needed is a question of individual behaviour and therefore standardized consumption values are used for planning. The most accurate values can be found in guidelines for dimensioning solar active systems. Large differences occur from one country to the other. Obviously, a key step is to specify appliances that have a minimal hot water demand. Appliances have a relatively short life span, however, so other measures are also important. Basically, two strategies are possible:

1 energy recovery; and
2 using renewables.

The heat in used hot water on the way down the drain is a tempting source for heat recovery. Such heat recovery systems tend to require an unacceptable amount of undesirable maintenance, however. Heat production from a renewable source (i.e. heat pump, biomass or active solar system) is likely to be a more attractive solution for the building owner.

2.2.4 Conclusions

Conservation is a very cost-effective means of achieving a high-performance house because the least expensive, most ecological kWh is the kWh which is never needed. In selecting the targets for conservation, priority has to be given in the order of the magnitude of energy end uses. While electricity for technical systems represents a small slice of the end-use pie, it is magnified when the primary energy needed to produce electricity is considered. Reducing the envelope heat losses by transmission and air leakage has added comfort benefits. Finally, heat recovery is an effective means of tapping a 'free' energy source – namely, energy which has already served a purpose in the house.

References

Feist, W., Pfluger, R., Kaufmann, B., Schnieders, J. and Kah, O. (2005) *Passivhaus Projektierungs Paket 2004*, Passivhaus Institut, Darmstadt, Germany, www.passiv.de

MINERGIE (2005) *Reglement zur Nutzung des Produktes MINERGIE®-P*, Geschäftsstelle MINERGIE® MINERGIE® Agentur Bau, Steinerstrasse 37, CH-3006, www.minergie.ch

2.3 Passive solar contribution in high-performance housing

S. Robert Hastings

2.3.1 Introduction

The deliberate use of sunlight transmitted through windows to provide warmth in winter is an ancient concept. The efficiency of this process has steadily improved as windows have improved. In the last decade, the glass industry brought products on the market with a fivefold improvement in insulation quality. During this same period, housing has entered the market with a tenfold decrease in heating demand. These two developments create new conditions for passive solar design. This section reviews this evolution and offers design advice regarding passive solar use in houses that require very little heating energy.

2.3.2 The evolution of passive solar heating

Gaining heat in winter from large south-facing openings was a strategy known to the ancient Romans. They had a primitive form of glass as far back as 100 AD. Wall openings of the hypocaustum, or the sweating room, of a thermal bath were glazed to trap the sun's heat.

Passive solar design received much attention in the early 20th century. Direct-gain houses were rediscovered by such renowned architects as Louis Kahn (see Figure 2.3.1). The topic fell dormant in

DESIGN ADVICE:	
Maximum window/façade ratio:	50% for south façades (+/–45° from south)
Window properties:	U ˉ 0.8–1.0 W/m^2K (average glass and frame) g-value ≥ 0.50
Window proportions:	As large and square windows as possible to reduce frame perimeter heat losses.
Solar usability:	Open-floor plan minimizes local overheating. Massive materials where sunlight falls. Fast-reacting auxiliary heating (air heat, not floor heating).
Realistic expectations:	Break even in temperate climates, middle latitudes, net heat loss in cold northern climates for high-performance housing.
Overheat protection:	As important in high-performance housing as conventional housing, ideally sun shading outside the window (adjustable blind).
Movable window insulation:	Relic of the past, not worthwhile given super glazings.

Source: Simon (1947) with permission from Pilkington North America

Figure 2.3.1 *A prototype direct gain house by Louis I. Kahn*

the middle of the century, when energy became unimportant. The technical breakthrough of central air conditioning de-coupled building design from climate as a form-giver.

The first oil crisis of 1972 revived interest in energy. By the middle of the decade the newly created US Energy and Research Administration (ERDA) launched a multi-million dollar programme. It included passive solar building design, and annual national conferences were held on this topic. By the early 1980s, passive solar housing was also springing up across Europe.

Today, in the 21st century, a tenfold reduction in heating energy demand compared with conventional housing is possible. This is achieved primarily by:

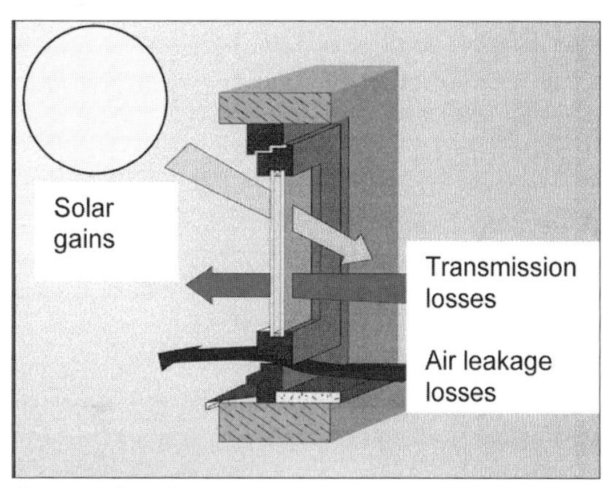

Source: Robert Hastings

Figure 2.3.2 *Window heat balance*

- reducing heat losses with highly insulated and air-tight construction;
- recovering heat from exhaust air; and
- producing heat very efficiently.

Each of these three strategies affects passive solar use. A planning mistake is more critical than before because the micro-heating systems in these houses have a very small capacity to compensate for errors.

2.3.3 The principle

Net passive solar gains occur when the solar input exceeds the heat losses of the window. High-performance windows achieve net gains more often than conventional windows. Although the glass coatings let less solar radiation into the house (g-value), this is more than offset by the reduced heat losses. Two examples follow of energy 'book-keeping' for a sunny and an overcast day with an average outdoor temperature of 0°C. The example uses a modern conventional window, not a super window (see Figure 2.3.3).

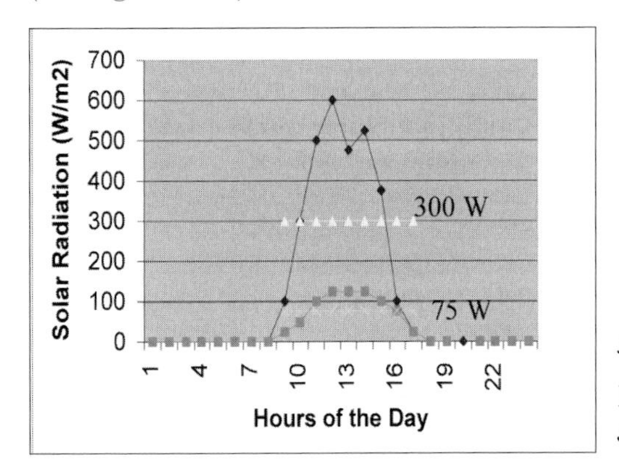

Source: Robert Hastings

Figure 2.3.3 *Vertical south solar radiation on a sunny (300 W) and overcast (75 W) day*

Income (solar radiation):

Sunny day:	G_{sol} = 9 h x 300 W average
Solar usability:	η = 85%
Overcast day:	G_{sol} = 9 h x 75 W average
Solar usability:	η = 100%
Direct + indirect energy delivered by window relative to total radiation striking the glass:	g = 0.6

Expenses (heat loss):

Window area:	A = 1 m^2
Insulation value:	U = 1.0 W/m^2K
Room temperature:	T_{room} = 20°C
Ambient temperature:	T_{amb} = 0°C
Temperature difference:	dT = 20 K

Balance:

$Q_{gain} = (G_{sol} \times g \times \eta) \times 9\ h$

$Q_{loss} = (U \times A \times dT) \times 24\ h$

Sunny:

$Q_{gain} = (300\ W \times 0.6 \times 0.85) \times 9h = 1377\ Wh$

$Q_{loss} = (1.0\ W/m^2K \times 1\ m^2 \times 20\ K) \times 24 = 480\ Wh$

Net gain = + 900 Wh

Overcast:

$Q_{gain} = (75\ W \times 0.6 \times 1.00) \times 9\ h \qquad = 405\ Wh$

$Q_{loss} = (1.0\ W/m^2K \times 1\ m^2 \times 20\ K) \times 24 = 480\ Wh$

$\text{Net gain} \quad = -75\ Wh$

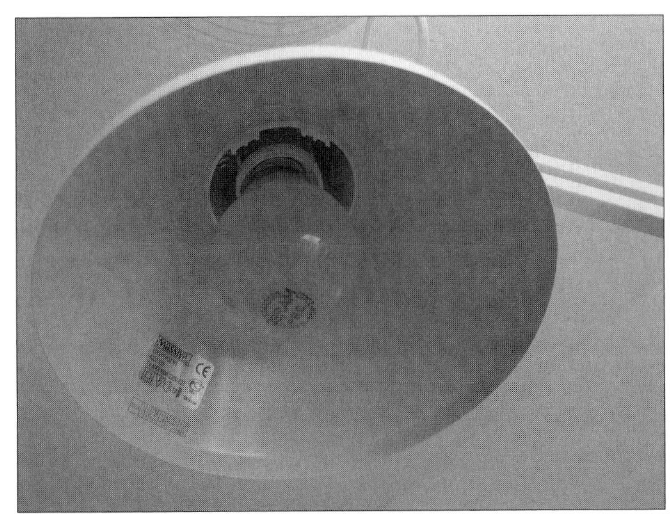

Source: Robert Hastings

Figure 2.3.4 *One-hour internal gains from a light bulb (75Wh)*

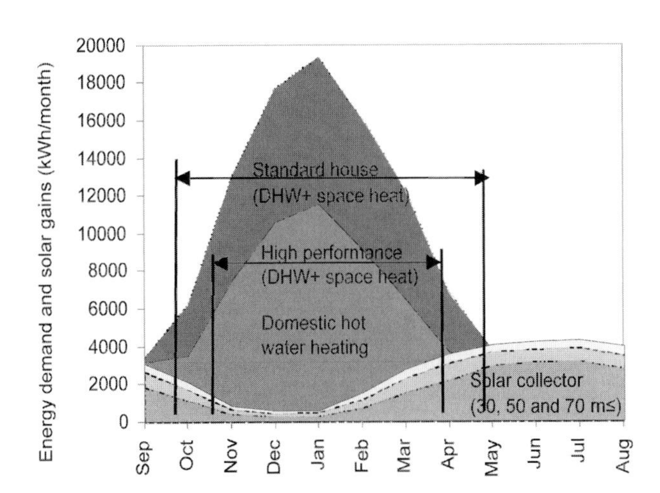

Source: Helena Gajbert, Lund University

Figure 2.3.5 *Heating demands and solar (south) per m² heated floor area*

Thus it can be seen that even a conventional south-facing window is able to offset its losses by the solar gains for days with an average outdoor temperature of 0°C. The balance for a high-performance window would be even better, so the critical issue is not the balance but if the gains occur when they are useable.

2.3.4 Direct gain in high-performance housing

The special conditions of a high-performance (hp) house limit the usability of passive solar gains. The net energy gains or losses of high-performance windows are small and the balance can easily switch from a 'profit' to a 'loss'. It is therefore useful to examine the 'accounting' more closely.

'Income' (solar gains)

Four factors affecting the 'income' are meteorology; glass transmittance (g); window area; and how effectively (η) the 'income' is applied to reducing the 'expenses'.

Meteorology (G_{sol}): given the very minimal heat losses, internal gains can keep the house comfortable without heating later into the autumn and starting earlier in the spring than standard houses. This reduces the heating season to the mid winter months when the days are shortest and solar radiation weakest (see Figure 2.3.5). Typically, passive solar gains are most useful in the spring when a heat demand still exists and the days are longer with stronger sun.

Glass transmittance (g): high-performance glass is called for in high-performance housing. The low emissivity coatings and typically three layers of such glazing reduce the amount of solar energy able to penetrate into the house. The g-value for commonly used heat insulating glazing ($U = 1.1\ W/m^2K$) in standard housing is in the order of 0.6. For high-performance glass ($U = 0.5\ W/m^2K$), the g-value can easily be as low as 0.4, or a one third reduction in the solar energy entering a room. A good value to look for in high-performance glazings is $\geqslant 0.50$.

Effectiveness (η): high-performance houses have one advantage regarding the usability of passive solar gains – their mechanical ventilation with heat recovery. Air and heat are extracted from the sunlit rooms and then used to heat ambient make-up air via the heat exchanger. This benefit should not, however, be overestimated. Consider the example of a 35 m^2 living room with ventilation extraction rate 0.45 room volumes per hour. If the sun warms the room from 20°C to 26°C, the exhaust air will provide 80 W more heating power to the heat exchanger.

The effectiveness of window gains in high-performance houses, as also in conventional houses, can be increased by:

- massive interior construction, ideally with the surface of the mass sunlit;
- an open-floor plan to allow better distribution of the solar heat; and
- a house concept with the largest window areas facing +/–45° south.

Overheating should be no more a problem in a high-performance house than in a standard house. Too large window areas without sun protection will also overheat a standard house. The high performance house, when window shading is used, can be cooler than the standard house. The increased roof and wall insulation are barriers to the summer surface heat build-up of sunlit exterior surfaces.

'Expenses'

Heat losses: a good U-value for windows (glass and framing) is 0.8 W/m^2K. Compared to a well-insulated wall with a U-value of 0.15 W/m^2K, the window heat loss rate is five times more heat than the wall. To maintain a room temperature of 20° C by an ambient temperature of –10°C, the heating system must deliver 19.5 W more power for 1 m^2 of window compared to a wall. Accordingly, the key limitation to window area is the demand on heating power.

2.3.5 Simulation results

Dynamic simulations of an apartment building in a temperate climate show that increasing window area need not increase heating energy demand. It may even slightly reduce demand. This can be seen in Figure 2.3.6 for the case of a top-middle apartment. This is most pronounced with the highly insulating glass in the lower curve. The extreme case is a middle mid-floor apartment. Even with the greatly reduced heating season (as a result of losses occurring from only two exterior surfaces), a break-even can be observed. With increasing window areas, increased heat losses are just offset by increased useable solar gains. Similar simulations in the cold climate indicate that window areas must be kept to the minimum needed for good daylighting. Solar gains during the dark winter months cannot offset the higher heat losses from colder ambient temperatures.

2.3.6 Monitoring results

Monitored data from an apartment building in Freiburg, Germany, confirm the extent to which direct solar gains reduce heating peak load (Voss et al, 2004). In Figure 2.3.7, the upper straight line represents the theoretic heating peak load as a function of ambient temperature with no solar radiation, no internal gains and a constant 20°C indoor temperature. The dashed line assumes a demand reduction resulting from 2.1 W/m^2 of internal heat gains (100 per cent usability). The points represent measured heating peak load, keyed by the intensity of solar radiation at the time. The diamonds indicate heating

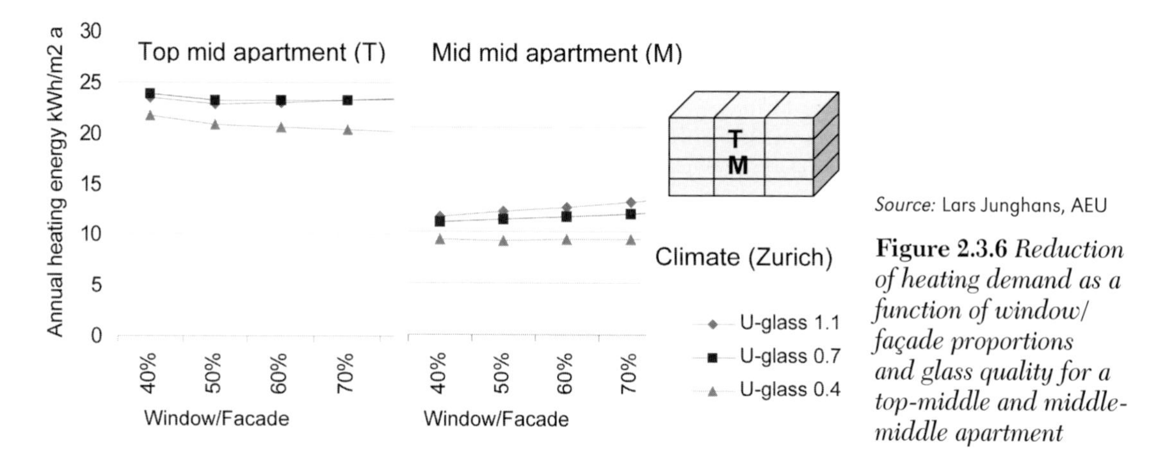

Source: Lars Junghans, AEU

Figure 2.3.6 *Reduction of heating demand as a function of window/ façade proportions and glass quality for a top-middle and middle-middle apartment*

peak load when solar radiation was less than 25W/m², triangles 25 to 90 W/m² and circles greater than 90 W/m².

From the data, two observations can be made:

1. The extent that heating peak load decreases with increasing solar radiation: this reduction would be even greater were the south-facing windows of this project not shaded by balconies and trees.
2. As the ambient temperature decreases, measured heating peak loads are furthest from the theoretical demand (dashed) line. This is explained by the fact that the 8 W/m² heating system capacity was unable to maintain the design 20°C. As the room temperature fell below the design temperature, the heating peak load decreased accordingly.

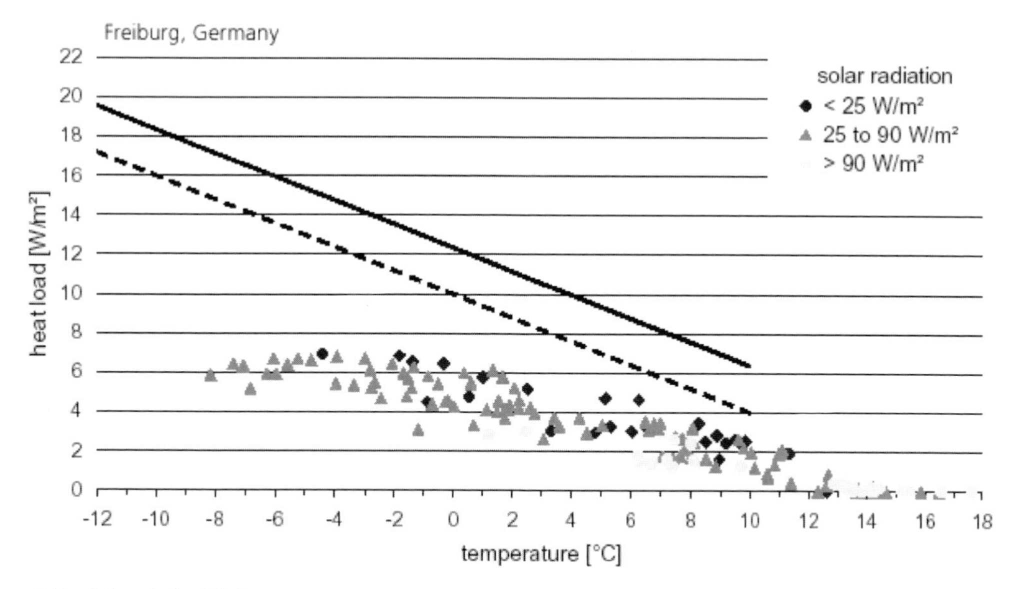

Source: K. Voss, Fraunhofer ISE, D

Figure 2.3.7 *Heating peak load versus ambient temperature for the apartment block living and working areas, Freiburg i.B*

2.3.7 Conclusions

In houses with very low heating demand, solar gains through windows are able to make a small but useful contribution. The contribution is small because of the shortening of the heating season as a result of thick insulation and ventilation heat recovery. The key issue becomes how much windows increase the needed heating power under design conditions.

Accordingly, the window area should not be over-dimensioned (greater than 50 per cent of the south façade area) and a highly insulating glass with good frames should be used (U_{window} 0.8–1.0 W/m^2K).

In cases where heating demand is extremely small – for example, a middle mid-floor apartment – useful solar gains are just able to compensate window heat losses. This, however, depends on the use of high-performance windows. Monitored data confirm the small but positive passive solar contribution in high-performance housing, even in this case of windows partially shaded by balconies and trees. In northern cold climates passive solar gains do not offset the heat losses, so the window area should be moderated to what is needed for daylighting.

Finally, it should not be forgotten that generous window areas, even if they only achieve an energy break-even, can be an important asset regarding daylight, view out and value of the real estate. With a good window construction and effective exterior sun shading, passive solar gains can modestly decrease auxiliary heating demand in winter without causing overheating in summer.

References

Simon, M. J. (1947) *Your Solar House*, Simon and Schuster Inc, New York

Voss, K., Russ, C., Petersdorff, C., Erhorn, H. and Reiss, J. (2004) *Design Insights from the Analysis of 50 Sustainable Solar Houses – Task 28/Annex 38 – D Sustainable Solar Housing*, Technical Report, Fraunhofer-Institut Solare Energie-systeme ISE, Gruppe Solares Bauen, Freiburg, Germany

2.4 Using daylight

S. Robert Hastings and Lars Junghans

Source: L. Junghans, AEU GmbH

Figure 2.4.1 *Computer-generated image of the reference room*

	DESIGN ADVICE:
Window size:	Maximum increase in daylight reached at 50 per cent window-to-façade ratio. Also less luminance contrast and glare at maximum.
Window sill height:	Low enough to allow view out if occupants are frequently seated.
Window head height:	High windows project daylight deeper into the room, give more uniform light distribution and make room appear brighter.
Window position:	Window in corner leaves other corner dark and gives impression of a dimly lit room.
Windows in two façades:	Ideal to have light sources from different directions.
Window frame:	Select window constructions with narrow frames.
Light transport:	Light tubes are effective and results are dramatic.
Glass:	Low iron glass increases daylighting by up to 6 per cent.
Room surfaces:	Obviously light colours preferred. Most critical are side walls and floor.
Glare/privacy control:	Curtains/blinds should be mounted to be pulled clear of glazing in order not to hinder daylight (especially at top of window).

2.4.1 Introduction

A window is a hole in the insulated envelope of a house that lets heat escape, but light enter. The first property increases heating costs, but the second property affects how quickly a house will sell or be rented, and the quality of life of the future occupants.

Upon entering a room, the quality of natural lighting strongly affects the first impression of the space. A well-lit, bright space without glare gives a good impression. Dark corners, a strong decrease of brightness over the room depth or glare leave a negative impression. A designer's first response is to plan very large windows. Yet, in highly insulated buildings, even good windows lose up to five times more heat than a highly insulated opaque wall. So the designer is confronted by a conflict: big windows for daylighting or small windows to conserve heat.

Fortunately, window size is not the only factor affecting the light quality of a space: many parameters influence human perception of light in a room, including the:

- window proportion;
- window position;
- cross-section of the window opening in the wall;
- treatment of room surfaces; and
- glare protection.

This section examines how these parameters affect the daylight quality of a reference room. Results are illustrated with computer-generated images of the room and luminance graphs. The conclusion offers design advice.

2.4.2 The reference room and conditions

The reference room is 3 m × 5 m × 2.5 m (width × depth × height). As a starting point, a conservative window area of 3 m^2 was chosen. This computes to a quite common window-to-floor area ratio of 20 per cent and a window/south façade ratio of 40 per cent. The visible transmission of the glazing is 62 per cent. The room surface absorptance of floor (70 per cent), walls (40 per cent) and ceiling (30 per cent) are typical for a home. The window opening is 40 cm deep, a not uncommon wall thickness for a highly insulated house. Performance was analysed for the three latitudes and climates (Stockholm, Zurich and Milan). A date of 21 September was selected for the analysis because it is not a seasonal extreme. Noon was chosen to give the greatest variations in the light penetration into the depth of the room. An overcast sky was considered because, with full sun, illuminance differences of the variations became obscured by the large absolute values. The program, Radiance (Minamihara, 2005) was used for the analysis.

2.4.3 Window size

Window size is a dominant factor affecting the first impression of a space. Window sizes ranging from 10 per cent to 100 per cent of the south-facing area were analysed. Logically, the illuminance increases with increasing opening sizes. This is, in fact, true for window-to-façade ratios up to 50 per cent. Increasing the window size much beyond this is not justified by improved daylighting, as seen in Figure 2.4.2 for the middle latitude. In northern latitudes, the curve is even flatter. In lower latitudes, however, increasing the window size beyond 50 per cent continues to increase the illuminance.

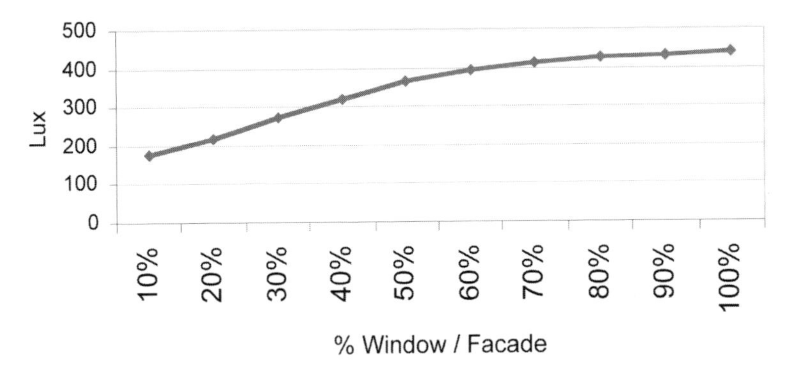

Source: Lars Junghans, AEU Ltd

Figure 2.4.2 *Daylight versus window percentage of façade*

Window areas up to a 50 per cent ratio not only increase the absolute amount of light, they also reduce the luminance contrast and resulting glare. The result is a better first impression on entering such a space.

2.4.4 Window proportions and position

The daylight qualities of the reference room were analysed considering different window shapes and proportions. In each variation, including the case of two windows, the total window area is held constant at 3 m^2. Figure 2.4.3 shows the illuminance values for the reference window as a basis of comparison.

A tall window (for example, a glass door)

This geometry offers occupants a large, vertical view angle from the sky to the ground. The first impression on entering the room is positive (see Figure 2.4.4). The height of the glass door is particularly beneficial in frequently overcast climates. In such conditions, the sky is up to three times brighter at the zenith than at the horizon, so the higher the window, the deeper the daylight will project into the room.

A horizontal window

This case has a horizontal window across the entire width of the room. In spite of the minimal light penetration into the room, the impression is pleasant (see Figure 2.4.5). Good light penetration into

Source: Lars Junghans, AEU Ltd

Figure 2.4.3 *Reference case*

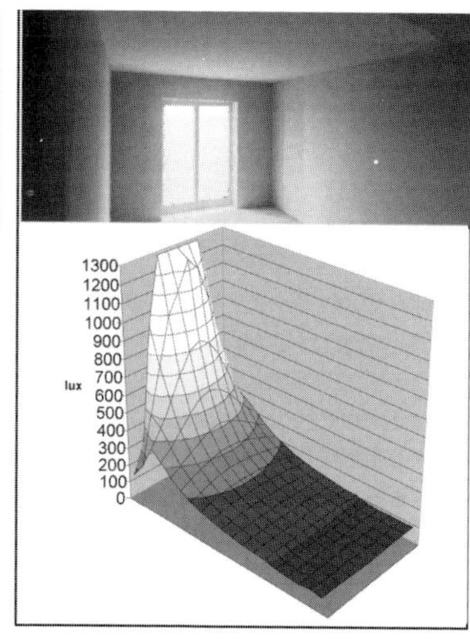

Source: Lars Junghans, AEU Ltd

Figure 2.4.4 *Glass door*

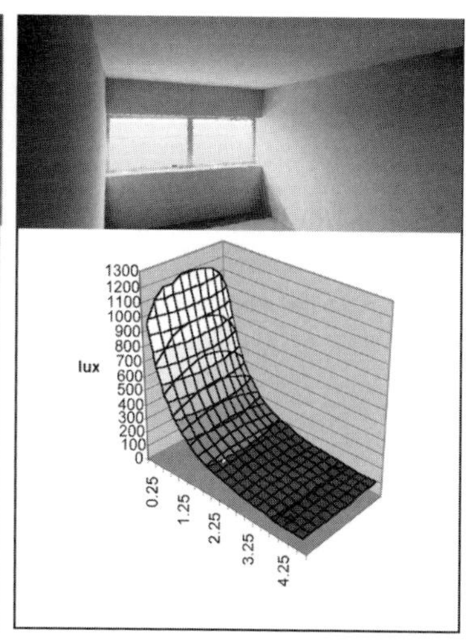

Source: Lars Junghans, AEU Ltd

Figure 2.4.5 *Horizontal window*

the room depth occurs, in any case, for east- and west-facing windows due to the low sun angles from these directions. A working place directly in front of a horizontal window has a special appeal. Ideally, the window sill should be low enough to provide a view out and down from desk height.

A high and low window combination

The function of the high window is to achieve good daylight penetration, while the low window offers a view to the outside. The light distribution into the room depth is slightly improved, but the human impression is unfavourable because of the too limited view to the outside and the dark other corner (see Figure 2.4.6).

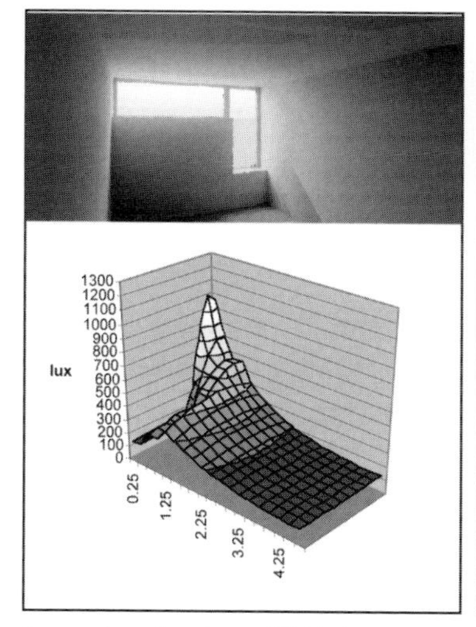

Source: Lars Junghans, AEU Ltd

Figure 2.4.6 *High and low windows (same total area)*

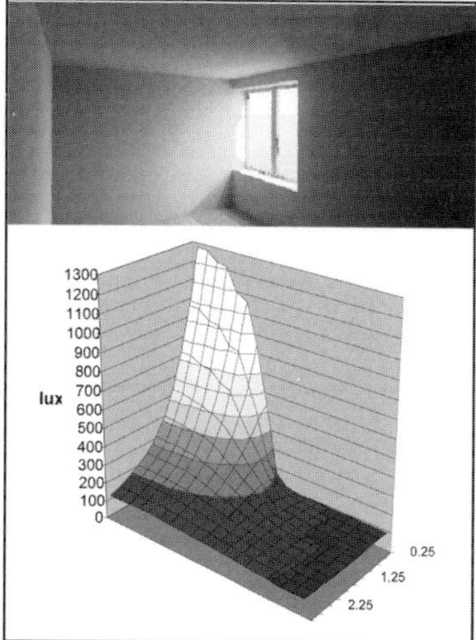

Source: Lars Junghans, AEU Ltd

Figure 2.4.7 *Corner window*

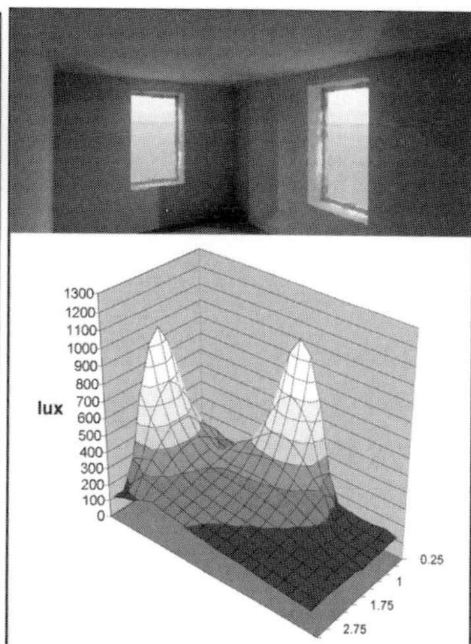

Source: Lars Junghans, AEU Ltd

Figure 2.4.8 *Windows on two sides*

A corner window

A window in one corner strongly illuminates the adjacent wall and corner. This minimizes contrast glare. The other corner, however, is left in the dark. The furniture arrangement and room use have to be adapted to this asymmetry (see Figure 2.4.7).

Windows on two walls

In this case a window is added to the west wall and the window in the south wall is reduced in size so that the total window area is the same as the standard case. This solution yields good daylighting – for example, to read a book in the daytime, for most locations in the room. Because the windows are small there is more glare (see Figure 2.4.8).

2.4.5 Window frame and glazing

The window frame can often 'consume' 30 per cent of the wall opening area. It lets no light in and has a worse U-value than today's high-performance glazings. So, a slim window frame construction improves both daylighting and thermal performance. The solution of insetting the window frame behind the exterior insulation to minimize heat losses is a good solution. A light colour for the room-facing side of the window frame reduces contrast glare.

The multiple coatings of high-performance glazings absorb light. This can be partially offset if a more expensive low iron content glass is specified. Such glass can admit up to 6 per cent more light than conventional glass.

Source: Lars Junghans, AEU Ltd

Figure 2.4.9 *Window flared into the room*

2.4.6 Window-opening cross-section

The cross-section of the opening in the thick walls of highly insulated structures strongly influences daylighting and visual comfort. The preferred position of the window is towards the exterior of the wall opening to avoid thermal bridges. This allows the flaring of the deep window cross-section into the room, as can be seen in the thick masonry walls of chateaus or monasteries.

Various combinations of the window head (top), jambs (sides) and sills (bottom) flared into the room were analysed. The resulting impression is that the window is much larger than it is in reality. Furthermore, visual comfort is greatly improved because the illuminated flared surfaces reduce contrast glare between the wall and window. East- or west-facing windows are special cases. If the south jamb of an east- or west-facing window is flared, the room will receive full sun later in the morning, or respectively earlier in the afternoon. In higher latitudes this benefit is even more evident. The window sill can be horizontal and even project out into the room. By making it a light colour, it can increase the daylight factor at the rear of the room by as much as 10 per cent (Lenzlinger, 1995).

2.4.7 Surfaces

How bright a room is depends very much on how much the room surfaces absorb light. Not all surfaces are equally important in this regard.

Walls

The walls to either side of a window are extremely important, as can be seen in Figure 2.4.10. By changing the colour of the walls from a dark colour with 80 per cent absorptance to white with 15 per cent absorptance, the illuminance increases by 100 lux. If it is preferable to paint one wall a darker colour, the wall on the back side of the room away from the window has only a small influence on the room illuminance. The wall with the window has the least effect on the luminance; but if it is too dark, the bright window will create contrast glare.

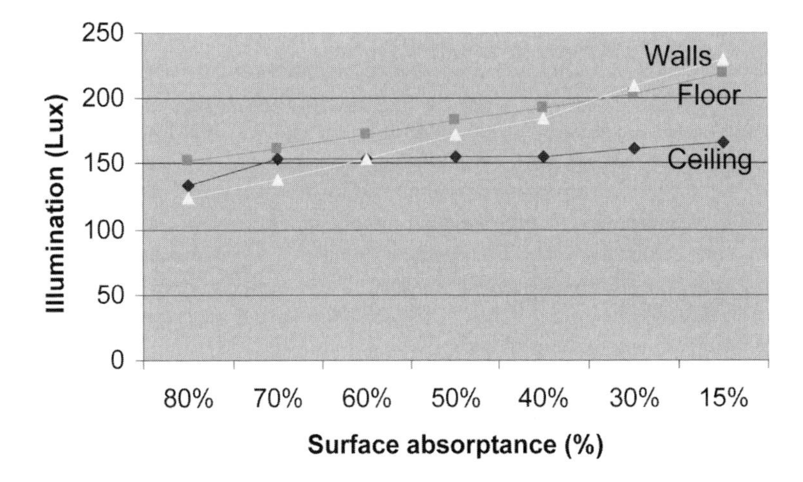

Source: Lars Junghans, AEU Ltd

Figure 2.4.10 *Effect of room surface absorptances on illumination*

Floors

The colour of floors also has a strong influence because much of the light from a window strikes the floor. Although a light-coloured floor makes the space appear brighter, dark floors are often preferred because they appear solid and stable. Furthermore, if the floor is massive – that is, tile, stone or brick – a darker colour will absorb the sunlight and store the heat. A good compromise is a floor colour slightly darker than the walls.

Ceiling

The ceiling colour has a smaller calculated influence than the floor or walls. The computed difference between an absorptance of 70 per cent and 10 per cent is only 15 lux. However, dark ceilings convey an oppressive feeling. In addition, ceilings are the room surface least likely to be furnished, so typically their full surface is available to reflect light.

2.4.8 Light transport

Skylights

In row houses and single family homes, 'skylighting' a room buried in the core of the house creatures an attractive feature. A skylight above the stair is a good example. A sky-lighted bathroom is particularly attractive as daylight and a sky view are provided without compromising privacy.

Tubular skylights

An alternative to a conventional skylight, which may not even be possible if there is a high attic, is a tubular skylight. While the absolute amount of light available is, of course, limited to the aperture area, such a light-transporting tube is highly efficient. The interior surface of the pipe has up to a 98 per cent reflectivity. When the light reaches the end of the pipe, a diffuser glass spreads the light into the room below. Compared to a standard 100 W incandescent bulb, which produces 1200 lumens, a 1.8 m long 250 mm pipe produces up to 4000 lumens. A 530 mm diameter light tube can deliver up to 18000 lumens (Hanley, 2005).

Source: Solatube Global Marketing, Inc., Carlsbad, CA, www.solatube.com

Figure 2.4.11 *A tubular skylight in Geneva*

Light wells

Too often, the only sources of natural light and air for a basement are small, high windows looking into narrow, half-round steel or cement culverts sunk into the ground. Such window wells are ugly and nearly worthless regarding daylighting. Yet, a basement is one of the most expensive spaces of a house to construct and will likely serve many activities that the builder did not anticipate. A small investment in better design can greatly enhance this space. The ideal solution if the basement is totally underground is to grade the earth down to the cellar window. The slope can be landscaped with light-coloured stone or planting to create a feature out of the utilitarian cellar window.

2.4.9 Glare control

Curtains and drapes

Curtains are effective in reducing glare, but they also reduce daylight levels. Short curtains on the upper part of the window impede daylight from the important sky zenith from projecting into the room. For side curtains, the tracks should extend beyond the window to allow the curtains to be pulled completely clear of the window opening.

Conventional Venetian blinds, even with white slats, obstruct much daylight in every tilt angle. Even if the slats are set horizontally, only a minimal amount of light is reflected to the ceiling. The room impression is then of a bright window and a dim room.

Fixed architectural shading, like roof overhangs, are problematic. With an overcast sky, the room is too dark. It can also be frustrating that the direct view to the sky zenith is permanently blocked.

2.4.10 Conclusions

Effective daylighting is important both for the marketability of a home and the quality of life of future homeowners. From detailed analysis with a light simulation program, the following design advice is offered:

- A window-to-south-façade ratio of up to 50 per cent is desirable (deep rooms can profit from a higher percentage and a higher ceiling). Window areas larger than 50 per cent of the façade do not increase the illuminance proportionally.
- The shape of a window affects the perceived and also real light distribution in a space. A tall window, such as a French door, projects light deep into the room and provides a good view from the sky to the ground. A horizontal window offers minimal light penetration into the room depth, but offers an ideal work space. A window in a room corner illuminates the adjacent wall, providing the least glare; but the other room corner is left dark. The ideal situation is where windows on two walls are possible, offering bi-directional and more uniform light distribution.
- A flared cross-section of the window opening in the wall improves both the light distribution and visual comfort. A flared window head improves room illuminance, especially by overcast skies. If the window head and sides are flared into the room, comfort is maximized because of the smaller luminance ratios and the room will appear brighter. A deep, light-coloured window sill also improved the illuminance.
- Light coloured room surfaces greatly increase the illuminance of a room, but some surfaces are more important than others in this regard. Most important are the floor and walls to the sides of a window. Less important in terms of illuminance is the ceiling. It is, however, the room surface most likely to be unobstructed and it strongly affects the perception of the brightness of a room.
- Daylighting spaces in the core of a house, like the stair or a bathroom, is very desirable. Light pipes have proven very effective where skylights are not possible. Improved daylighting of a basement is also very desirable, given the cost of its construction and the many future uses it can serve.
- Curtains should be mounted so that they can be moved fully out of the aperture area of the window. Fixed architectural shading, such as a roof overhang, blocks valuable zenith sky light, which is the strongest light direction in overcast skies.

References

Baker, N., Fanchiotti, A. and Steemers, K. (1993) *Daylighting in Architecture: A European Reference Book*, James and James Ltd, London

Fontoynont, M. (1999) *Daylight Performance of Buildings*, James and James Ltd, London

Hanley, B. (2005) Information from Brett Hanley at Solatube Global Marketing Inc, Carlsbad, CA, www.solatube.com

Lenzlinger, M. (1995) 'Regeln für Gutes Tageslicht', *Diane Projekt Tageslichtnutzung*, no 805.165d, Bern, Germany

Minamihara, M. (2005) *Radiance: Building Technologies Program*, Publications Coordinator, Lawrence Berkeley National Laboratory, Berkeley, CA, mminamihara@lbl.gov, www.radsite.lbl.gov/radiance/HOME.html

2.5 Using active solar energy

Helena Gajbert

DESIGN ADVICE

The amount of energy to heat domestic hot water is a large fraction of the total energy needed in high-performance housing, and an active solar system can easily cover up to 50 per cent of this demand. Here are example rules of thumb from simulations (see also Part II in this volume).

Suitable dimensions for an example solar system in northern or temperate climates:

An apartment building (approximately 1600 m^2, 16 apartments or 48 occupants):

- Tank volume: 3000–5000 l/building or 60–100 l/occupant.
- Collector area: 30–80 m^2/building or 0.6–1.75 m^2/occupant.

A single family house (approximately 150 m^2, 4 occupants):

- Tank volume: 400–600 l/house or 100–150 l/ occupant.
- Collector area: 5–10 m^2/house or 1.3–2.5 m^2/occupant.

Collector slope: For domestic hot water (DHW) systems, approximately 40°–50° slope. For combi-systems, the slope is preferably steeper, although a larger collector area is then required.

Economics: If water-based heat distribution is preferred (i.e. a low temperature floor heating system), a solar combi-system with direct injection of solar heat to the floor heating system is a good approach.

2.5.1 Introduction

Heat captured by an active solar heating system can cover a significant part of the very low energy demand of high-performance houses. Because the space heating demand of such houses is very low, the year-round energy demand for domestic hot water (DHW) becomes relatively important. An active solar system can cover a large part of this energy demand, often more than 50 per cent, since the demand also occurs in summer. The question then is: if there is to be a solar system, would it not be good to enlarge it to also supply some space heating (a 'combi-system')? A key issue, then, is estimating the space heating contribution realistically, given the shorter heating period of high-performance housing.

2.5.2 Solar thermal system designs

If the client desires a water-based heating system, this is a good situation to argue for a solar combi-system. It provides flexibility and there are many interesting system designs on the market, varying by region and the locally used auxiliary energy sources (Weiss, 2003). The storage tank can serve both the DHW and the space heating system, and it can be heated by both the auxiliary and the solar systems, depending on conditions. The system illustrated in Figure 2.5.1 uses a wood pellet boiler and an electric resistance heater as auxiliary energy sources to back up the solar circuit. It is equipped with a device to enhance thermal stratification in the tank. The DHW is heated in an external heat exchanger.

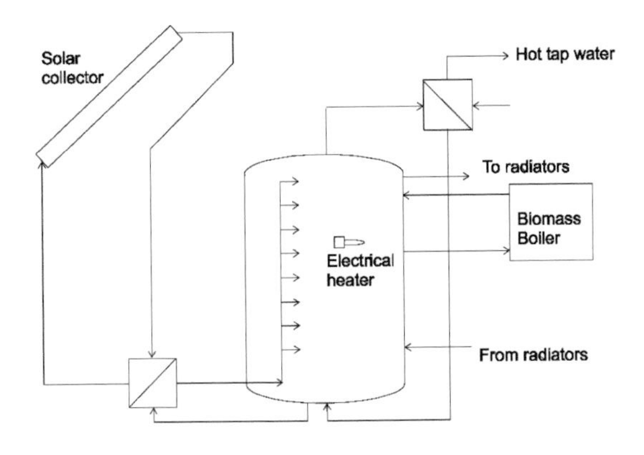

Source: Helena Gajbert, Lund University SE

Figure 2.5.1 *A solar combi-system with a joint storage tank for the domestic hot water (DHW) and space heating systems*

A critical design condition is the case of minimum heat demand by maximum solar gains, as often occurs in summer. To minimize the system overheating in such instances, the solar system should be dimensioned to just cover the summer energy demand, accepting a small solar coverage in winter. Nonetheless, an important advantage is that the combustion backup heating can be shut down during the summer. As a result, short cycling of the back-up system in summer can be eliminated, increasing the life expectancy of the system.

The shorter heating season of a high-performance house results in a more pronounced seasonal mismatch between solar gains and space heating demand than is the case for conventional houses. This is illustrated in Figure 2.5.2, with the case of a reference apartment building in the cold (Stockholm) climate. The DHW demand is here assumed constant over the year, although it is often slightly lower in summer – for example, due to vacations.

The overheating problem could be reduced by installing a larger storage tank than the system design would otherwise require. However, this approach is expensive, takes up more house volume and leads to higher tank heat losses.

There are also good control solutions available – for example, using partial evaporation to empty the collector during stagnation. Active removal of energy transported via the steam during stagnation is also possible with a small volume heat sink. Alternatively, the tank can be cooled down by circulating the heat transfer medium through the collectors at night or by using an air cooler (Hausner and Fink, 2002; Weiss, 2003). However, the more cautious design strategy is to not over-dimension the collector area.

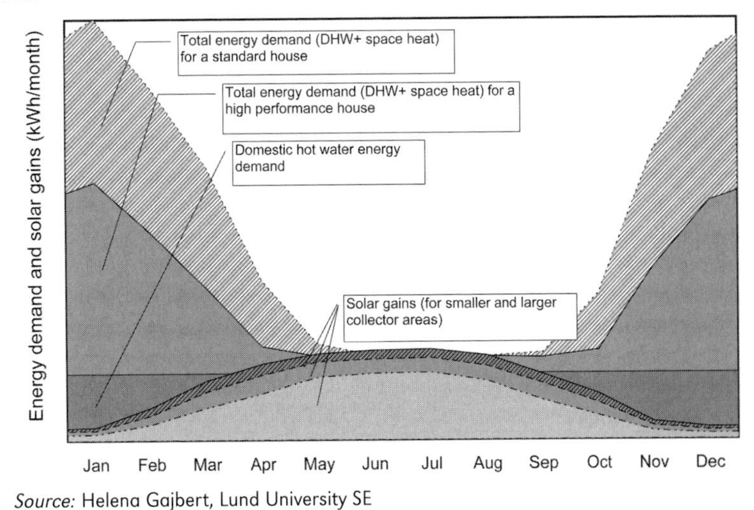

Source: Helena Gajbert, Lund University SE

Figure 2.5.2 *Seasonal variations in solar gains and space heating demand in standard housing versus high-performance housing*

An advantage of a combi-system is that during the heating season, when energy for space heating is withdrawn from the tank, the operating temperature in the collectors can be lowered, increasing the collector efficiency. This typically occurs in autumn and spring. This gain in efficiency can be maximized if the solar collectors supply heat directly to low temperature surface heating, as shown in Figure 2.5.3. This requires a good control strategy. Such systems are common in France, Denmark and Germany (Weiss, 2003).

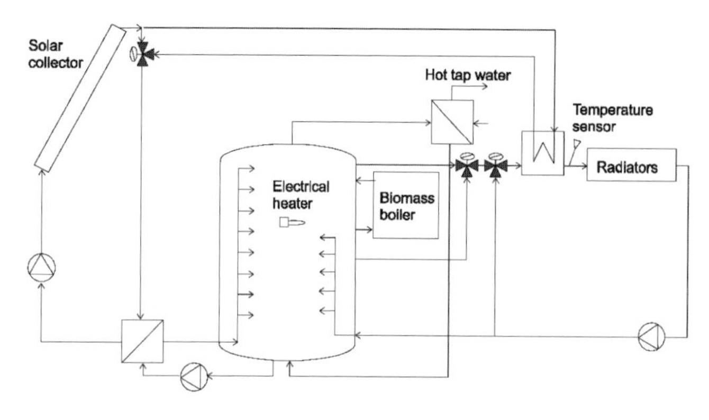

Source: Helena Gajbert, Lund University SE

Figure 2.5.3 *A solar combi-system with the possibility of delivering solar heat directly to the heating system without passing through the tank first*

One way to better balance the solar gains to the heating demand is to mount the collectors vertically on the façade or to mount them with a high tilt. This positioning makes better use of low winter sun angles and suppresses summer overheating. This results in a more even distribution of solar gains over the year; but a larger collector area is required to cover the summer domestic hot water demand. Figure 2.5.4 shows how the solar fraction increases with increased collector area for collectors tilted 40° and 90°. The results are from Polysun simulations of a flat-plate combi-system coupled to a single family house in the cold climate.

A larger collector area is possible without risking overheating if the collector tilt angle is increased in relation to the optimal tilt angle, which approximately lies between 35° in southern latitudes (latitudes around 45°) and 50° in northern latitudes (latitudes around 60°). The suitable collector area to cover 95 per cent of the summer hot water heating demand with minimal system overheating is illustrated in Figure 2.5.5. The results are from Polysun simulations of a flat-plate combi-system coupled to an apartment building in the cold climate.

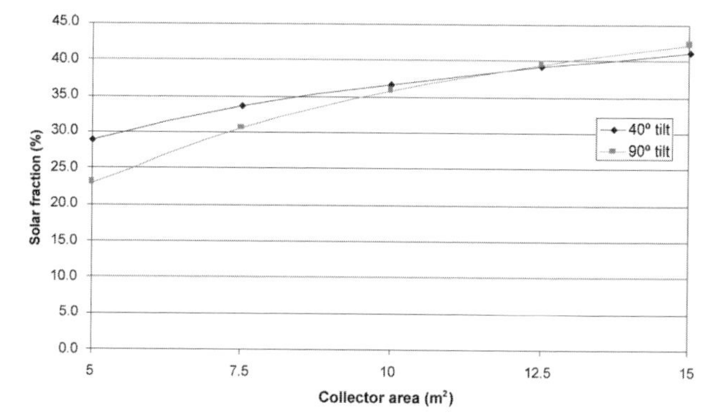

Source: T. Boström

Figure 2.5.4 *Effect of collector tilt and area on solar fraction*

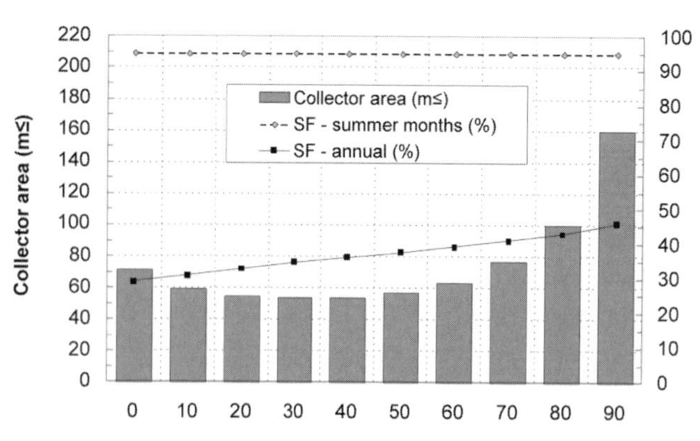

Source: Helena Gajbert, Lund University SE

Figure 2.5.5 *Suitable collector areas at different tilt angles for a collector dimensioned to cover 95% of the summer demand; solar fractions for the year and for the summer are also shown*

Where houses are closely spaced, a micro-heating grid with a central solar system may be plausible. Collectors on the houses then can share a large semi-seasonal solar storage tank enabling some summer solar gain excess to be stored for the autumn. The large tank volume has an advantageous surface-to-volume ratio reducing tank heat loss. A shared system can also even out peak demands or supply excesses. Disadvantages are the relatively high heat losses of the pipe grid relative to the very small space heating demand, and the investment and administrative costs for meeting such a small demand. Such micro-heat distribution grids for row houses have proven effective in demonstration projects in Germany (Russ, 2005).

2.5.3 Collector types and placement

There are many reliable and efficient collectors and types on the market today.

Flat-plate collectors have been highly optimized with selective coated absorbers, low iron anti-reflective coated glazing, durable frames and gaskets, and effective back insulation.

Evacuated tube collectors, with their extremely good insulation, can operate at higher temperatures with only very small heat losses. Because the losses are so small, evacuated tube collectors can deliver heat even by very low irradiance levels occurring in winter and during overcast periods. They are thus very suitable for high-performance houses with their shortened heating season. The individual tubes can be rotated to an optimal absorber tilt angle, allowing great flexibility in where they are placed. However, the separation between tubes must avoid one shadowing the next. Heat pipe collectors need to be mounted with a slight tilt to allow for the gravity flow. A critical issue is snow retention on the tubes. A façade-mounted system minimizes this problem (Kovacs and Pettersson, 2002).

Concentrating flat-plate collectors can also be envisioned, although such systems are not commercially available for housing. Prototype systems have been built and tested – for example, in Sweden, the UK and Australia. A low-cost reflector increases the irradiance on the absorber. The system geometry can be optimized for spring and autumn solar angles. While this suppresses the summer performance, the solar gains are better balanced to the heating demand over the year. Using an evacuated tube collector as the target of the reflector is a promising solution.

Solar air collectors, while interesting for conventional houses or renovations, are difficult to justify economically for high-performance houses. The heat exchanger of the ventilation system competes with the solar air collectors. One system or the other must operate with decreased efficiency. A plausible configuration would be an entire roof slope covered with a low cost sheet-metal solar air collector. During summer an air-to-water heat exchanger should easily cover all or most of the water heating energy demand. In the autumn and spring, the system could supply much of the space heating. Pre-cast hollow-core concrete floor planks or large core masonry walls can also serve for heat storage and distribution (Morhenne, 2000).

2.5.4 Tank location

The tank location is an important design decision. On the one hand, it is sensible to locate it inside the insulated envelope of the building. Heat losses from even a very well-insulated tank are therefore not lost to the ambient, but remain within the house (although this is a disadvantage in summer). Short pipe runs keep circulation heat losses small. A good location for the tank could be in or next to the bathroom. The warm tank surface improves the comfort in the bathroom. Comfort here is critical because this is often where exhaust air for the whole living space is extracted, and the occupant may be feeling cold after showering. Also, the pipe run is minimized between the tank and one of the biggest points of demand. This decreases pipe heat losses and offers the convenience of immediate hot water at the faucet. An upper floor bathroom is ideal, as the pipe run to roof collectors is also minimized. However, the insulated building interior is a very expensive volume to sacrifice for a tank, so the basement may be the preferred location. This otherwise cold space can then be slightly tempered, improving comfort for the many uses found for a basement. Pipe runs from the tank should be within the heated house envelope as much as possible.

2.5.5 Regional design differences

It is easiest to market a solar combi-system in regions where homeowners prefer hot water radiant heating, particularly low temperature systems. Regions where oil or wood pellet heating is preferred have proven to be good markets for combi-systems, often even large systems. Solar combi-systems have been widely sold in Germany, Austria, Switzerland, Sweden, Denmark and Norway. It seems more difficult to market combi-systems in regions where mainly direct electricity or gas is used for heating – for example, in The Netherlands (Weiss, 2003).

2.5.6 Conclusions

A solar thermal combi-system can be a valuable component of sustainable housing. The dimensioning and economics of the system have to reflect the very low space heating demand and short heating season. For a high-performance house, a solar domestic hot water system is more easily argued than a solar combi-system, given the more constant demand for hot water over the entire year and the cost of a water-based radiant heat distribution system. If the client desires a radiant heating system, a combi-system is suitable. The DHW and the space heating system can then share a storage tank, which can be heated both by the solar collector and by the auxiliary heat sources. In such an application, a higher collector tilt improves the balance between the space heating demand and seasonal solar gains, although a larger collector area is required to cover the summer demand. Façade-integrated collectors are very suitable and make the system less prone to summer overheating. Capturing and using energy from the sun is a very natural and marketable concept for sustainable low energy housing.

References

Hausner, R. and Fink, C. (2002) *Stagnation Behaviour of Solar Thermal Systems: A Report of IEA SHC – Task 26*, AEE INTEC, Austria, www.iea-shc.org/

Kovacs, P. and Pettersson, U. (2002) *Solvärmda Kombisystem: En Jämförelse Mellan Vakuumrör och Plan Solfångare Genom Mätning och Simulering*, SP rapport 200220, SP Swedish National Testing and Research Institute, Borås, Sweden

Morhenne, J. (2000) 'Controls', in *Solar Air Systems: A Design Handbook*, James and James Ltd, London, Chapter 5

Russ, C. (2005) *Demonstration Buildings: Design, Monitoring and Evaluation*, IEA, Task 28, Subtask D, www.iea-shc.org.

Weiss, W. (ed) (2003) *Solar Heating Systems for Houses: A Design Handbook for Solar Combisystems*, IEA Task 26, James and James Ltd, London,

2.6 Producing remaining energy efficiently

S. Robert Hastings

2.6.1 Introduction

How the remaining energy needed by a high-performance house should be supplied is an important issue. Remaining energy refers to the energy still needed after conservation measures and optionally active solar heat or photovoltaic electric production.

This section provides an overview of a sample of systems and their appropriateness for high-performance housing. Not included are solar systems that were mentioned in section 2.5 of this chapter, which require a very large investment or are very location dependent (such as geothermal systems), or systems not yet readily available (such as fuel cells). More details on the systems in this overview are provided in the companion volume to this book: *Sustainable Solar Housing: Exemplary Buildings and Technologies, Volume II.*

A key selection criterion for the heat-producing system is that the energy supply source should have minimal fixed costs. To take the example of gas, while condensing gas furnaces can achieve a nominal efficiency over 100 per cent, the very small quantity of gas consumed results in the fixed costs exceeding the energy costs, in the following example (Table 2.6.1), by 235 per cent!

DESIGN ADVICE

Important characteristics of the equipment supplying the remaining energy for high-performance housing include:

- *Simplicity*: because so little energy is needed and the required heating power is very low, the system investment cost should reflect this. Simple systems are also more likely to be reliable and less prone to being falsely set during installation or by the occupants.
- *Fast reacting*: because high-performance housing reacts very quickly to passive solar or occupant heat gains, the heating system should be fast responding, not continuing to supply heat when no more heat is required (i.e. plate radiators and piping with minimal water content, or radiant surface heating with minimal mass).
- *Well insulated*: the heat-generating equipment should be well insulated to minimize start-up heat losses, tanks well insulated to minimize standing heat losses and pipes, and valves, etc. well insulated to minimize circulation losses.
- *Less is more*: if a tank can be avoided (by using a flow-through heater) and pipe and duct runs kept short, investment costs are reduced, heat losses minimized and less parasitic power is needed to move the water or air.
- *Low parasitic power*: electricity consumed by fans supplying combustion air and expelling flue gas and pumps or fans for circulating the produced heat – given the many hours of operation and the primary energy factor of electricity – comprise a significant fraction of the total primary energy of such housing. A 60 W circulation pump running 12 hours a day adds over 3 kWh/m^{2}a of primary energy for a 1200 m^2 row house. The primary energy used for the ventilation system, de-icing of the heat exchanger, circulation pumps, fuel injection and operation of the solar system can approach 9 to 10 kWh/m^{2}a (Feist, 2002).
- *Low base costs*: because so little energy is needed for space heating, a supply source with minimal fixed costs is preferable, unless the heating plant can be shared by several households.

Table 2.6.1 *Example costs for heating a high-performance house with gas*

Assumptions	
Heated floor area	150 m^2
Heating energy	15 kWh/m^{2}a
Gas furnace efficiency	98%
Total space heating	2295 kWh/a
Winer gas price (2005)	€0.024/kWh
Variable costs for gas	
Variable cost	€55
Fixed costs	
Base price	€5.00
Meter fee/year	€70.00
Flue gas check	€20.00
Service subscription	€40.00
Total	€135.00
Total price	**€190**
Variable costs	29%
Fixed costs	71%

2.6.2 Combustion systems

Fossil-fuel combustion systems

Given the very small heating capacity needed by high-performance housing, a gas or oil-fired heating plant should ideally be shared by a row of houses. For a single house, the fixed costs (connect charge, meter reading, chimney cleaning and combustion gas checks) will be disproportionate to the actual cost of consumed gas or oil.

Today, condensing gas plants (but now also oil-heating plants) are common. By recovered heat from the flue gases, the annual system efficiency can be increased. Gas systems achieve a greater benefit from the condensation heat recovery than oil (approximately 11 per cent versus 6 per cent), and the condensation is less acidic, so in many locations it can be directly discharged into the house drain. The advantage of oil is that there is a reserve in the house, versus gas, which depends completely on the uninterrupted supply from the utility.

Biomass combustion systems

This term can include anything from gas out of sewage sludge to sunflower oil to wood. The discussion here addresses only the latter. Although wood embodies only about half the energy per kg (approximately 5 kWh/kg) as oil, it is appealing because it is 'CO_2 neutral'. The CO_2 from combustion originates from CO_2 the tree extracted from the atmosphere. Less often stated is that wood combustion also emits fine particulates. This contributes to the serious air pollution problem that can occur in built-up areas by temperature inversions.

Three types of wood combustion are common:

1 Wood stoves are now available with carefully regulated combustion air supply, well-engineered heat exchangers (to air or water) and with a slow burn (i.e. vertical placement of only two to three pieces of firewood) appropriate to the small heating power requirement of high-performance housing (see Figure 2.6.1). The stove can be in the living space, with a window to add 'atmosphere' to the room. But these stoves must be hand fed and ashes manually removed. In mechanically ventilated air-tight housing, the fire box must be gas tight and 100 per cent supplied by outside air. Such high-performance wood stoves have a high price for the energy they produce, but other benefits make them popular.

2 Pellet stoves have the big advantage that they are fully automated and the heating power output can be regulated to match the demand. The stove can also be located in the living space with the advantage that heat radiated and convected from the stove – even if well insulated – is still within the heated volume of the house and not lost. A disadvantage is that the ignition is electric, so frequent cycling should be avoided by a rational control strategy.

Source: TOPOLINO, www.twlag.ch

Figure 2.6.1 *A high-efficiency woodstove*

3 Wood chip furnaces have been popular for larger projects (i.e. apartment buildings or central systems for whole neighbourhoods with district heating). The disadvantage is obviously the volume, which must be transported and stored compared to pellets of compressed waste wood or firewood.

Combined heat and power systems

The heat produced by combustion can be used to generate electricity. For example, gas or propane can fuel an internal combustion engine that drives an electric generator, and the 'waste' from the motor is then used for space heating and water heating. Disadvantages are maintenance costs, the need to isolate noise and vibration and high investment costs. Finally, the system should serve at least 20 housing units to be sensible.

Source: Biotech Hugler AG, www.huggler-technik.ch

Figure 2.6.2 *A wood pellet central heating system*

2.6.3 Heat pumps

Ambient air heat source

Given the primary energy factor of electricity, here assumed to be 2.354, reflecting the non-renewable energy needed in electricity production, on average, across Europe, a heat pump makes sense if its coefficient of performance (COP) is greater than this. Typically, heat pumps using ambient air as a heat source have a COP in the order of 2.5 to 2.75, so other heat sources for a heat pump are more attractive.

Ground/surface water heat source

The seasonal storage effect of the ground at depths of 1.5 m to 2 m ensures that this heat source will be a constant temperature of approximately the average annual temperature. This greatly improves the COP of a heat pump.

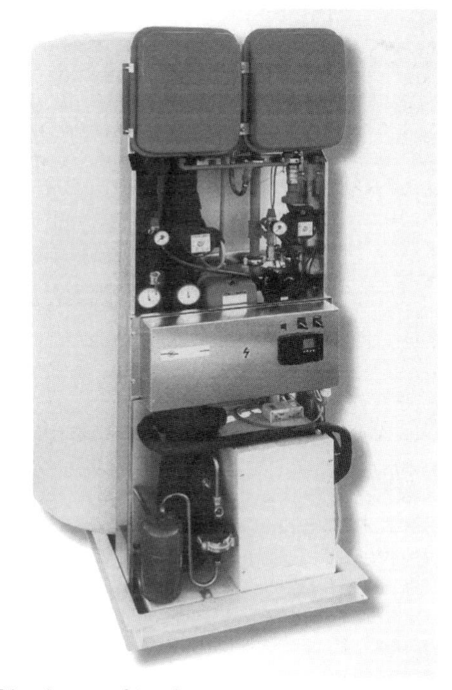

Source: Friap Ag, www.friap.ch

Figure 2.6.3 *A compact heat pump-combined heating water and ventilation system*

Exhaust air heat source

So-called 'compact heating, ventilation and hot water systems' use the air exhausted from the ventilation heat exchanger as a heat source (see Figure 2.6.3). As a result, a COP of 3 or better is possible; but it is important that the temperature demand of the system not be too high (< 60°C). A typical system has a grid connection of only 500 to 700 W, which can supply 1.5 to 2.0 kW of heating power. To prevent the heat exchanger from freezing up, the supply air is preheated in a pipe buried 1.5 m to 2 m underground.

2.6.4 Direct electric heating

Given the very minimal amount of heating needed by high-performance houses, it is tempting to consider using direct electrical heating. For space heating, base board electric heating allows individual room controls. For domestic water heating, a simple electric element is all that is needed in the storage tank. Achieving a very low primary energy consumption is, however, difficult with this solution unless a very beneficial primary energy factor for electricity is used reflecting a high fraction of generation by renewable energy. This argument is weakened today, given the international connection of electrical grids.

2.6.5 Conclusions

The key issue in selecting the system to produce the needed remaining energy is probably not the cost of the energy source – whether it is gas, oil or wood. Rather, it is the total cost, including the fixed costs of connection and administrative fees from the utility and the cost of amortizing the investment in the equipment. Ideal, and most economical, is a solution that requires the connection of one type of energy carrier. Since electricity is needed, in any case, in a house, the logical conclusion is then the ideal choice – for example, a heat pump. The high primary energy (and exergy) value of electricity may make other alternatives, such as biomass, more attractive. System performance and efficiency improve if the heat can be delivered at a lower temperature – as is possible with radiant surface heating. This must ideally be achieved, however, in a manner that allows the heat delivery to be quickly cut off when passive solar and internal gains have already achieved the needed room temperature.

References

Feist, W. (1998) *Das Niedrigenergiehaus – Neuer Standarf für Energiebewusstes Bauen*, C. F. Müller Publishers, Heidelberg, Germany

Feist, W. (ed) (2000) *Arbeitskreis Kostengünstige Passivhäuser, Protokollband no 6, Haustechnik im Passivhaus, Protokollband no 20 Passivhaus*, Versorgungstechnik Passivhaus Institut, Darmstadt, Germany, www.passiv.de

Feist, W., Pfluger, R., Kaufmann, B., Schnieders, J. and Kah, O. (2005) *Passivhaus Projektierungs Paket 2004*, Passivhaus Institut, Darmstadt, Germany

Hoffmann, C., Hastings, R. and Voss, K. (2005) *Wohnbauten mit geringem Energieverbrauch – 12 Gebäude Planung, umsetzung und Realität*, C. F. Müller Publishers, Heidelberg, Germany

3

Ecology

3.1 Introduction

Carsten Petersdorff

In houses consuming very little energy to maintain comfort, the energy needed to build the houses represents an important part of the energy consumed over the whole life cycle of the building. Thus, it can happen that in the path to reducing energy over the house life cycle, at some point selecting a construction or component with very little embodied energy may be more effective than a measure which reduces heating energy. Another important factor to consider in selecting components is the flow of materials that occurred in their production. This includes extracting raw material, processing, fabrication and installation, through to demolition and disposal/recycling. The question quickly arises: what is sustainable?

In 1987, Dr Gro Harlem Brundtland, the former prime minister of Norway, as chair of the World Commission on Environment and Development, introduced us to the word 'sustainability' in her report, *Our Common Future*, (WCED, 1987) to the United Nations. The report, known as the Brundtland Report, defines sustainability as 'development that meets the needs of the present generation without compromising the ability of future generations to meet their own needs'.

Generally, sustainability is interpreted as addressing *economic, environmental and social aspects* all together. Depending on the background of the researchers, the emphasis varies: economists tend to focus on sustainable economic development; ecologists emphasize sustainable interactions with natural systems; while sociologists focus on quality-of-life issues.

This volume presents four selected methods to assess the sustainability of buildings:

1. Cumulative energy demand (CED) includes the whole life cycle of buildings (production, construction, use, demolition and disposal). It focuses on quantitative results for the used energy over the whole life cycle.
2. Life-cycle analysis (LCA) focuses on ecological aspects (such as pollution).
3. Total quality assessment (TQA) addresses ecological aspects, as well as economical and social aspects, and quantifies their influences.
4. The methodology of architecture towards sustainability (ATS) explores how a building performs in environmental, social, economic and political dimensions, or 'the milieu', in order to improve quality of life.

Numbers 1, 2 and 4 are examined in this chapter, while TQA is covered in Chapter 5. Not all methods address all three aspects (ecology, economy and social aspects of buildings) equally well. The methods tend to have a particular focus (see Figure 3.1.1). The scope, required input and type of results for

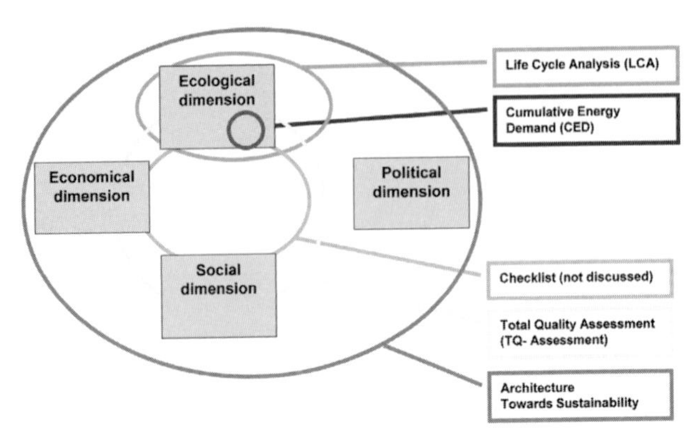

Source: Kristel de Myttenaere, University Catholique de Louvain

Figure 3.1.1 *Aspects covered by the different methodologies*

Table 3.1.1 *Characteristics of the different methods presented*

Tool	Scope	Data need	Results	Possible user	Potential	Limitations
Cumulative energy demand (CED)	Ecological assessment over the whole life cycle of buildings, considering only the cumulative energy demand. No time and site-specific assessment.	Inventory data of all building components and energy use. Cumulative energy factors for all components and energy systems.	Quantitative.	Specialist.	Quantitative comparison possible. Base for energetic optimization of buildings. Information on energetic impacts.	Only energetic aspects included. Different databases and boundary conditions.
Life-cycle analysis (LCA)	Ecological assessment over the whole life cycle of buildings. Environmental impacts on a global, regional and local level. No time and site-specific assessment.	Inventory data of all building components and energy use. Impact assessment factors for all components and energy systems.	Quantitative. Interpretation and sensitivity analysis of the results are very important.	Specialist.	Quantitative comparison possible. Base for optimization of buildings. Information on various ecological impacts.	Lack of consistent databases. Complexity limits the inclusion of aspects. Uncertainties in early stage of design process. Only ecological aspects considered.
Total quality assessment (TQA)	Quantitative indicators for different criteria of sustainability in buildings, including ecological, social and financial aspects.	Building data used for design and quality control (materials, energy). Material inventories and weighting factor are included in the calculations.	Quantitative results for nine different aspects. Quality certificate for building. Documentation for building owner.	Architect and planner.	Quantitative comparison possible. Applicable for planning phase and prior to handing over of the building. Inclusion of different aspects of sustainability.	User behaviour is not included. Some ecological impacts omitted due to missing experience.
Architecture towards sustainability (ATS)	Qualitative integration into particularities of the milieu. Environmental, social, economical and political dimensions.	–	–	Architect and planner.	–	–

the different methods are presented in Table 3.1.1. The list is not comprehensive; the methods discussed are those which were used in this International Energy Agency (IEA) work. For a comprehensive overview of methods or tools see:

- IEA BCS Annex 31: 'Energy Related Environmental Impact of Buildings' (www.uni-weimar.de/scc/PRO/TOOLS/index.html); and
- the project, Greening the Building Life Cycle, of Environment Australia (www.buildlca.rmit.edu.au/matrix.htm).

References

WCED (World Commission on Environment and Development) (1987) *Our Common Future*, The Brundtland Report, Oxford University Press, Oxford

3.2 Cumulative energy demand (CED)

Carsten Petersdorff

3.2.1 Introduction

In the design or building permit process, energy consumption for one year of operation of the building is typically estimated. Due to improved insulation and the subsequent reduction in energy needed for heating, the proportion of energy consumed during the building process becomes more important than by conventional buildings and should therefore also be considered.

Cumulative energy demand (CED) is the entire primary energy demand over the whole life cycle of a product or a service. It is a good indicator to assess the ecological balance of a building because it is the sum of energy used both in producing materials and components, as well as in their operation over their lifespan.

The idea behind cumulative energy is that every technical process needs energy. The supply of this energy is linked to different environmental issues. Therefore, the amount of energy used reflects the environmental impact of a product. This allows a fair comparison of products and services with respect to energy over their life cycle. For comparability among forms of energy, whether they are thermal, mechanical, chemical or electrical, all forms of energy are converted into the unit of primary energy.

In assessing the CED, the following parameters are considered:

- amount of materials used;
- energy used in the processing of the materials;
- service life of the materials;
- energy required over the lifetime of the materials through to the disposal or recycling of parts, components or materials; and
- emissions related to energy conversions during production, use and disposal.

3.2.2 Overview and boundaries

To quantify the CED, a standardized unit is used according to the VDI guideline 4600 (VDI 1997). This guideline provides a useful basis for prioritizing energy-saving opportunities throughout design, production, use and disposal.

The CED states the entire primary energy demand during the production (CEDP), the use (CED_U) and the disposal (CED_D) of the product:

$$CED = CED_P + CED_U + CED_D \quad [3.1]$$

The boundary for calculating the CED of a product extends from the raw material at its place of origin to its final disposal or storage. Calculating the CED requires an unambiguous definition of these boundaries and a quantification of the material and energy flows crossing them. The boundaries are set according to local, time and technological criteria and are often interdependent. This makes the analysis complex.

3.2.3 General remarks on the methodology of balancing

To balance the CED of a product or a service, either the whole process chain is investigated, or input and output analyses are used:

- The input–output analysis usually relies on national economic statistics and energy-use data. It is a technique used in economics for tracing resources and products within an economy. Producers and consumers are divided into different branches, which are defined in terms of the resources they require as inputs and what they produce as outputs. The quantities of input and output are usually expressed in monetary terms. However, because of the high degree of aggregation and the reliance on monetary values, this method is not directly suited to determine the CED of a product. Nevertheless, due to the inventory of energy, materials and products as input and output, it allows estimations on the cumulative energy demand of a product.
- The whole process chain (see Figure 3.2.1) can be analysed as micro- or macro-analysis. In macro-analysis, data on energy use of a higher degree of aggregation are processed (i.e. at least of entire plant groups) and therefore usually only provides a blurred picture. In micro-analysis, the data show little or no aggregation. Here, the production sequence is broken down into individual processes and examined accordingly. This method, however, is only feasible if large amounts of data are to be processed. To determine the CED, a successive approximation by the combination of micro- and macro-analyses is recommended.

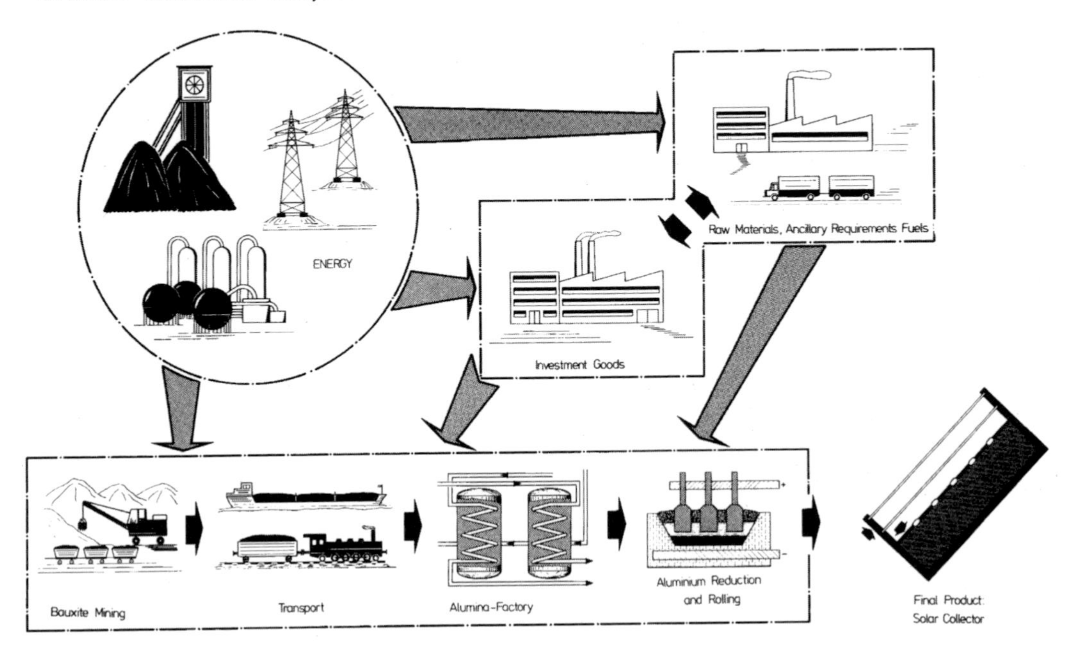

Source: VDI (1997), http://www.vdi.de

Figure 3.2.1 *Example of a process chain*

3.2.4 Building specific aspects

The CED can be applied at different scales for a building project:

- For a given building material, the production and disposal phase can be assessed. Product-specific values, taken from databases, are helpful in choosing materials.
- Functional elements (for example, 1 m^2 exterior wall with specific characteristics such as a defined U-value) can be assessed by the CED values to help in planning decisions. Not only the average demands for production, but also secondary aspects such as transport or expanses during construction, are considered. The lifetime and maintenance for the elements are inputs for this assessment.
- The whole building assessment requires careful selection of the criteria to be used. Are functional units (per square metre living area) or lifestyle aspects (per dwelling, per person) more relevant? Results typically include energy demand for construction and demolition, energy demand during use for heating and DHW, and energy demand for maintenance.
- Urban planning can also make use of the CED and can assess both the buildings and their infrastructure.

3.2.5 Potential and limitations

It is not always possible to determine the CED without gaps for all parallel stages of the process chain trees. Process chains are often complex and the energy relevance of individual parts is often of subordinate significance. Because of this inevitable uncertainty, it is important to indicate the gaps in the results, the criteria used to set the boundaries and the method of analysis used to assess the process chains. The assumed service life of the buildings and components often differs widely, so these assumptions must also be reported.

The CED itself is a comparably easy method to calculate the environmental impact of buildings and to give an overview of critical points in the lifetime of buildings. Results of a CED evaluation are comparable to other evaluations.

3.2.6 Recommendations

General guidelines for conducting a CED are as follows:

- The energy demand for space heating and DHW dominates the CED over the life of standard buildings. In the case of very low energy-demand buildings, the energy used in the operation of the building over its lifetime is so reduced that, indeed, energy used for construction is of the same magnitude. Hence, if the goal is to reduce energy consumption over the building lifetime, the selection of certain components for the construction may be just as effective as the energy performance of systems.
- In the construction and maintenance phase, the building envelope dominates energy demand. Compact buildings (for example, apartment buildings) reduce the building envelope and therefore the energy demand for construction, as well as for heating and cooling.
- Houses in light construction generally have a lower cumulative energy demand for construction and maintenance. Within lightweight constructions, massive wood has a 25 per cent lower value than plywood. Treatment of the wood should not cause unwelcome emissions. For massive buildings, masonry units baked at low temperature (for example, limestone) show the best values. Usually hard units (for example, bricks) need more energy for production.
- The use of reinforced concrete should be minimized because both the concrete and reinforcing steel are energy intensive to produce and dispose of at the end of the building life.
- The level of building insulation yields a positive effect even taken as far as the Passivhaus standard. Insulation produced out of recycling material (for example, cellulose) yields the best values.

- The gas used in the gap of insulating gas (for example, krypton) strongly influences the construction energy balance; but the overall benefit over the building lifetime is positive.
- It should be considered whether energy-intensive constructions can be replaced by light components (for example, a garden shed instead of a concrete basement or a carport instead of a full garage).
- Last but not least, choosing components with a long lifespan may lead to large energy savings over the lifetime of the building.

References

Boermans, T. and Petersdorff, C. (2003) 'KEA als Entscheidungsparameter in Solarsiedlungen – Analyse der Solarsiedlung Koldenfeld' in *Rahmen des Projektes Anwendung und Kommunikation des kumulierten Energieaufwandes (KEA) als praktikabler Entscheidungsindikator für nachhaltige Produkte und Dienstleistungen hinsichtlich Reduzierung des Ressource und Energieverbrauchs*, Ecofys, Köln, Germany

Fritsche, U. R. (2003) *Global Emission Model for Integrated Systems*, Manual, Ökoinstitut, Darmstadt, Germany

VDI (1997) *Kumulierter Energieaufwand: Begriffe, Definitionen, Berechnungsmethoden (Cumulative Energy Demand: Terms, Definitions, Methods of Calculation)*, VDI-Richtlinie 4600, VDI-Gesellschaft Energietechnik, Düsseldorf, Germany

Wagner, H.J., Schuchardt, R., Siraki, K., Petersdorff, C. and Boermans, T. (2002) *Ökologische Bewertung im Gebäudebereich – KEA: Untersuchung der Solarsiedlung Gelsenkirchen*, Endbericht, AG Solar NRW, Universität GH Essen, Essen, Germany

WCED (World Commission on Environment and Development) (1987) *Our Common Future*, The Brundtland Report, Oxford University Press, Oxford

3.3 Life-cycle analysis (LCA)

Alex Primas

3.3.1 Introduction

Life-cycle analysis (LCA) is increasingly used to assess the environmental impact of buildings. As the name indicates, LCA studies the whole life cycle of a product or a process – in our case, a building – from raw material extraction, production and the use phase, to disposal and/or deconstruction of the product. For the different stages of the life cycle, an inventory is made of the energy and material consumption, and of the emission into the environment. This makes it possible to identify products or processes where environmental aspects can be improved. Due to the variety of environmental impacts within building processes, it is impossible to give as a result one simple number. The results need to be interpreted with regard to the uncertainty of the data and method used.

3.3.2 Overview

An LCA is structured in three steps:

1. definition of the goal and scope;
2. inventory and life-cycle impact assessment; and
3. interpretation of results (see Figure 3.3.1).

These steps are executed in an iterative process. Changes in the first phase may be necessary after results are known from the last phase, so it is usually an iterative process.

The LCA process is, to a large extent, specified and defined by:

- The Society of Environmental Toxicology and Chemistry (SETAC) (Consoli et al, 1993);
- CML (Heijungs et al, 1992);
- the *Nordic Guidelines on Life-Cycle Assessment* (Lindfors et al, 1995); and
- the International Organization for Standardization (ISO,) which standardized the framework within the series ISO 14040.

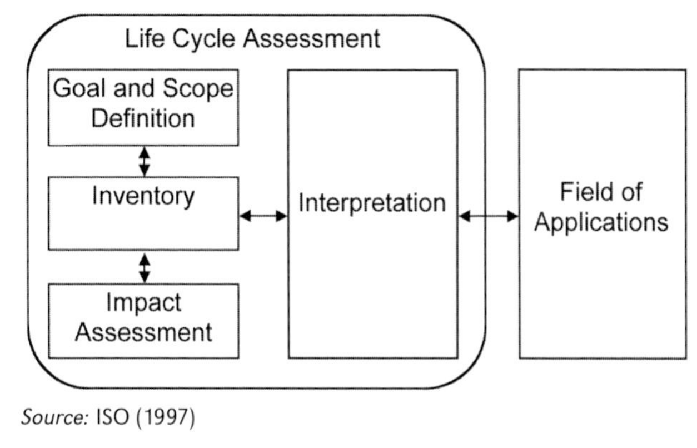

Source: ISO (1997)

Figure 3.3.1 *Structure of a life-cycle assessment*

Definition of goal and scope

This step includes defining the goal and scope of the study, the life cycle of the product, the system boundaries, the functional unit, and required data quality. The functional unit has to be clearly identified and measurable since it is the basis of comparing alternatives.

Life-cycle inventory (LCI)

Inventory analysis consists of data collection and calculation procedures to quantify the inputs and outputs (in physical units) of a product system. Besides used resources and emissions, these data include products or processes employed (for example, steel, electricity or transports). Inventory data have to be related to reference flows for each unit process in order to quantify and normalize input and output in terms of the functional unit chosen for the study. Inventory data for building materials, building processes and energy-supplying processes have to share common system boundaries to ensure the consistency of the data base.

In various cases a process results in different products. Allocation splits up the environmental loads to the different products. The procedure of allocation is described in ISO (1997); but no methodology to define the allocation key can be preferred definitely. In some cases, it is also possible to avoid the allocation step by the expansion of the system boundaries.

The inventory table consists of a huge number of data of used resources, from nature (as oil, ore or land use) and emissions to nature (to air, water and soil) in different physical units.

Life-cycle impact assessment (LCIA)

The large amount of inventory data can hardly be interpreted; thus, data must be aggregated. The effects of the resource use and generated emissions are grouped and quantified to a limited number of impact categories, which may then be weighted for importance. Heijungs (1992) introduced an impact-oriented classification of inventory data in five steps:

1. Definition of the impact categories (for example, greenhouse gases; non-renewable resources, etc.).
2. Classification: in the classification step, all substances are sorted into classes according to the effect that they have on the environment. Certain substances are included in more than one class.
3. Characterization: not all substances have the same affect on the environment, so they have to be weighted in relation to a reference substance.
4. Normalization: each effect is benchmarked against the known total effect for this class.
5. Evaluation: the different normalized effect scores are multiplied by a weighting factor representing the relative importance of the effect.

There are different methods to weight the environmental impacts. Eco-indicator 99 (Goedkoop and Spriensma, 2001) offers many aggregating methods where all of the environmental impacts are weighted. The result is one number that indicates the total environmental impact. Three types of environmental damages are differentiated within Eco-indicator 99: human health, ecosystem quality and depletion of resources.

There are different LCA-based tools to evaluate the ecological performance of buildings. A list of the different instruments and tools with a description of their application and performance is presented in IEA (1999) and RMIT (2001).

Interpretation

After each step of an LCA, the data can be interpreted. This may include sensitivity analysis or a systematic evaluation of opportunities to reduce the environmental impact of the building life cycle.

Result

The ecological impact of the studied process or product on a global scale is reported. The results can help to obtain an ecological optimal solution, while also showing the impact of different solutions on the ecosystem.

3.3.3 Building specific aspects

The functional unit for the LCA of a building may be residents, apartments, m^2 usable or m^2 heated floor area. If the scale for the comparison is the urban structure, the unit 'resident' or 'apartment' may be more appropriate than 'area'. On the other hand, these units make a comparison between single buildings difficult. If m^2 usable floor area is used as the functional unit, the comparison prioritizes the differences in material use. If m^2 heated floor area is used as the functional unit, the comparison prioritizes the differences in operation energy use. In this book the functional unit is defined as m^2 heated floor area (interior dimensions of all room space or the equivalent if all floor area were carpeted) and years of operation, which includes assumptions on the lifetime of the building.

For long-lived products, such as a building, the life cycle incorporates assumptions about the functional service lifetime, use and maintenance scenarios, repair and replacement of components, major refurbishment or renovation, and demolition and recycling scenarios.

The disposal scenario of a building built today is speculative, at best, and therefore should be reported and displayed separately in order to keep the results transparent.

LCA can be used during the design of new buildings or when refurbishing existing buildings. Further LCA may provide useful insights for the ecological improvement of the buildings' portfolio of companies working according to ISO 14000. LCA can also be applied to generate data sets on the ecological performance of components or construction types as a guideline for practitioners.

3.3.4 Potentials and limitations

Potentials

An LCA provides information on the ecological performance of building concepts in a readily accessible format to help during planning decisions. Quantitative results allow a comparison among different solutions. This simplifies the otherwise difficult task of identifying problematical steps within the life cycle.

Limitations

- A consistent and peer-reviewed international level databases of life-cycle inventories of building-related products is lacking. In September 2003, the Swiss Centre for Life Cycle Inventories

published a central web-based LCA database with consistent data sets (Ecoinvent, 2003). As this evolves, it can fill this information gap, at least regionally.
- Early in the design, the planner has defined very little; yet an LCA requires detailed construction data.
- It is difficult to reflect the impact of the building's locations with regard to travel and transportation. Furthermore, LCA investigates only ecological impacts, not social or economical aspects. The ecological impacts are considered mainly on a global scale and may not be suitable for the assessment of the impacts on regional or local scales.
- The amount of work needed for an LCA is high compared to other methods such as simple checklists.

3.3.5 Conclusions and recommendations

High-performance houses can reduce the total environmental impact over the lifetime of a building by as much as 50 per cent compared to a conventional building. Energy savings during operation have the largest impact in conventional buildings; but due to the low operation energy demand of high-performance buildings, the construction and renewal of the building become increasingly important. A large savings potential lies in the basic concept of the building and includes the following points:

- A compact building design leads to a low energy demand and needs less material. Therefore, the lowest surface-to-volume ratio is preferable.
- Use wood instead of concrete. Due to the lower impacts over the life cycle, a wooden building structure rates better. If a massive floor is needed for thermal reasons, the ideal material would be native stone slabs over a dry floor construction in combination with a wooden under-floor structure. In wood construction, however, the local wood should not be chemically treated.
- Use long-lasting and recyclable building materials.
- Avoid composite materials. If materials with high embodied energy are used (plastic, metals) these should be separable to facilitate recycling.

Key energy aspects during the operational life of a house include the following:

- Energy conservation has highest importance. The choice of insulation materials is unimportant with regard to the total impact and therefore a high insulation level is advantageous.
- Solar collector areas exceeding 0.1 m^2 per m^2 net heated floor area do not show a corresponding lessening of the life-cycle energy demand.
- Ventilation system performance must also reflect electricity demand for the fans and, possibly, backup heating. The materials choice is less important.
- Electricity demand by household appliances can be sharply reduced if the builder specifies best efficiency class-rated equipment.
- Electricity mix (for houses with a heat pump) will show a better LCA if part of the electricity comes from a renewable source – such as photovoltaics or wind.

References

Consoli, F., Allen, D. Boustead, I., Fava, J., Franklin, W., Jensen, A.A., de Oude, N.,Parrish, R., Perriman, R. Postlethwaite, D., Quay, B., Seguin J. and Vigon, B. (1993) *Guidelines for Life-Cycle Assessment: A Code of Practice*, Society of Environmental Toxicology and Chemistry (SETAC), Brussels

Ecoinvent (2003) *Ecoinvent Database*, Swiss Centre for Life Cycle Inventories, Duebendorf, www.ecoinvent.ch/en/index.htm

Goedkoop, M. and Spriensma, R. (2001) *The Eco-indicator 99: A Damage Oriented Method for Life Cycle Impact Assessment*, Methodology Report, third revised edition, PRé Consultants B.V., Amersfoort, The Netherlands

Heijungs R., Guinée, J. B., Huppes, G., Lankreijer, R. M., Udo de Haes, H. A., Wegener Sleeswijk, A., Ansems, A. M. M., Eggels, P. G., van Duin, R. and de Goede, H. P. (1992) *Environmental Life Cycle Assessment of Products – Guide*, Centre of Environmental Science, Leiden, The Netherlands

IEA (1999) *IEA Annex 31: Energy Related Environmental Impact of Buildings*, www.uni-weimar.de/scc/PRO/TOOLS/index.html

ISO (1997) *Environmental Management – Life Cycle Assessment: Principles and Framework*, ISO/FDIS 140401997 (E), International Organization of Standardization, Geneva

Lindfors, L. G., Christiansen, K., Hoffman, L., Virtanen, Y., Junttila, V., Hanssen, O.J., Rønning, A., Ekvall, T. and Finnveden, G. (1995) *Nordic Guidelines on Life-Cycle Assessment*, Nord 199520, Nordic Council of Ministers, Copenhagen.

RMIT (2001) *Greening the Building Life Cycle: Life Cycle Assessment (LCA) Tools in Building Construction*, Environment Australia Centre for Design, RMIT University, Australia, www.buildlca.rmit.edu.au.

3.4 Architecture towards sustainability (ATS)

Kristel de Myttenaere

3.4.1 Introduction

Sustainable architecture is more than energy efficient or zero-emission architecture. It must adapt to and respect its environment in the broader context of 'milieu'. This encompasses the natural, ecological, bio-economic, cultural and societal setting. A successful solution must address the following principles, taken from the 1992 Rio Declaration (WCED, 1992):

- *Shared but differentiated responsibilities*: we, citizens of the Earth, are all responsible for the future of the planet, and Westerners still more so than others. This shared responsibility is exercised among individuals, institutions and countries, and the ecosystems that surround them. These responsibilities inevitably bind all people.
- *Intra- and intergenerational equity*: we all have a right to a quantitatively and qualitatively healthy environment, both today and tomorrow, all over the world. A shared responsibility for both today and tomorrow means that we are the stewards of a healthy environment and that future generations have this right as well.
- *Integration of the components of sustainable development (environment, society, economy and politics)*: to achieve sustainable development, protection of the environment must be an integral part of the development process and cannot be considered on its own; it is important to have a cross-sector approach. Rethinking the ties between the environment, society, politics and economics as it applies to development requires engaging multiple stakeholders in various sectors. This successful approach is both timely and effective.
- *Precaution and acknowledgement of scientific uncertainty*: we must limit hypothetical or potential risks. We are responsible for the consequences of our current actions. A cooperative, integrated approach is a tool that helps to balance the various components of development without risking a loss for future generations.
- *Participation and good governance*: we owe it to ourselves to be well informed in order to take position and act in full knowledge of the local and global issues. Access to education and information is essential. Our shared responsibility is universal. Everyone's participation gives us all a choice and a voice.

3.4.2 Applying the principles to planning

Shared responsibility

All spaces are a collective good for which society, and particularly architects, are responsible. If the architect makes an undertaking personally to a client, he also makes an undertaking to society. The impact of his constructions exceeds the boundaries of his project. How can the architecture responsibly respond to the various scales of the public space?

Intra- and intergenerational equity

What type of architecture can provide housing to all, with respect to both individual and collective well-being? A minimal density is necessary to ensure access for everyone to health facilities, schools, the work place, cultural events, trade, points of transportation, etc. How should the built environment be articulated to solve this in both the dimensions of space and time? Our heritage, both natural and cultural, handed down from generation to generation must be respected in this process.

Integration of components of sustainable development

Sustainable development entails environmental, social, economic and political dimensions. Sustainable architecture must ensure individual comfort, while also preserving ecosystems in natural areas. The architecture must operate at individual levels (functionalism) and community levels (mixed functions).

Precaution and acknowledgement of scientific uncertainty

We cannot anticipate the individual and collective spatial needs of future generations any better today than previous generations did earlier. Therefore, our architecture should be adaptable and designed to meet the needs of future occupants. Many of the buildings we have inherited from previous generations incorporate natural materials, daylighting, passive cooling techniques and a rational use of passive sources of energy. These existing buildings can often be utilized to offset the impact of new structures

Participation and good governance

Here, the idea is to improve public awareness of environmental, social, economic and political problems at the scales of neighbourhood, town, territory, country or the planet. At issue is the articulation between the individual and the community. The challenge is to develop viable public infrastructures (housing, schools, hospitals, shops, cultural areas, natural areas, etc.) to promote individual development, participation and an enhanced community life.

3.4.3 A case study

To illustrate how these principles can be addressed, a project in Vienna is analysed. It responds to many aspects, which, of themselves, may be contradictory. It attempts to profit from positive elements of its environment, while protecting itself from the negative attributes and, at the same time, limiting its own negative impact on the environment.

Project description

The project is located in the second ring of Vienna along a major boulevard linking the town centre to the suburbs. It consists of a strip apartment block along a busy and noisy boulevard. From the sheltered rear of the apartment block, lower row houses extend back perpendicularly. The development includes 215 housing units from 60 to 130 m^2, a daycare centre with three units, a playground, an office building, a restaurant and a car park with 215 spaces. The project was constructed from 1993 to 1996.

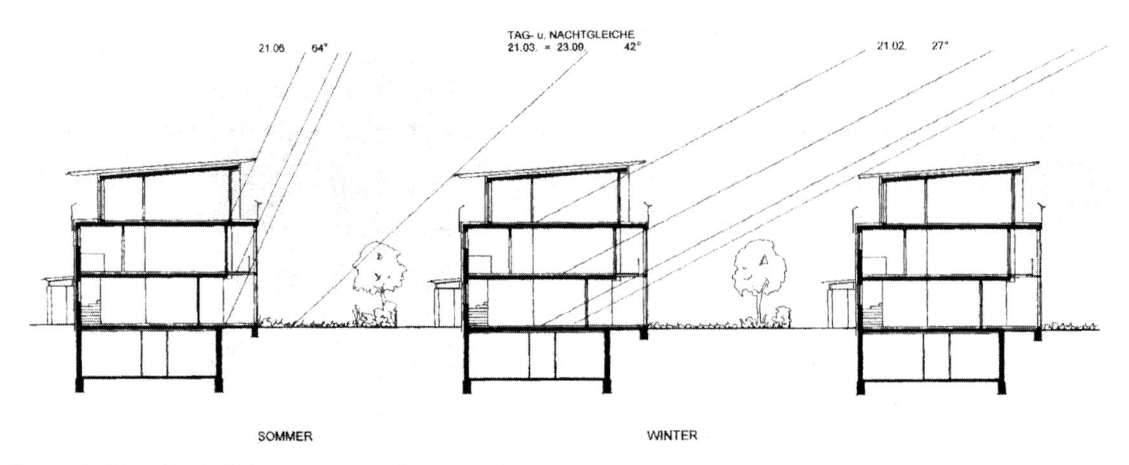

Source: Architect Martin Treberspurg, www.treberspurg.at/

Figure 3.4.1 *Case study of the Hirschenfeld housing development on Brunnerstrasse in Vienna*

Analysis at the town planning scale

The environmental dimension: the project is consistent with the density of the area; therefore, there is reduced dispersion and sprawl into the surrounding countryside.

The social dimension: the housing is designed to meet the needs of various social classes. The project offers a diversity of types of living spaces and opportunities for tenants to customize their individual space and to provide community space.

The economic dimension: the location of the project within the city, rather than in the suburbs, reduces the cost of roads, public transport, electric grid, water and sewers, etc. from the standpoint of the project and of the town.

The political dimension: the diversity of dwelling types should attract persons of different ages, incomes and social structures, and promote a social cohesion.

Analysis at the neighbourhood scale

The environmental dimension: the 'wall' of apartments along the boulevard shields the interior of the block from traffic noise, pollution and the dangers associated with an urban thoroughfare. The neighbourhood to the east of the project also profits from

Source: Kristel de Myttenaere, University Catholique de Louvain

Figure 3.4.2 *Photo impressions of the milieu of Brunnerstrasse in Vienna*

this shielding. This concept of protection can be observed at various scales. For example, within the building, spaces are organized in successive strata from the noisiest to the quietest (public hallways, technical walls separating the hallways from the housing, 'service' areas and 'living' areas). The façade is also layered: a grid with vegetation, a glass façade and masonry walls.

The social dimension: the project provides a structural social transition between the urban space to the west of the boulevard that is dense and is comprised mainly of social housing, and the low density space to the east with single family homes.

The economic dimension: the project houses mixed uses, including daycare, a restaurant and offices, as well as residences. The infrastructure is laid out to interconnect these elements. Such multi-functionality creates a certain dynamic and increases the likelihood that some people can live, work and make use of services without travel.

The political dimension: inner pedestrian ways, green spaces, play areas and the restaurant provide opportunities for people to meet casually to discuss issues and build relationships.

Analysis at the building scale

The environmental dimension: a glazed buffer space provides access to the apartments while also shielding the project from noise, traffic exhaust and crime.

The project makes use of passive solar and conservation principles, including opening the row houses to the south, glazed verandas, entry air locks and a good insulation with minimal thermal bridges.

The social dimension: semi-public, semi-private and private areas establish an effective interface between individuality and community. Transition areas such as vestibules, terraces, and balconies and verandas create differentiated degrees of separation physically but also symbolically. Social interaction can occur in common areas such as laundry rooms, bicycle parking, storage areas, play areas for children and paths crossing the development.

The economic dimension: the living spaces are simple, well structured, rational and flexible, making the project adaptable for different occupancies today and tomorrow. The project is open to changes in the future.

The political dimension: the different types of housing can accommodate different life styles and ways of living, reinforcing social cohesion. The simplicity and occupant friendliness of the living spaces appeal to all types of individuals. Transition areas between indoor and outdoor spaces reinforce the sense and security of community.

Analysis at the scale of materials and systems

The environmental dimension: diverse solar and conservation strategies were included in the design:

- high thermal insulation;
- passive solar gain coupled with massive construction;
- solar thermal collectors on south-facing roofs;
- heat recovery from exhaust air; and
- combined heating and ventilation to reduce the need for direct ventilation from the noisy side of the project.

The social dimension: the heating systems are regulated collectively but can also be adjusted for each individual housing unit. Energy consumption therefore varies considerably among the units.

The economic dimension: state-of-the-art technologies were used, so there was little additional cost. There is no need for sophisticated understanding of the systems to use them well.

The political dimension: the occupants of each housing unit control their own consumption. However, a public display of the energy consumption of the different housing units would have made inhabitants more energy conscious.

Learning from the project

Today, certain technical improvements would be possible. The ventilation and heating system could be more efficient, effective and better regulated. Savings could also have been increased if the occupants were made more aware of their energy consumption. It would be interesting to study the collection of rainwater and recycling/disposal of wastewater at this scale in detail. It is impressive how many issues were addressed effectively by the architecture and planning. It is a good example of a multidimensional and cross-sector approach to sustainable development.

3.4.4 Conclusions

Assessing the probable economic, political, social and environmental impact of a development is a complex enterprise due to the diversity of the issues dealt with. Designing sustainable architecture requires anticipating these interactions within its natural, ecological, societal and cultural setting – or its 'milieu'. It is difficult to define a list of criteria and to establish a weighting system in order to make objective planning decisions. Qualitative criteria are also important and are even more difficult to define and to rank. A person perceives himself both as an individual and as a member of a community (family, nation, culture, etc.). Design must also respond at these levels. As the design evolves, paradoxes inevitably arise. These paradoxes can only be solved through a system of compromise and collaboration that no longer targets attaining a maximum for any one criterion, but an optimum for all criteria. There is no absolute truth (scientific, economic, etc); a solution must find the right proportions within understood limits. A design ethic should respect the principles of the Rio Declaration:

- shared but differentiated responsibilities;
- intra- and intergenerational equity;
- integration of the components of sustainable development;
- precaution and acknowledgement of scientific uncertainty; and
- participation and good governance.

References

United Nations Programme (1993) *Agenda 21 Earth Summit: Action from Rio*, United Nations, www.un.org/esa/sustdev/documents/agenda21/index.htm

WCED (1992) *27 Principles of the Rio Declaration*, Oxford University Press, Oxford, www.un.org/cyberschoolbus/peace/earthsummit.htm

Websites

Belgian Federal Plan Bureau: www.plan.be/fr/welcome.stm
International Institute for Sustainable Development: www.iisd.org
United Nations Division for Sustainable Development: www.un.org/esa/sustdev

4

Economics of High-Performance Houses

Berthold Kaufmann

4.1 Introduction

This chapter shows that an extra investment in thermally optimizing a building can yield substantial added value for the owner and occupants. The higher first costs of construction are easily paid back within the building's lifetime. Indeed, given the recent escalation in energy prices, the payback appears shorter than ever.

The thermal optimization of housing within economic constraints also results in renewable energy being able to cover a greater fraction of this reduced demand, whether the energy source is solar thermal, biomass, photovoltaics (PV) or wind power (Kaufmann et al, 2003).

The economic savings from improving the thermal performance of a building can be assessed by comparing the return on investment of alternative solutions – for example, insulation; high-performance windows; mechanical ventilation system with heat recovery; and high efficiency heat supply, controls and distribution systems. A good measure is the cost of a saved kWh of heating. This can be easily compared to actual prices for the purchased or end energy. In this chapter an average price of €0.055/kWh end energy is assumed to be realistic for the next ten years. Saved energy costs based on reduced heating demand have to be amplified by the inefficiency of the heating system (90 per cent assumed) and the cost for electricity to power pumps, fans and controls (€0.034/kWh assumed). An additional cost is the capital cost of the heating system. All factors added together result in an overall cost for heat production of €0.08/kWh based on the end energy price of €0.055/kWh (Feist, 2005c; Kaufmann, 2005). The price for conventional electricity in the year 2005 is taken as €0.17/kWh. A higher price for 'renewable' electricity is assumed.

In some studies the payback period needed for an investment is used as an indicator. For high-performance components, the payback from saved energy may not fall within the life expectancy of the product. This can happen, for example, with highly insulated windows and with a mechanical ventilation system with heat recovery. These results are correct, but cannot really help in decision making because a comparison of costs among alternatives is not possible. A better measure for comparison is the cost of a saved kWh.

If the extra costs for an energy saving action such as an optimized thermal insulation thickness are lower than the actual costs for delivered heat, the financial benefit is obvious. If, on the other hand, the saved kWh costs are higher than heating costs (€/kWh) it can be helpful to compare this

Table 4.1.1 *General data for cost calculation*

Annual rate of interest	3.5%	Real rate, adjusted by inflation
Calculation time of interest	20 years	With constant rate of interest
Lifetime of components	20 years	Mechanical ventilation, domestic hot water, etc.
Lifetime of components	30 years	Window
Lifetime of components	50 years	Thermal insulation of wall and roof
Price for end energy	€0.055/kWh	Oil, gas, district heat, not electricity
Total costs for supply of heat	€0.080/kWh	Explanation see text, not electricity
Price for electricity	€0.170/kWh	End energy
Climatic region		Temperate (Central Europe)

energy saving measure to other measures. Using this method any 'extra benefit' of a measure can be given a value and included in the calculation.

The calculations in this chapter are based on the dynamic method for calculations of capitalized values or present cash values, respectively. The boundary conditions for the calculations are defined as follows (see Table 4.1.1). The medium (real) annual rate of interest for an investment in the last decade was about 3.5 per cent where an inflation rate of about 1.5 per cent is already taken into account. The financial calculation time with a fixed rate of interest is assumed to be 20 years, which is shorter than the lifetime of most building components used in middle Europe today. This assumption results in a residual financial value at the end of the calculation period, which decreases the actual cash value of the investment costs significantly (Feist, et al, 2001; Feist, 2004).

4.2 Cost assessment of high-performance components

Two questions serve as an introduction:

1 What are the costs of doing 'nothing' with respect to the thermal performance of buildings, especially providing no thermal insulation for a new building?
2 What are the costs for a simple new surface treatment (ie patching and painting?)

Doing nothing may be an expensive option. The running costs resulting from 1 m^2 of poorly insulated wall over its lifetime are significantly higher than the investment costs for insulation.

4.2.1 Insulation layer outside the wall versus new surface treatment only

The total costs of the needed investment and cost of energy losses for adding insulation are presented in Table 4.2.1. Costs for the 'anyway action' of a new surface treatment only are excluded, so only costs that are related to the energy saving investment are compared. The optimum layer thickness is 23 cm. All costs are area-specific (€/m^2 wall area).

Only the energy-related costs have to be taken into account when calculating the total costs of an energy saving action. For a new building, a surface treatment only of the outer wall is necessary in any way. So, these 'anyway costs' (€30/m^2) can be subtracted from the investment costs. The energy losses and the related running costs can be calculated directly by using the U-value of the construction times the number of degree days × 24 for the relevant climatic region. The added insulation layer is assumed to have a lifetime of 50 years. The economic optimal insulation thickness is about 23 cm, resulting in costs of about €0.030 per saved kWh. But the cost function in Table 4.2.1 and in Figure 4.2.1 is rather flat around the minimum. Therefore the 'future scenario' with a layer thickness of 30 cm, which is suitable for passive houses, has only slightly higher costs of €0.031 per saved kWh. This is more than 40 per cent below the actual price for end energy of €0.055/kWh.

For comparison: a layer thickness of 12 cm with a U-value of about 0.3 W/(m^2K) was required by the local building code in Germany in February 2002.

Table 4.2.1 *Total costs (investment plus energy losses) of insulation added to the exterior wall*

Variations of several measures	Anyway action: new rendering for wall	Thermal insulation (6 cm)	Thermal insulation (12 cm)	Thermal insulation (23 cm): 'Optimum'	Thermal insulation (30 cm): 'Future'
Economical boundary conditions and building data					
Thickness of insulation layer (cm)	0.00	6.00	12.00	23.00	30.00
Old U-value without any action (W/m^2K)	1.41	1.41	1.41	1.41	1.41
New U-value (W/m^2K)	1.41	0.45	0.27	0.15	0.12
Energy losses through component (kWh/m^2a)	106.00	34.00	20.00	12.00	9.00
Investment costs					
Investment costs for action (€/m^2)	30	90	95	105	111
Minus the anyway costs (€/m^2)	30	30	30	30	30
Total investment costs (€/m^2)	0	60	65	75	81
Residual value of investment costs (€/m^2)	**0**	**24**	**26**	**29**	**32**
Investment costs minus residual value	0	36	40	45	49
Present cash value of total costs (see Figure 4.2.1)					
Present cash value of investment (€/m^2)	0	36.00	40.00	45.00	49.00
Present cash value of energy costs (€/m^2)	120.00	39.00	23.00	13.00	10.00
Present cash value of total costs (€/m^2)	120.24	75.02	62.50	58.37	59.16
Possible energy savings (end energy) by action (kWh/m^2a)	0.0	81.4	96.9	106.7	109.5
Annual capital costs for saved end energy (€/kWh)	–	0.0315	0.0287	0.0298	0.0313
Annual rate of total costs					
Annual rate for total investment costs (€/a)	0.00	2.56	2.78	3.18	3.43
Energy costs per area of component (€/m^2a)	8.46	2.72	1.62	0.93	0.73
Annual rate of total costs (€/m^2a)	8.46	5.28	4.40	4.11	4.16

As can clearly be seen in Figure 4.2.1, insulating is economical; doing nothing will cost more! The energy consumption costs and the investment costs are summed up and are depicted as a function of insulation layer thickness (Feist, 2005c). All thermal insulation scenarios have lower total costs than the one with no thermal insulation.

4.2.2 Insulation layer for roof (anyway action: new roof tiles and slats)

Roof insulation is by far the most economic measure. Taller roof rafters in many cases can be specified with only moderate extra costs (Feist, 2005c). Table 4.2.2 shows the costs for a highly insulated roof. The costs for the greater insulation thickness are low and the economic benefit is apparent, even for the 'future scenario'.

Figure 4.2.2 shows the total investment costs as a function of the insulation thickness. Since the costs for energy consumption decrease rapidly with the thickness (1/d), but the construction costs only increase linearly with layer thickness, the function for total costs has a minimum determined by energy price. For the assumptions given above, the actual cost minimum lies between 20 and 50 cm of insulation. Taking into account the cumulative primary energy embodied in the insulation (for example, mineral wool), the optimum layer thickness from the ecological point of view would be between 50 and 225 cm (Feist, 2005c). It should be noted that the minimum of the function is very flat. Therefore, it is sensible to choose a thickness at the lower boundary of the ecologic optimum range where it overlaps with the economic optimum range.

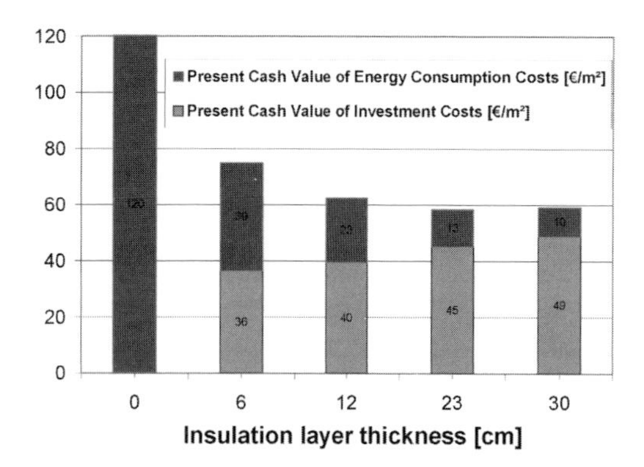

Source: Feist (2005c)

Figure 4.2.1 *Total costs for a compound thermal insulation layer in the wall (area of €/m² relates to the wall area)*

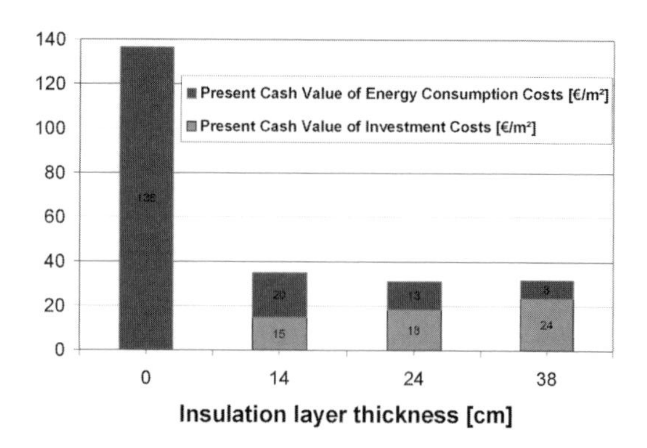

Source: Feist (2005c)

Figure 4.2.2 *Total costs for a thermal insulation layer between roof rafters*

Table 4.2.2 *Total costs for a thermal insulation layer in the roof; all area-specific numbers (€/m²) here are related to the roof area*

Variants of several actions Economical boundary conditions and building data	Anyway action: Roof tiles and slats only	Insulation		
		Between rafters (14 cm)	Between and on top rafters 'Optimum' (24 cm)	Between and on top rafters 'Future' (38 cm)
Thickness of insulation layer (cm)	0.00	14.00	24.00	38.00
Old U-value without any action (W/m²K)	1.6	1.6	1.6	1.6
New U-value (W/m²K)	1.60	0.24	0.15	0.10
Energy losses through component (kWh/m²a)	120.00	18.00	11.00	7.00
Investment costs				
Investment costs for action (€/m²)	90	114	120	129
Minus the anyway costs (€/m²)	90	90	90	90
Total investment costs (€/m²)	0	24	30	39
Residual value of investment costs (€/m²)	0	10	12	15
Investment costs minus residual value	0	15	18	24
Present cash value of total costs				
Present cash value of investment (€/m²)	0.00	15.00	18.00	24.00
Present cash value of energy costs (€/m²)	136.00	20.00	13.00	8.00
Present cash value of total costs (€/m²)	136.44	35.04	31.13	31.86
Possible end energy savings by action (kWh/m²a)	**0.0**	**115.8**	**123.3**	**127.7**
Investment costs for saved end energy (€/kWh)	–	0.0090	0.0105	0.0130
Annual rate of total costs				
Annual rate for total investment costs (€/a)	0.00	1.04	1.30	1.65
Energy costs per area of component (€/m²a)	9.60	1.43	0.89	0.59
Annual rate of total costs (€/m²a)	9.60	2.47	2.19	2.24

Source: Feist (2005c)

Roof insulation thicknesses of between 40 and 50 cm are suitable for high-performance houses. Table 4.2.2 shows the optimum insulation thickness to be about 24 cm. A 40 cm thick insulation is available at very low extra costs.

In Figure 4.2.2 the energy consumption costs and the investment costs for the insulation layer between roof rafters are summed up and depicted as a function of insulation layer thickness. The area (€/m^2) is related to roof area.

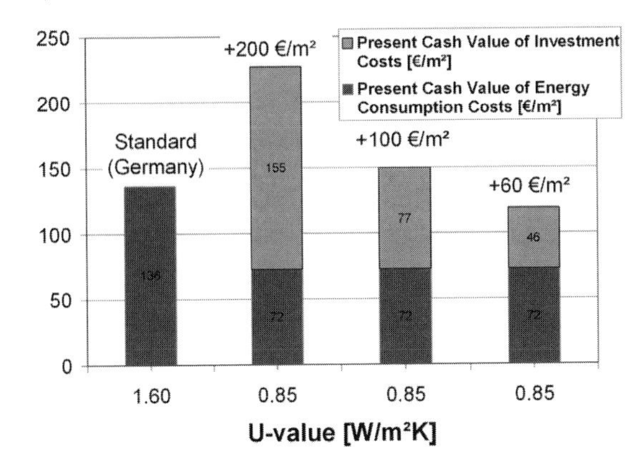

Source: Schöberl and Hutter (2003)

Figure 4.2.3 *Total costs for high-performance windows (€/m^2 window area)*

4.2.3 High-performance windows

Windows are one of the valuable and, hence, high cost components of high-performance houses. But it is worth comparing costs in detail. Table 4.2.3 shows that the present cash value of the energy loss costs for a standard window with U_w = 1.6 W/(m^{2}K) are about €136/m^2. Replacing this by high-performance windows with U_W = 0.85 W/(m^{2}K) can drop the energy losses and the related costs to about €72/m^2.

Comparing the total costs, including energy losses and investment expenses, for the various scenarios shows that the extra costs for high-performance windows are moderate. The first developments of so called 'passive house windows' had investment costs of about 450 Euros/m^2 (scenario '+200 Euros'), which were due to high development costs and low production numbers. Currently, such windows are available for about €350/m^2 (scenario '+€100'), which leads to a present cash value for the total costs of €150/m^2. This is only 14 Euros/m^2 higher than for a standard window.

In Figure 4.2.3 the energy consumption costs and the investment costs are summed up and graphed as a function of roof insulation thickness. All area specific numbers (Euros/m^2) here are related to the construction area.

Going one step further: scenario '+€64' leads us to how much a high-performance window (U_W = 0.8 W/m^{2}K) may cost in order to be cost effective at the actual energy prices. The result is not surprising: extra investment costs of about €64/m^2 compared to a standard window (U_W = 1.6 W/m^{2}K) are acceptable. The extra capital costs are balanced by the lower energy expenses during lifetime of the window.

These results lead to prices for the optimized window of about €310/m^2. This may be realistic for the market in the near future. The window costs can even be reduced by placing it in the wall opening in such a manner that wall insulation covers much of the window frame. Thus, a less expensive window frame can be specified as much of it is covered. In an Austrian project window alternatives were investigated in 2003 for a large social housing project (2778 m^2 living area). Their absolute numbers are significantly smaller (€250/m^2 for plastic windows; €280/m^2 for timber/aluminium constructions) than those given in Table 4.2.3 because of the scale of the project. They found high-performance windows cost about 30 per cent to 35 per cent more than standard windows. This is an indication of how mass production can lower costs. As this price reduction occurs, the present cash value for the extra costs will diminish (Schöberl et al, 2003).

Table 4.2.3 *Total extra costs for high-performance windows*

Variants of several actions	Anyway action: New standard windows	Passive windows '+€200'	Passive windows '+€100'	Passive windows '+€60'
Economic boundary conditions and building data				
Old U-value without any action (W/m²K)	1.60	1.60	1.60	1.60
New U-value (W/m²K)	1.60	0.85	0.85	0.85
Energy losses through component (kWh/m²a)	120.00	64.00	64.00	64.00
Investment costs				
Investment costs for action (€/m²)	250	450	350	310
Minus the anyway costs (€/m²)	250	250	250	250
Total investment costs (€/m²)	0	200	100	60
Residual value of investment costs (€/m²)	0	45	23	14
Investment costs minus residual value	0	155	77	46
Present cash value of total costs				
Present cash value of investment (€/m²)	0.00	155.00	77.00	46.00
Present cash value of energy costs (€/m²)	136.00	72.00	72.00	72.00
Present cash value of total costs (€/m²)	136.44	227.03	149.76	118.85
Possible end energy savings by action (kWh/m²a)	0.00	63.80	63.80	63.80
Investment costs for saved end energy (€/kWh)	–	0.1706	0.0853	0.0512
Annual rate of total costs				
Annual rate for total investment costs (€a)	0.00	10.87	5.44	3.26
Energy costs per area of component (€/m²a)	9.60	5.10	5.10	5.10
Annual rate of total costs (€/m²a)	9.60	15.97	10.54	8.36

The total extra costs for high-performance windows, assuming a 30-year lifetime, are presented in Table 4.2.3. Areas (€/m²) are related to the window area.

In addition to the purely financial arguments, high-performance windows are essential to the performance of high comfort, low energy housing. Low U-values ensure warm inner glass surface temperatures and, hence, comfort. Because there are no downdrafts along the glass, there is no need for a radiator beneath the window, also saving additional capital costs.

4.2.4 Combined heat pump systems versus direct electric heating

The investment costs and the resulting running costs for a combined system (including hot water boiler, heat pump and mechanical ventilation with heat recovery) are compared to the same system but with direct electric resistance heating instead of the heat pump (assumptions are given in Table 4.2.4). For this study, the seasonal performance factor (SPF) of the heat pump is conservatively assumed to be only 2.5, although there are actually systems available with an SPF of 3.0 or even higher. The economic performance is summarized in Table 4.2.4.

The investment costs for the direct electric heating system are nearly 50 per cent lower than those for the combined systems with heat pump. Costs may differ for various systems; but this assumption seems good enough to describe the actual market situation.

The running costs include (in order of importance) the electric energy consumption, maintenance and the basic fees for the electric power supply, which are higher for a high power connection. The price of electricity is assumed to be €0.17/kWhel, as mentioned above.

To compare investment costs and running costs, both are spread to an annual rate by the calculation of interest method (dynamical cost calculation method). The assumptions for this calculation are summarized in Table 4.1.1 and Table 4.2.4. The annual rate of interest is assumed to be 3.5 per cent. The inflation is included in this rate (Feist, 2005c). The financial calculation time equals the lifetime

Table 4.2.4 *General data for cost calculation with respect to electric heating and heat pump systems*

Annual rate of interest	3.5 %	
Calculation time of interest	20 a	
Lifetime of components	20 a	
Actual price for delivery of electricity	€0.17/kWh	Including fixed costs (e.g. connection fee)

of the equipment of about 20 years. The lifetime of the compressor of the heat pump may be shorter (15 years) as it is for fans of the ventilation system and pumps of a heating system. On the other hand, the lifetime of the non-movable parts – as, for example, air ducts – is significantly longer (30 years and more). For simplicity of the exemplary calculation given here, the lifetime of the whole system is assumed to be 20 years.

In Table 4.2.5, all annual costs are drawn back to the present to be summarized with the actual investment costs so that the present cash value of the total costs can be compared. It can be seen that the combined system with heat pump is, with respect to the total costs, 5 per cent more expensive than the direct electric heating solution. The annual rate of total costs is shown at the bottom of Table 4.2.5.

From the annual rate of the extra investment costs and the amount of electricity saved per year, it is possible to calculate the costs per saved kWh. The costs of about €0.2/kWh for the heat pump system can be compared with the actual price for electricity of €0.17/kWh.

This result is obvious from column 4 and 5 in Table 4.2.5, where the price for electricity ('green power') is assumed to be €0.20/kWh. The present cash value of the two variants is then almost the same.

As some experts assume that the price for 'green' or even conventional electricity will increase in the near future to €0.25/kWh, total costs are calculated in column 6 and 7 to check the financial consequences. The present cash value of the heat pump system then amounts to 8 per cent more cost effective than the direct electric heater. This result confirms the striking argument that electricity is far too valuable to be simply used directly for heating. It can be seen clearly that direct electric heating is even more expensive from a financial point of view.

As shown in column 2 of Table 4.2.5, the optimization of the seasonal performance factor (SPF) is beneficial both in terms of reduced environmental impact and economy. If SPF = 3 is assumed, the extra investment costs for the saved kWh decrease significantly by €0.02/kWh. The resulting costs per kWh is not much higher than the actual price for electricity of €0.17/kWh.

Better performance of the system results in significant higher economic advantage and is therefore economically relevant even at the actual price level for electricity.

4.2.5 Optimal design of ground plans

Optimizing the design of ground plans can help to save running costs and building costs. Short hot water circuits with good pipe insulation from the boiler to the points of use reduce hot water heat losses; and short duct runs reduce duct heat losses and the fan power needed to supply the air. Short runs also reduce construction costs. Therefore, bathroom, kitchen and toilets should be arranged to share a common wall, where air ducts, hot and cold water, and waste plumbing can be bundled.

The mechanical ventilation system can also be clustered with the hot water boiler either in the bathroom or kitchen. A compact combined system with heat pump needs less than 1 m^2 of floor area. If the supply system is placed in the cellar or on the top floor beneath the roof, it should be located directly above or below the wet rooms to minimize pipe and duct runs.

Air jet diffusers that project the ventilation air far into a room eliminate the need to run ducts to the exterior perimeter of a room to ensure good air circulation. With such diffusers, room air inlets can be placed over living or bedroom doors from the corridor, as shown in Figure 4.2.4, saving some 10 m of duct length (Kaufmann et al, 2004).

Table 4.2.5 *Investment and running costs for a combined system with a heat pump compared to a system with direct electric heating*

	Actual price (€0.17/kWh)			Green power (€0.20/kWh)		Green power (€0.25/kWh)	
Equipment variations	**Combi-system with heat pump 1**	**Combi-system with high efficiency 2**	**Direct electric heat 3**	**Combi-system with heat pump 4**	**Direct electric heating 5**	**Combi-system with heat pump 6**	**Direct electric heating heat pump 7**
Technical data of building and supply							
Heating demand (kWh/m²a)	15	15	15	15	15	15	15
Domestic hot water demand (kWh/m²a)	20	20	20	20	20	20	20
Sum of heat demand (kWh/m²a)	35	35	35	35	35	35	35
Net floor area (m²)	100	100	100	100	100	100	100
Total annual heating demand (kWh/a)	3500	3500	3500	3500	3500	3500	3500
Heat pump							
Seasonal performance factor (SPF)	2.5	3.0	1.0	2.5	1.0	2.5	1.0
Resulting electrical demand (kWhel/a)	1400	1167	3500	1400	3500	1400	3500
Price for end energy delivery (€/kWhel)	0.17	0.20	0.25				
Investment costs							
Air heater with control unit			500		500		500
Flow water heater (18 kW) with controller			400		400		400
Mechanical ventilation with heat recovery			2500		2500		2500
Air duct system	1500	1500	1500	1500	1500	1500	1500
Cost for tradesmen for installation	1000	1000	1000	1000	1000	1000	1000
Combined heat pump, boiler and ventilation (compact units available as low as €6300)	9000	9000		9000		9000	
Total investment costs (€)	11,500	11,500	5900	11,500	5900	11,500	5900
Present cash value of total costs							
Present cash value of investment (€)	11,500	11,500	5900	11,500	5900	11,500	5900
Present cash value of annual expenses (maintenance, etc.) (€)	2132	2132	1421	2132	1421	2132	1421
Present cash value of basic power supply (€)	426	426	853	426	853	426	853
Present cash value of energy costs (€)	3383	2819	8456	3979	9949	4974	12,436
Present cash value of total costs (€)	17,441	16,877	16,630	18,038	18,123	19,033	20,610
	(3)-(1)	(3)-(2)		(5)-(4)		(7)-(6)	
Possible savings end energy (kWh/a)	2100	2333		2100		2100	
Additional investment costs for combined heat pump system (€/a)	414	414		414		414	
Additional investment cost per saved kWh end energy (€/a)	0.197	0.177		0.197		0.197	
Annual rate of total costs							
Annual rate for total investment costs (€/a)	809	809	415	809	415	809	415
Maintenance and other annual expenses (€/a)	150	150	100	150	100	150	100
Basic power supply costs (€/a)	30	30	60	30	60	30	60
Costs for electric consumption (€/a)	238	198	595	280	700	350	875
Annual rate of total costs (€)	1227	1187	1170	1269	1275	1339	1450

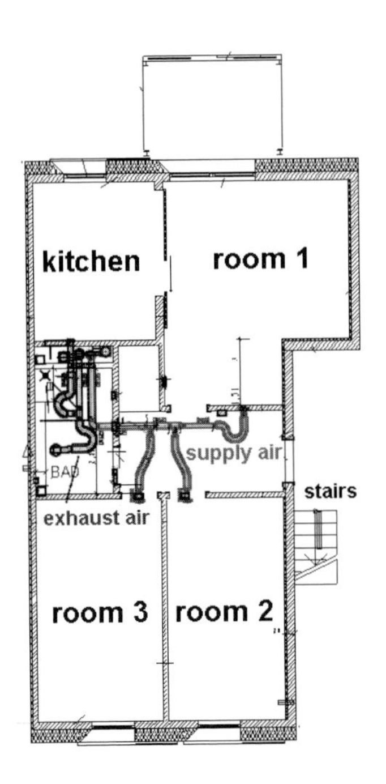

Source: Passivhaus Institut, Darmstadt, Germany, www.passiv.de

Figure 4.2.4 (Left) *Optimized ground plan of an apartment in a social house project in Kassel, Germany;* (right) *a view of the air ducts*

4.2.6 Detailed design without thermal bridges

Avoiding thermal bridges does not imply higher material costs; it primarily depends on the ingenuity and care of the planner. Avoiding thermal bridges is especially important in buildings where much effort has been given to achieving a high level of insulation. In the worst case, the thermal bridges can increase the heat losses by 5 to 15kWh/(m^2a). This represents from 30 per cent to 100 per cent of the heating energy demand of a high-performance building.

The thermal separation of bearing walls above the cellar or the ground slab requires special materials to minimize the thermal bridge effect. Most thermal bridges can be avoided by careful detailing and avoiding steel or aluminium anchors penetrating the insulation layer.

4.2.7 Air-tight envelope

The building envelope of high-performance houses has to be air tight, but tight connections at corners need not be really costly. This is again a question of careful planning and quality control at the building site. Taking time to train and motivate the craftsmen and tradesman is a good investment.

4.2.8 Building costs: Expenses for living

This section has shown that the added costs for high-performance components are easily justified when total costs, including reduced operating costs, are considered. Clearly, in the construction, upfront costs are decisive for homebuyers; but this is a question of priorities. High quality envelope construction and systems have to be marketed, just as marble floors, granite kitchen countertops or designer faucets are very successfully marketed.

4.3 Additional expenses

4.3.1 Expenses to fulfil an additional benefit for the user

The additional benefit from a high-performance envelope can quickly justify the added first costs.

Superior insulation levels without thermal bridges covers the whole building envelope and is the first line of defence against cold, wind and rain. Shelter is the reason for living in a house, so it makes sense to optimize how well the house serves this function. Good insulation not only reduces energy costs, it also improves comfort because the interior room surfaces are warmer and there are no drafts. Furthermore, the risk of condensation, structural decay and mould can be ruled out with a well-built envelope (tight, minimal thermal bridges and overall U-values lower than 0.15 W/m²K). Finally, the energy savings from the superior envelope are so high that the total costs of construction and heating are less than the total costs of a standard envelope.

High-performance windows with U-values less than 0.85 W/(m²K) also provide superior thermal comfort. On winter nights the minimum inner surface temperatures can be kept higher than 17°C without a radiator or convector beneath the window (also a cost saving).

Mechanical ventilation with heat recovery can cut heat losses by about 70 per cent compared to window ventilation, while substantially raising the room air quality. The air change rate necessary for good indoor air quality (30 m³/h and person) can be provided by a small and highly efficient ventilation unit day and night. The carbon dioxide (CO_2) concentration is then far less than in standard residential buildings.

The above examples illustrate that the extra benefit of high-performance building components are often the decisive factors in decisions and therefore have to be calculated into the economics. The additional costs for superior construction of the whole building, including technical systems, can be 7 per cent to 15 per cent (Schöberl et al, 2003; Berndgen-Kaiser et al 2004). Basic and added construction costs for high thermal performance components are presented in Table 4.3.1. The area-specific numbers (€/m²) in this table are related to the net floor area of the building, assumed to be 120 m².

Investigations of Schöberl et al (2003) in Vienna, Austria, and Berndgen-Kaiser et al (2004) in Nordrhein-Westfalen (NRW), a federal state of Germany, show that extra costs for high thermal performance components are not extraordinarily high. The extra cost numbers given by these authors are 7% (social dwellings in Vienna) and 10 per cent to 15 per cent (single family houses in NRW).

Extra investment costs of 10 per cent and 15 per cent are compared to the running costs, especially energy costs in Table 4.3.1. Assuming a realistic energy saving potential compared to the local building code, these energy saving costs can nearly compensate for the extra expenses. The price for a saved kWh is calculated to be €0.059/kWh for the scenario '+10 per cent' extra costs. The scenario '+15 per cent' extra costs leads to a price of €0.089/kWh of saved energy. The present cash values and annual rates of the investment costs, respectively, are calculated in Table 4.3.2, with subtraction of residual values as before.

Table 4.3.1 *Basic and added construction costs for high thermal performance components*

Technical data of building and supply	Standard	High performance (+10%)	High performance (+15%)
Heating demand (kWh/m²a)	100	15	15
Domestic hot water demand (kWh/m²a)	30	20	20
Sum of heat demand (kWh/m²a)	130	35	35
Energy saving potential (kWh/m²a)		95	95
Net floor area (m²)	120	120	120
Total annual heating demand (kWh/a)	15,600	4200	4200
Energy saving potential (kWh/a)		11,400	11,400
Basic building costs (€/m²)	1200	1200	1200
Extra costs for high thermal performance components (%)		10%	15%
Extra costs for high thermal performance components (€/m²)		120	180
Investment			
Investment 1: construction costs (€)	144,000	144,000	144,000
Investment 2: extra costs of high thermal performance components (€)		14,400	21,600
Investment costs 1 + 2 (€)	144,000	158,400	165,600
Investment costs 1 + 2 minus residual value (€)	95,836	105,420	110,211

Table 4.3.2 *Annual rate of total costs*

Energy costs (end energy) (€/kWh$_{el}$)		0.055			0.10	
Annual rate for investment 1 (€/a)	6743	6743	6743	6743	6743	6743
Management, taxes, etc., running costs for investment 1 (€/a)	1000	1000	1000	1000	1000	1000
Annual rate for investment 2 (€/a)	0	674	1011	0	674	1011
Maintenance expenses running costs for investment 2 (€/a)	150	150	150	150	150	150
Costs for heating (end) energy (€/a)	858	231	231	1560	420	420
Annual rate of total investment costs 1 + 2 (€/a)	**8751**	**8798**	**9136**	**9453**	**8987**	**9325**
Costs for saved kWh (€/kWh)	-	**0.0592**	**0.0887**	-	**0.0592**	**0.0887**
Present value of total costs						
Present value of investment 1 + 2 (€)	95,836	105,420	110,211	95,836	105,420	110,211
Present value of running investment costs 1 + 2 (€)	16,344	16,344	16,344	16,344	16,344	16,344
Present value of energy costs (€)	12,194	3283	3283	22,171	5969	5969
Present value of total costs (€)	124,375	125,047	129,839	134,352	127,733	132,525
Additional costs (€)		672	5464		-6619	-1827
Percentages		0.5%	4.4%		-4.9%	-1.4%

4.4 Summary and outlook

High-performance housing is construction with very good insulation, with negligible thermal bridges, low U-value windows, air-tight envelope construction and mechanical ventilation with heat recovery. These high-quality components are essential to achieve an extremely low heating load (10 W/m^2) compared to standard buildings, and this results in some added costs. The benefits from this investment are easily offset by the economy of a smaller heating system (thanks to the reduced heating load), drastically reduced operating costs and the value of improved comfort.

Nonetheless, the extra costs must compete with other features typically offered as extras for a home. Unlike marble floors, investing in better construction is rewarded by lower total costs; a marble floor does little to reduce operating costs. Furthermore, the needed high quality components are increasingly catalogue items and therefore the cost difference is diminishing.

For home buyers whose purchase is at their borrowing limit, the drastically reduced energy costs protects them from being financially squeezed by the volatility of imported energy prices, and from one oil crisis after the next.

Purchasing a high-performance house therefore turns out to be a win–win story for all groups of people involved:

- the customer who gets a comfortable sustainable and, hence, affordable home;
- the people producing the buildings and related products: architects, craftsmen, industry;
- the national economy which profits from an investment locally versus exporting capital to energy exporting countries for decades ahead, and this at steadily increasing rates as energy prices rise; and
- future generations, who will profit from a larger remaining supply of non-renewable resources and from a cleaner environment.

References

Berndgen-Kaiser, A., Fox-Kämper, R., Reul, J. and Helmerking, D. (2004) *Passivhäuser in NRW Auswertung, Projektschau Wohnerfahrung*, Institut für Stadtentwicklungsforschung und Bauwesen (ILS), Aachen, Germany

Bundesgesetzblatt (2002) 'Verordnung über den Energiesparenden Wärmeschutz und Energiesparende Anlagentechnik bei Gebäuden', Energieeinsparverordnung – EnEV, verkündet am 21 November 2001, 1 February, no 59

Feist, W. (2001) *Wärmebrückenfreies Konstruieren*, Passivhaus Institut, Arbeitskreis Kostengünstige Passivhäuser, Protokollband no 16, 2, Auflage, Darmstadt, Germany

Feist, W. (2002) *Architekturbeispiele Wohngebäude*, Passivhaus Institut, Arbeitskreis Kostengünstige Passivhäuser, Protokollband no 21, first edition, Darmstadt, Germany

Feist, W. (2004) *Einsatz von Passivhauskomponenten für die Altbausanierung*, Passivhaus Institut, Arbeitskreis Kostengünstige Passivhäuser, Protokollband no 24, 1, Auflage, Darmstadt, Germany

Feist, W. (2005a) *Hochwärmegedämmte Dachkonstruktionen*, Passivhaus Institut, Arbeitskreis Kostengünstige Passivhäuser, Protokollband no 29, 1, Auflage, Darmstadt, Germany

Feist, W. (2005b) *Lüftung bei Bestandssanierung*, Passivhaus Institut, Arbeitskreis Kostengünstige Passivhäuser, Protokollband no 30, 1, Auflage, Darmstadt, Germany

Feist, W. (2005c) *Zur Wirtschaftlichkeit der Wärmedämmung bei Dächern*, in Protokollband no 29, Arbeitskreis Kostengünstige Passivhäuser (AKKP), 1, Auflage, Darmstadt, Germany

Feist, W., Baffia, E. and Sariri, V. (2001) *Wirtschaftlichkeit ausgewählter Energiesparmaßnahmen im Gebäudebestand*, Studie im Auftrag des Bundesministeriums für Wirtschaft, Abschlussbericht 1998, Passivhaus Institut, 3, Auflage, Darmstadt, Germany

Kaufmann, B. (2005) *Das Passivhaus – der Entwicklungsstand ökonomisch betrachtet*, Proceedings of the International Passive House Conference 2005 in Ludwigshafen/Rhein, Germany

Kaufmann, B., Feist, W., John, M. and Nagel, M. (2002) *Das Passivhaus – Energie-Effizientes-Bauen*, Informationdienst Holz, Holzbau Handbuch, Reihe 1, Teil 3, Folge 10, DGfH, München, Germany

Kaufmann, B., Feist, W. and Pfluger, R. (2003) *Technische Innovationstrends und Potenziale der Effizienzverbesserung im Bereich Raumwärme*, Studie im Auftrag des Institut für Ökologische Wirtschaftsforschung (IÖW), Berlin, Germany

Kaufmann, B., Feist, W., Pfluger, R,. John, M. and Nagel, M. (2004) *Passivhäuser erfolgreich planen und bauen, Ein Leitfaden zur Qualitätssicherung im Passivhaus*, Erstellt im Auftrag des Instituts für Stadtentwicklungsforschung und Bauwesen (ILS), Aachen, Germany

Peper, S., Feist, W. and Sariri, V. (1999) *Luftdichte Projektierung von Passivhäusern, Eine Planungshilfe*, CEPHEUS Projektinformation no 7, Fachinformation PHI-1999/6, Passivhaus Institut, Darmstadt, Germany

Reiß, J. (2003) *Ergebnisse des Forschungsvorhabens Messtechnische Validierung des Energiekonzeptes einer großtechnisch umgesetzten Passivhausentwicklung in Stuttgart-Feuerbach*, Passivhaustagung Hamburg, Fraunhofer IBP, Stuttgart, Germany

Schöberl, H., Hutter, S., Bednar, T., Jachan, C., Deseyve, C., Steininger, C., Sammer, G., Kuzmich, F., Münch, M. and Bauer, P. (2003) *Anwendung der Passivhaustechnologie im sozialen Wohnbau, Projektbericht im Rahmen der Programmlinie Haus der Zukunft*, Bundesministerium für Verkehr, Innovation, Technologie, Wien, Austria

Steinmüller, B. (2005) *Passivhaustechnologie im Bestand – von der Vision in die breite Umsetzung*, Proceedings of the International Passive House Conference 2005 in Ludwigshafen/Rhein, Germany

Wärmeschutz und Energieeinsparung in Gebäuden (2001) DIN V 4108, Teil 7, Luftdichtheit von Gebäuden, Anforderungen, Planungs und Ausführungsempfehlungen sowie beispiele, Deutsches Institut für Normung e. V., Berlin, Germany

World Energy Outlook 2000 (2000) *Highlights/IEA*, International Energy Agency, OECD, Deutsche Ausgabe, Weltenergieausblick, Paris

5

Multi-Criteria Decisions

5.1 Introduction

Sustainable building design involves a wide range of complex issues within fields of building physics, environmental sciences, architecture and marketing. An integrated approach to sustainable building design would address these diverse aspects in a holistic way. Sustainable building design views the individual building systems not as isolated entities, but as closely connected and interacting with the rest of the building. In an integrated approach, all the diverse design criteria need to be traded off against each other in order to make decisions for the 'overall goodness' of the design.

Unfortunately, 'sustainable' building measures have often been 'optimized' without taking the whole building performance into account. For example, a design is optimized for its energy performance within economic constraints, with too little regard for the other important dimensions that must be considered. There are numerous examples of how comfort issues, environmental issues and aesthetics have not been properly weighted in decisions.

Close cooperation between the design team and the client is imperative for the success of sustainable building design, as in any design. It is important that environmental considerations are included in the early design stages because the decisions taken here are essential for achieving a well-integrated sustainable building.

In this complex decision-making context, it may be useful to follow a structured approach in order to ensure proper consideration of all important issues, as well as to formalize the evaluation process and to make the value judgements as consistent and transparent as possible.

5.2 Multi-criteria decision-making (MCDM) methods

Inger Andresen and Anne Grete Hestnes

5.2.1 Introduction

Multi-criteria decision-making (MCDM) methods are systematic approaches to solving problems with a range of attributes (criteria) that may have different units of measurement and may be conflicting. The aim is to help organize conflicting and complex information in a way that facilitates making a decision. In addition, the methods strive to help decision-makers be conscious of their own value system and to learn of the values of others. MCDM methods have been successfully included in various 'green building' assessment methods – for example, the BREEAM method (Prior, 1993) and the LEED method (USGBC, 1999).

The MCDM approach described in this section is based on Andresen (2000) and Balcomb et al (2002). Within the International Energy Agency (IEA) Task 23, a computer tool called MCDM'23 was developed to support this process (see www.iea-shc.org/task23/).

5.2.2 How to use MCDM in the design of solar sustainable buildings

Several phases of building design, particularly during the early stages of design, tend to be iterative or cyclical in nature. Any such design cycle might benefit from an MCDM approach, both for structuring the design work and as part of the evaluation phase. It is recommended to use a trimmed-down or simplified version of the method during the early phases of design and then to use a fuller, more comprehensive version later in the process.

The application of the MCDM method may be separated into seven main steps:

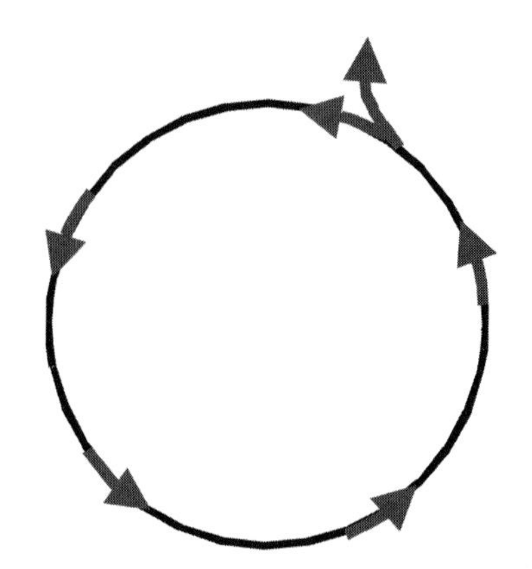

Source: SINTEF Building and Infrastructure, Norway, www.sintef-group.com

Figure 5.2.1 *The cyclical process of decision-making in design*

1. Select main design criteria and sub-criteria.
2. Develop measurement scales for the sub-criteria.
3. Weight the main criteria and sub-criteria.
4. Generate alternatives.
5. Predict performance.
6. Aggregate scores.
7. Present and discuss results, and make decisions.

Step 1: Selection of main design criteria and sub-criteria

The client is the ultimate arbiter of criteria; but it will normally be necessary for the design team to discuss and interpret the client's priorities and to add needed additional criteria before design begins, preferably at their initial meeting. Typically, many priorities are defined in the brief and the team priorities need to reflect those of the client. The importance of having the team specify criteria is to foster the development of a common mission for the design team, and to have an agreed-upon reference for evaluating the performance of design.

Although most of the work of criteria selection is done in the programming stage, criteria may be added, removed or reformulated as the design proceeds. The number and nature of the criteria will vary from case to case. Checklists of criteria may be used to help the search and to ensure that no important issues have been overlooked. Some criteria will be quantifiable, such as annual energy use. Others will be qualitative, such as architectural expression.

In order to have a manageable number of criteria, the number of main design criteria and sub-criteria should not be too large. A maximum of six to eight main criteria and a maximum of six to eight sub-criteria underneath each main criterion is recommended. A good procedure is to develop an exhaustive list first and then refine it by:

- eliminating criteria of little importance;
- grouping those that remain into main categories;
- selecting the titles for the main design criteria; and
- refining the sub-criteria.

The approach recommended is to start out wide with general, strategic criteria, and then narrow in, proceeding to specific criteria until a level is reached that is reasonable. More than 30 sub-criteria would probably be too many – for example:

- main goal (for example, sustainable building);
- main criteria (for example, resource use);
- sub-criteria (for example, annual fuels);
- indicators (for example, kWh/m^2a).

Table 5.2.1 shows a list of main design criteria and sub-criteria for solar building design that has been proposed by the IEA Task 23 participants. It is representative, but not necessarily comprehensive, for solar sustainable housing design.

The exact selection of criteria will be context and design phase dependent. Even if the main design criteria are the same, the sub-criteria may differ from design phase to design phase.

In the pre-design phase, the criteria need to be quite general. These can be criteria such as volume, shape and orientation, functionality, resource use and environmental loading. At this point it may be appropriate to keep the discussion of architectural concepts as a separate discussion.

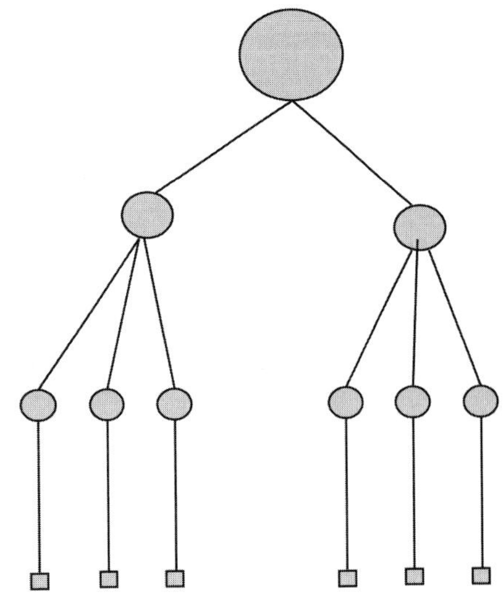

Source: Inger Andresen, SINTEF, www.sintef-group.com

Figure 5.2.2 *The hierarchical structure of design criteria*

Table 5.2.1 *Example of main design criteria and sub-criteria*

Main design criteria	Sub-criteria
Life-cycle cost	Construction cost Annual operation cost Annual maintenance cost
Resource use	Annual electricity Annual fuels Annual water Construction materials Land
Environmental loading	CO_2 emissions from construction Annual CO_2 emissions from operation SO_2 emissions from construction Annual SO_2 emissions from operation NO_x emissions from construction Annual NO_x emissions from operation
Indoor climate	Air quality Lighting (including daylight) Thermal comfort Acoustic
Functionality	Functionality Flexibility Maintainability Public relation value
Architectural expression	Identity Scale/proportion Integrity/coherence Integration in urban context

In the concept design phase, when certain decisions already have been made, more specific criteria may be considered. These can be criteria related to the structure and the systems of the whole building, such as cost (life-cycle cost), functionality (multi-functionality, modularity, flexibility), indoor climate, resource use, and compatibility with the (already chosen) architectural concept.

In the design development phase, yet more specific criteria may be applied. Nevertheless, some of the headings may be the same.

Step 2: Developing scales for the sub-criteria

A scale for each sub-criterion is necessary to be able to measure the performance. A measurement scale is a way of converting a value into a score. A value can be a number or a phrase, depending on whether the criterion is quantitative or qualitative. Quantitative values are used for criteria that can be measured directly with numbers, such as annual energy use, life-cycle cost or carbon emissions. Qualitative values are words or phrases that can be used to characterize how well a building scheme rates against a particular criterion where the rating is more a matter of judgement, not normally subject to quantification. These are quality issues, such as architectural expression or functionality. Some criteria can be characterized either way, such as indoor air quality, which can be either qualitative or rated based on a numerical value.

The common measurement scale need not have a large number of intervals. This is because fine gradations do not make sense for the qualitative criteria; these criteria can only be described verbally, and humans have a limited vocabulary to express qualitative gradations. Furthermore, for the quantitative criteria, it does not make sense to have a very fine scale, partly because of the large uncertainties of the performance predictions in the early design phase, and partly because they should be compatible with the qualitative ones. A nine- or ten-point scale seems to be the maximum usable gradation. The ten-point scale is usually reduced to a seven-point scale because the performances that are assigned a score lower than four are so poor that they cannot be compensated elsewhere. The seven-point scale has also been widely used within behavioural sciences. Thus, a seven-point scale, with scores ranging from four to ten, seems to be appropriate.

Table 5.2.2 *The common measurement scale*

Score	Judgement
10	Excellent
9	Good to excellent
8	Good
7	Fair to good
6	Fair
5	Acceptable to fair
4	Marginally acceptable

The upper end, a score of ten, means that the building rates as 'excellent'. To be more exact, the ten means that the building is the 'best reasonably attainable' with regard to the particular criterion. This is a bit softer than saying that it is the best theoretically attainable.

The lower end, a score of four, means that it is just marginally possible to construct a building that scores so poorly. For example, the maximum building energy use allowed by regulation could be the lower bound. One is not legally allowed to construct a building that performs worse than the regulation.

The next step is to create a measurement scale for each of the criteria, indicating the assessment of the merit of achieving particular scores. The scale should be divided into intervals that are felt to be equal (i.e. the utility of a unit step on the scale must be the same whether it is at one or the other end of the scale).

The process of creating measurement scales should generate much discussion, causing participants in the process to focus on the interpretation of the criteria that they have defined in addition to assessment of options. It will often come to light that the same words have different meanings for different individuals, which can lead to a restructuring of the model. The process of setting end points on the scales can lead to an active search for alternative options, spurred by the feeling of: 'Can we not do better than that?'

Scaling the objectives of a problem in this manner not only helps the design team arrive at uniform measurement scales, but is also a way of defining the general nature and context of the problem. The process of defining and constructing these measurement scales involves the collective participation of the entire team and allows each team member to express his or her own values and expertise to the group as a whole.

Table 5.2.3 *Example of measurement scales for a qualitative criterion (flexibility) and a quantitative criterion (energy use)*

Score	Judgement	Flexibility	Annual energy use (kWh/m^2)
10	Excellent	Different clients without changes	80
9	Good to excellent	Different clients by: • moving adjustable partitions; *or* • adding installations prepared for	100
8	Good	Different clients by: • moving adjustable partitions; *and* • adding installations prepared for	120
7	Fair to good	Different clients by rebuilding: • non-load bearing partitions; *or* • some installations	140
6	Fair	Different clients by rebuilding: • non-load bearing partitions; *and* • some installations	160
5	Acceptable to Fair	Different clients by rebuilding: • some load bearing partitions; *or* • several installations	190
4	Marginally acceptable	Different clients by rebuilding: • some load bearing partitions; *and* • several installations	250

Step 3: Weighting the main criteria and sub-criteria

The main design criteria weights reflect the central priorities of the project. Although the client may be ultimately responsible for selecting the final scheme, he or she deserves help from the design team. It is useful for the team to evaluate schemes for presentation to the client and to make a recommendation. In order to do this, priorities are necessary.

There are different ways of eliciting weights. The grading method works with the weights directly. The criteria weights are determined on a ten-point scale similar to the one used for scoring the performances. The decision-maker expresses the importance of criteria in grades on the scale of ten to four. The most important criterion receives a grade of ten. All of the other criteria are compared to this – for example, if a criterion is felt to be somewhat less important than the most important one, it receives a grade of eight.

A useful tool is to graph the weights in a chart. One can then visualize the results. Participants who respond better to graphs than numbers (almost all of us) will find this attractive.

Table 5.2.4 *The weighting scale*

Grade	Relative importance (compared with the most important criterion)
10	Of equal importance
9	
8	Somewhat less important
7	
6	Significantly less important
5	
4	Not important

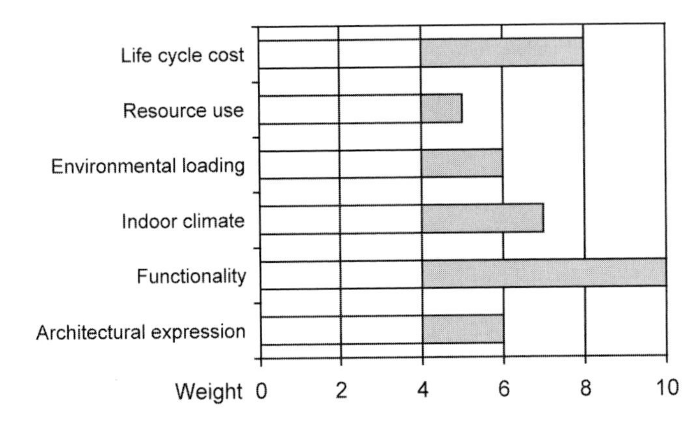

Source: SINTEF Building and Infrastructure, Norway, www.sintef-group.com

Figure 5.2.3 *Graphical presentation of the weights*

Step 4: Generate alternatives

The design team then generates alternative solutions that they think will answer well to the specified performance criteria and the priorities of the client. Since the generation of alternatives is mainly a craft, little formal guidance can be given. Each designer will have developed their own approach. It is important that the alternatives are generated keeping in mind the criteria and their relative importance. It may be wise to start out wide to test the extremes of the criteria and to be sure that a wide range of possibilities has been considered.

Step 5: Predict performance

The levels of predicted performance of the proposed solutions with respect to the criteria are determined. The performance prediction may be based on computer simulations, databases, rules of thumb, experience or expert judgement. The level of detail should be chosen based on an estimation of the available time and resources, and the accuracy required.

Step 6: Aggregate scores

The simple additive weighting model is used to aggregate the scores into one score based on the criteria weights (total score = sum of normalized criterion weight × criterion score):

$$S = \sum_{j=1}^{m} w_j s_j \qquad [5.1]$$

where S is the total score, m is the number of criteria, w_j is the normalized weight of the criterion, and s_j is the score for the criterion. The weights in the sum are first normalized by dividing the individual weight with the total sum of weights.

This is used first at the sub-criterion level to obtain the criteria scores and again at the main level to calculate the total score.

This step can be done using an Excel worksheet or the MCDM-23 software (Tanimoto and Chimklai, 2002). The procedure is used to perform both the aggregation of sub-criteria scores into a main criteria score and to aggregate the main criteria scores into one score that represents the overall performance of the building.

Blind faith in a single final score will invariably mask the process and judgements that went into developing a total score. One important value of the method is that it can be used as documentation of the selection process. This could be particularly important in the case of a public building, where it is important to clearly document the process and results.

Step 7: Present and discuss results, and make decisions

A star diagram is recommended for presenting the overall performance of an alternative (see Figure 5.2.4). The generation of the star diagram is a standard feature of Excel and is included in the MCDM-23 software. In this diagram it is possible to show multiple dimensions; thus, all the individual performance measures can be gathered into one picture. Each 'finger' represents the scale for one criterion. The performance on each criterion is plotted on each 'finger'. The centre of the star usually designates the minimum score for each criterion. The outer unit polygon represents the maximum score for each criterion. Although the star diagram may be used to give an indication of the overall performance of an alternative, it should be used with caution. This is because the main criteria are shown as if they are all equally important, whereas in most cases some criteria are more important than others (different weights). A bar diagram such as the one in Figure 5.2.5 may be used to show the products of weights and scores for each design scheme.

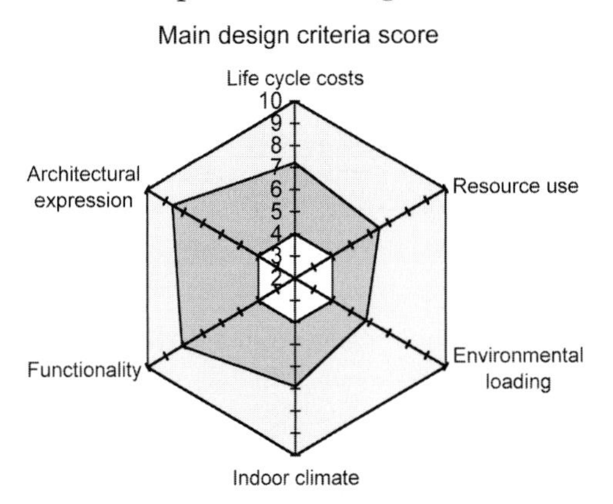

Source: SINTEF Building and Infrastructure, Norway, www.sintef-group.com

Figure 5.2.4 *A star diagram showing the scores for each criterion*

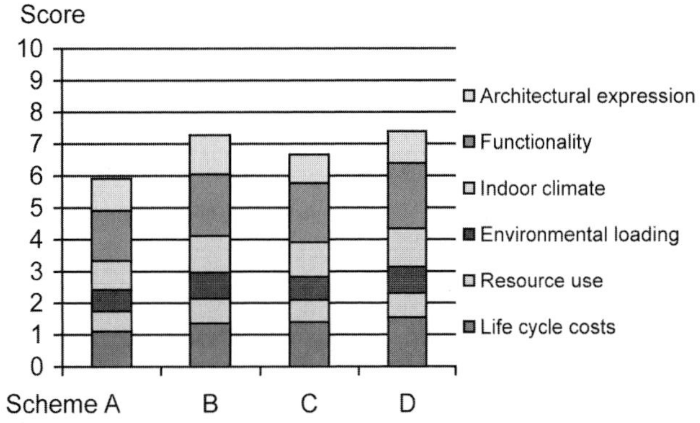

Source: SINTEF Building and Infrastructure, Norway, www.sintef-group.com

Figure 5.2.5 *A bar diagram showing the total weighted scores for each alternative design scheme*

The team should study their results, come to a conclusion regarding their recommendation, and present this to the client for his or her final decision. If the presentation and logic are clear, and if the team and the client were working towards commonly agreed goals, the conclusion will usually be evident.

It may be that at this point a new scheme should be developed that combines the best features of the leading scheme while eliminating some of its problems.

The most important use of the method is to structure the discussions and to help the design team reach a common understanding of the problem at hand and of the value of the various solutions.

Thus, the design team will be able to make a better recommendation to the client, and the client will have a better basis for their decisions.

References

Andresen, I. (2000) *A Multi-Criteria Decision-Making Method for Solar Building Design*, PhD thesis, Department of Building Technology, Faculty of Architecture, Planning and Fine Arts, Norwegian University of Science and Technology, Trondheim, Norway

Balcomb, D., Andresen, I., Hestnes, A. G. and Aggerholm, S. (2002) *Multi-Criteria Decision-Making: MCDM-23. A Method for Specifying and Prioritizing Criteria and Goals in Design*, International Energy Agency, Solar Heating and Cooling Programme, Task 23 Optimization of Solar Energy Use in Large Buildings, www.iea-shc.org

Prior, J. (1993) *Building Research Establishment Environmental Assessment Method (BREEAM)*, Building Research Establishment, Garston, UK

Tanimoto, J. and Chimklai, P. (2002) *MCDM'23: IEA Task 23 Multi-Criteria Decision-Making Tool*, Kyushu University, Japan

USGBC (1999) *LEED Green Building Rating System: Leadership in Energy and Environmental Design*, US Green Building Council, San Francisco, CA

5.3 Total quality assessment (TQA)

Susanne Geissler and Manfred Bruck

In Austria, a total quality building assessment and certification system was developed in 2000 to encourage the construction of user-friendly, environmentally benign and cost-efficient buildings. Builders like the system because it is both a quality management and marketing instrument. Furthermore, it requires only a small effort to collect and analyse data. In 2002 and 2003, five projects were partly financed by the Austrian government as pilot tests of this assessment method. By mid 2003, 15 buildings were under assessment, of which half achieved the certificate prior to construction. In spring 2004, the first buildings were awarded the second certificate on completion of construction.

5.3.1 Scope

Total quality assessment (TQA) aims to provide the information necessary to design a high-performance ecological building and to confirm this performance both prior to construction and on completion of construction during commissioning. TQA does not assess architectural quality; it is limited to technical issues addressing:

- ecologically relevant aspects (i.e. energy use, CO_2 emissions and water consumption);
- economically relevant aspects (i.e. investment costs, operational costs and external costs); and
- socially relevant aspects (i.e. thermal comfort in summer and winter, green spaces and accessibility for disabled people).

The TQA framework uses a life-cycle approach in that assessment criteria consider the impacts caused during both construction (building materials) and operation (energy supply systems). Based on these data, total life-cycle energy use and CO_2 emissions of buildings are assessed. This assessment is limited to the technical performance of the building. Impacts from a specific type of user behaviour during the operation of the assessed building are not taken into account. TQA has been revised to meet the requirements of the European Building Performance Directive in January 2006 (Directive 2002/91). TQA was developed in cooperation with construction companies, building owners, architects and engineers. The documentation of this process is available at www.e3building.net. The main requirements in developing the process were:

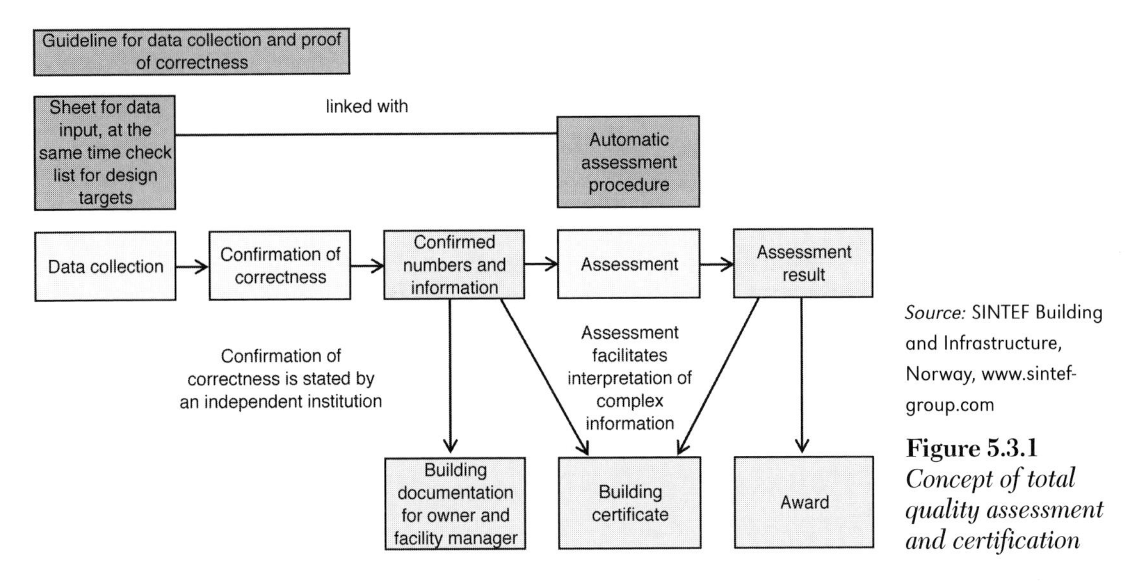

Source: SINTEF Building and Infrastructure, Norway, www.sintef-group.com

Figure 5.3.1 *Concept of total quality assessment and certification*

- low effort for data collection;
- transparency;
- easy and time-saving assessment; and
- usefulness of TQA result for marketing.

To achieve these goals:

- the assessment uses data derived from planning and quality control measures necessary during construction;
- the system is a computer program easily checked by the user; input data remain visible beside the assessment results; input data must be confirmed by hand calculations, drawings, etc. in order to allow independent experts to verify the TQA file and issue a TQA certificate; and
- the programme has several automatic calculations to save the user time.

The TQA system has three components:

1 guidelines on which criteria are used, which data are needed and how to improve the design;
2 a computerized data framework with an automatic assessment procedure; and
3 a procedure for building certification based on the results of the building TQA. This impartial assessment can then be useful for marketing. Figure 5.3.1 describes the TQA concept.

The TQA criteria are organized in the following categories (for detailed information see www.tq-building.org):

- resource consumption;
- harmful impacts on humans and the environment;
- comfort;
- longevity;
- safety and security;
- design/planning quality;
- quality control during construction;
- quality of amenities and site; and
- economics.

5.3.2 Assessment scale and weighting

The TQA system is based on design targets: for each criterion there is an assessment scale, consisting of either:

- 8 steps, each rated from –2 to +5; or
- 6 steps, each rated from 0 to 5.

The best score is five. Negative scores indicate such poor performance that the building will fail the assessment. In order to sum up results, weighting factors are defined that are based on experts' discussions. These factors are kept constant to ensure comparable results (see Table 5.3.1).

Each step on the scale corresponds to a design target. For each assessment criteria, designers and clients define the design targets from the assessment scales. This serves as a reminder of which aspects are most important to consider. Performance-oriented targets are preferable in order not to limit the design team to few measures and/or technologies, but rather to allow freedom to develop the best solution under the given circumstances. Table 5.3.1 illustrates assessment scales and weighting factors.

Table 5.3.1 *Weighting factors for energy-use criteria under the category of resource consumption*

Categories and criteria	Scores	Weighting factor	Weighted scores
1 Resource use		0.16	
1.1 Energy use of the building	3.25	0.30	0.98
1.1.1 Primary energy use for building materials	5.00	0.25	1.25
1.1.2 Heating energy use	5.00	0.25	1.25
1.2.3 Share of renewable energy carriers to cover heating energy use	2.00	0.25	0.50
1.1.4 Solar energy for DHW	1.00	0.25	0.25

5.3.3 Calculation of external cost as an optional function

The TQA system provides the option to calculate external costs of energy and materials use as an alternative method of presenting results. External costs result from production and service processes. Usually the polluter pays principle applies; but for political reasons some costs are left unpaid and must be borne by taxes or, still worse, by those affected. Estimating external costs are a form of risk management. In the construction sector, external costs are caused by airborne emissions from energy used to produce building materials and component and energy used during operation. The latter are the most important external costs of buildings. For electricity, oil and gas used to operate buildings, external costs amount to €0.018/kWh to €0.021/kWh. This amounts to 12 per cent of electricity costs and 45 per cent to 50 per cent of oil and gas costs. For wood fuels, external costs account for €0.0013/kWh to €0.0017/kWh, or 4 per cent to 8 per cent of fuel costs (Bruck and Fellner, 2001).

Reducing external costs by investing more money for energy saving building concepts and CO_2-reduced heating systems is cost effective. Measures that save energy during building operation not only reduce external costs, but also reduce direct operational cost. Usually, the payback time is very short.

5.3.4 Recommendations

During pre-design, the TQA criteria, weighting factors and performance targets make it evident for the building owners what aspects must be emphasized in the project.

TQA encourages the integrated optimization of a building by providing assessment criteria that are independent from each other. Heating energy consumption is not reduced at the expenses of embodied energy of building materials because embodied energy of building materials is part of the assessment, too. The TQA method supports the design team in reducing heating energy consumption, as well as embodied energy of materials.

The requirement to calculate follow-up costs prevents the design team from choosing options characterized by low investment costs but high life-cycle costs. Finally, the TQA explicitly requires the provision of daylight and sunshine in winter as constraints for the optimization.

5.3.5 An example of total quality assessment: The Wienerberg apartment project

One of the winning projects of the builders' competition was the concept of a multi-storey brick house in passive house quality with 97 dwellings (sizes between 54 m^2 and 102 m^2). Special attention was given to ecological and economic criteria.

Due to east–west orientation and shading by neighbours, it was not possible to maximize the solar gains by large windows. Therefore, the architectural concept addressed minimization of transmission losses (i.e. small openings in a large volume in order to achieve the passive house standard).

Source: GEBOES Builders, Vienna

Figure 5.3.2 *Wienerberg city apartment building*

A flexible ground plan guarantees that between two flats one room can be assigned to either of the adjacent dwellings. The external walls of concrete structure with face brick are highly insulated with ecologically recommended materials. Ventilation losses are minimized by a mechanical ventilation system with heat recovery, combined with a ground-coupled heat exchanger. Backup heating is provided by a central gas furnace. Hot water heating demand is partly covered by roof-integrated solar collectors.

The Wienerberg city project was assessed with the TQA method (version 2.0) by a representative of the builder and an external expert, who collected data, filled out the assessment tool and compiled the supporting documents for submission for certification.

The tool used for the project was 'TQA version 2.0'. This tool is based on the following categories and criteria (for details on sub-criteria and indicators, see www.tq-building.org):

1 *Resource consumption:*
   - energy consumption of the building;
   - quality of the soil;
   - consumption of potable water; and
   - use of building materials.
2 *Harmful impacts on human beings and the environment:*
   - airborne emissions;
   - solid waste;
   - wastewater;
   - individual car transport;

- human toxicity and eco-toxicity of building materials;
- avoidance of radon;
- electro-biological installation; and
- avoidance of mould.

3 *Comfort:*
- indoor air quality;
- thermal comfort;
- daylight (daylight coefficient in the largest room of a typical unit);
- winter sun (hours of sun on 21 December in the largest room of a typical unit);
- sound protection; and
- building automation.

4 *Longevity:*
- flexibility of the structure with respect to changing user needs; and
- maintenance and performance.

5 *Safety and security:*
- natural risks (avalanches, earthquakes, etc.);
- others: high voltage;
- fire protection; and
- no barriers.

6 *Design (planning) quality.*

7 *Quality control during construction:*
- construction supervision; and
- final inspections;

8 *Quality of amenities and site:*
- access to public and other services; and
- amenities of building.

9 *Economic performance:*
- construction costs;
- operation costs;
- life-cycle costs; and
- external costs.

As a result of this analysis, a TQA certificate was issued prior to construction. This certificate confirms that all data submitted for assessment are correct. The next step, not described here, is the submission of data and supporting documents as the condition for achieving the second certificate prior to handing over the building. The second certificate proves that the building was constructed the way it was planned – or not. It is possible that a building achieves the first certificate prior to construction and fails in the second assessment after construction is completed.

The certificate consists of a briefcase including all relevant information in short form (four pages) with approximately 30 pages of computer printouts containing detailed information. All input data used for assessment are presented, complete with additional information that might be relevant for potential users of the building.

Table 5.3.2 lists some of the most important indicators that account for the results of the assessment categories in the TQA certificate.

Table 5.3.2 *Performance of selected indicators of total quality assessment categories*

TQA categories	Performance of selected indicators
Resource consumption (2.5 points of 5)	11.75 kWh/m^2 a primary energy for space heating Share of renewable energy for space heating: 21.8% Thermal solar energy for domestic hot water planned
Environmental loadings (3.8 points of 5)	Emission of 1.45 kg CO_2 equivalent/m2a from space heating Waste management concept for the building site available Sealing of soil on building site: 54.5% Separate billing of potable water planned; no water saving appliances and no rainwater use planned
Indoor quality (3.2 points of 5)	Concept of how to avoid indoor air pollution from materials available Efficient mechanical ventilation with heat recovery planned High comfort in summer for the majority of the flats proven High comfort in winter for all flats proven 95% flats with 1.5 hours of sunshine in winter proven
Quality control design (4.0 points of 5)	Design targets defined; alternative design options evaluated; all types of consequential costs calculated; no building information system and no building management system planned
Quality of amenities and site (3.0 points of 5)	Hobby room and playground provided for children Access to a common green space for all inhabitants available Balcony or loggia for almost all flats (four flats without) planned Nearby shops for daily needs (< 300 m)

Summary and conclusions from the example

It was the builder's goal to construct a multi-family structure to the Passivhaus standard. This decision is reflected in 5 points out of 5 in the assessment sub-category 'Heating energy use' under the category of 'Resource consumption'. However, the lower mark of 2.5 in the category 'Resource consumption' reflects the share of renewable energy needed to cover the remaining energy demand. It is low because, for example, water saving measures are insufficient, more ecological materials could have been used in the construction and much of the ground is sealed.

The assessment result in the category 'Indoor quality' (3.2 points out of 5) is based on the mark of 4 for the ventilation system, 4 for comfort in summer and 5 for comfort in winter. Furthermore, 4 was given for sunshine in winter and good marks were given for noise isolation. These very good marks are offset due to the fact that daylight availability achieves only 1 point. This reduces the overall score to 3 out of 5 points for the category of 'Indoor quality' as a whole.

The results reflect where the builder gave priority. Energy savings were more important than other criteria. In order to encourage better buildings, the assessment scales of TQA are very demanding and, currently, it is very difficult to achieve the mark of 5 points at reasonable costs.

The TQA approach had a valuable effect on the building performance of Wienerberg city, on the builder's staff and on the design team. The list with TQA criteria was used as a checklist as the project evolved. In the end, the builder improved the design to achieve an overall performance of 3.44 points out of 5. Due to the demanding TQA scales, not one of the buildings assessed until now has achieved 4.5 or more points. Typical scores range between 2.5 and 3.9 out of 5 points. Builders who use the process and realize that during the preliminary examination their building achieve less than 2 points either upgrade the design targets or stop the TQA.

Discussion of TQA criteria educated the design team, created awareness and increased the likelihood that a team member will adopt the TQA system in a new design team. Training for the builder's sales staff was essential in order to enable them to use the TQA result – the TQA certificate – as a marketing tool.

Acknowledgements

Development of the total quality assessment method was funded by the ministries of agriculture, forestry, environment and water management, and by the ministries of economic and labour, and transport, innovation and technology. For more information, contact susanne.geissler@arsenal.ac.at or bruck@nextra.at.

References

Bruck, M. and Fellner, M. (2001) *Externe Kosten: Referenzgebäude und Wärmeerzeugungssysteme, Band III*, Wien, Austria, January 2002, bruck@ztbruck.at

Directive 2002/91 (2002) EU-Richtlinie 'Gesamtenergieeffizienz von Gebäuden', Richtlinie 2002/91/EG des europäischen Parlaments und des Rates vom 16 Dezember 2002 über die Gesamtenergieeffizienz von Gebäuden, Amtsblatt der Europäischen Gemeinschaften L1/65, Germany

Geissler, S. and Bruck, M. (2001) *ECO-Building: Optimierung von Gebäuden. Entwicklung eines Systems für die integrierte Gebäudebewertung in Österreich*, Ergebnisbericht, www.hausderzukunft.at

Geissler, S. and Bruck, M. (2004) *Total Quality (TQ) Planung und Bewertung von Gebäuden*, Ergebnisbericht, www.hausderzukunft.at

Geissler, S. and Tritthart, W. (2002) *IEA Task 23 Optimization of Solar Energy Use in Large Buildings*, Berichte aus Energie und Umweltforschung, 23/2002, Herausgegeben vom BMVIT, Wien, Austria, www.ecology.at

Hestnes, A. G., Löhnert, G., Schuler M. and Jaboyedoff, P. (1997–2002) The Optimization of Solar Energy Use in Large Buildings: IEA Task 23: Energetische, ökologische und ökonomische Optimierung von Gebäuden unter besonderer Berücksichtigung der Sonnenenergienutzung, Instrumente für den integrierten Planungsprozeß, www.task23.com

Websites

TQ Building www.tq-building.org
Arbeitsgemeinschaft IS wohn.bau www.iswb.at
Haus der Zukunft www.hausderzukunft.at
Austrian Institute for Applied Ecology www.ecology.at

6

Marketing Sustainable Housing

6.1 Sustainable housing: The next growth business

Edward Prendergast, Trond Haavik and Synnove Aabrekk

6.1.1 Main driving forces

Sustainable housing is currently finding its place in the housing market. As a result of global developments, public awareness and policy decisions, the housing industry is realizing that sustainability is an important market. Consumers begin to favour sustainable solutions for many reasons:

- direct savings from lower energy costs: experience shows potential savings as high as 75 per cent, with a payback as fast as two years for many features of such houses;
- these saving will be greater as non-renewable energy prices rise; and
- there are non-energy benefits:
  - better air quality, reducing asthma discomfort;
  - higher thermal comfort levels because of better insulation;
  - convenience (i.e. windows are easy to open, also making cleaning easier);
  - a better house, which will be easier to resell; and
  - taking responsibility for the environment.

Studies by Skumatz Economic Research Associates both in the US and in New Zealand show that homeowners of such houses value non-energy benefits, on average, more than twice as much as the energy savings (ACEEE, 2004).

6.1.2 Market position

Companies striving to be a market leader (whether local, regional, national or international) must differentiate their product from that of their competitors. Leaders cannot be identified as another 'follower'. This requires a consistent marketing concept and a product that fulfils the expectations created through the marketing.

Sustainable housing, until now, has been in an initial phase in the market. As it now enters into its growth phase, it is the right time for companies to define a leadership position. Delaying will make it more difficult to move away from being another follower. The main competitive tool of followers is limited to price. Successful market leaders promote sustainable housing with its non-energy benefits and added values. Brand building on sustainability is a decisive opportunity to become a market leader. Sustainability in housing renovation has at least the same market potential as sustainable new housing.

6.1.3 Marketing trends

Trends that can be observed in marketing today include the following:

- *Moving from national to international business*: most products are introduced on the domestic market first and the strategies used are adapted to penetrate international markets. The traditional thinking has been that each of the markets is different and has to be treated accordingly regarding product, marketing and communication. Because trends are crossing borders, businesses increasingly develop international communication strategies in order to build a strong international position.
- *Moving from 'product' to 'concept'*: traditional marketing has focused on product function. In order to show the 'added value' of the product, modern marketing focuses on 'concepts'. One such concept could be how the customer contributes to 'making the world a better place in which to live'. However, the message must be convincing and company philosophy must be consistent with this concept.
- *Shorter product life cycles*: a product being one of a kind does not last long. Companies are rapidly attracted to profitable markets. With an increasing number of competitors, the pressure to lower prices increases and profitability decreases. As a result, profitable periods are becoming shorter and newer products rapidly replace the market share of an existing product.
- *Immediate response to market changes*: dramatic events such as international conflicts and natural catastrophes strongly influence national markets. Companies are also affected (positively or negatively) by changes in demand as consumer reaction to these events. Energy efficient technology, for example, has benefited from higher energy prices and the need for security.
- *Branding*: as big companies launch more products and enter new markets, they use different brand names to differentiate themselves. There is now a trend towards fewer brands, with each brand associated with certain attitudes and values (conceptual marketing). The brand is used to take a defined position in the market. From that position, a broad range of products and services related to the brand can be marketed.
- *Changing consumer groups*: the populations of industrialized countries are ageing, and new housing is needed for this segment. Many older consumers have a relatively high income level and distinct requirements regarding comfort and security. Seniors are also often willing to invest in sustainability.

6.1.4 Marketing strategies

Marketing aims to communicate a company's products, services and philosophy to potential consumers. Successful marketing is consistent marketing, which requires a structured marketing campaign. These questions are summarised with the four 'Ps':

1. *Product (or service)*: what is to be sold?
2. *Price*: what are the competing products? How should the product be positioned in this market?
3. *Place*: where is the product to be sold? What distribution channels can be used?
4. *Promotion*: how will the product be communicated to customers?

A typical mistake is to rush directly to promotion. Such marketing is purely operational and not founded on strategic decisions. Marketing is not just making a nice brochure; it is linking the four 'Ps'. The four Ps are strongly interrelated and connected to the target group. Marketing identifies the target group and defines the product accordingly.

References

ACEEE (2004) *Summer Study on Energy Efficiency in Buildings*, 2000 Sustainable Building Conference, Asilomar, CA, August 23–27

6.2 Tools

Edward Prendergast, Trond Haavik and Synnove Aabrekk

6.2.1 The six-step process

This process can help companies to plan business development within sustainable housing:

1 *Information-gathering*: good decision-making is based on facts, not anticipation. It is a continuing challenge to differentiate between important and unimportant information (see section 6.3).
2 *Analysis*: based on the collected information, the position of the company relative to its competitive environment is analysed. Several market analysis tools are available for this purpose. The analyses help to identify the strategic marketing options (see section 6.3).
3 *Setting goals*: defining the objectives for the business includes setting qualitative and quantitative targets. Measurable targets are necessary to benchmark the degree of success. Measured interim performance influences the setting of new goals as strategies are implemented. Experience shows that only what is measured gets attention.
4 *Strategies*: good strategies clearly describe which products or services will be sold to whom, at what price and how. The main issue in planning strategies is setting priorities. For example, once it is decided to engage strategic partners for distributing a product, it is difficult to later change systems. It is therefore important to ensure that all strategic issues are addressed and that there are no inconsistencies.
5 *Action plans*: for each strategy, concrete action plans define what to do in order to achieve the goals. Good action plans state who is responsible, what the budgets are (external and internal costs) and a time schedule. A Gant diagram is useful for this purpose. As the company proceeds, new actions are added to the action plan. The important control question is then 'Which strategy is the proposed action meant to support?' Not being able to easily answer this question suggests that the strategy has been added without considering its purpose.
6 *Control*: as the company moves forward, progress must be measured to determine if the strategy is successful. By measuring the right factors, it is possible to isolate which actions have had an impact. Ongoing feedback can be helpful in revising or defining new strategies to adapt to market realities.

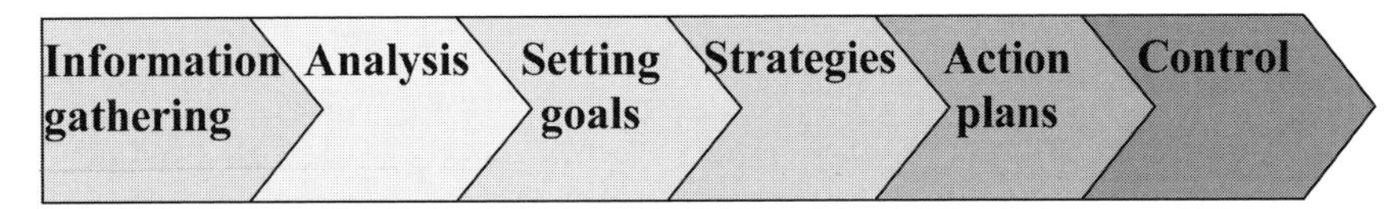

Source: Anliker AG, CH-6021 Emmenbrücke, www.anliker.ch

Figure 6.2.1 *The six-step process*

6.2.2 Information-gathering and analysis

Good marketing research consists of identifying and collecting information relevant for further decision-making. Collecting irrelevant information can be a distraction and must be avoided. 'Buzz words' for information-gathering and market analysis include:

- BI: business intelligence (has a wider scope than the two others);
- CI: competitive intelligence (focuses on competitive arena, commonly used in the US); and
- MIS: marketing intelligence systems (similar to CI: focus on continuous monitoring).

A worldwide association for professionals in this subject area is found at www.scip.org.

Before starting to gather information it is important to define the information that is needed and what types of analyses can be effective. There are two levels of analysis: PEST (political, economical, social and technological) and SWOT (strengths, weaknesses, opportunities and threats) analyses.

PEST analysis

PEST analyses examine political, economic, social and technological factors. Such factors indirectly affect business through their impact on the competitive arena. Before defining goals and strategies, it is necessary to identify which of the PEST factors are relevant for the actual business case to be analysed. Figure 6.2.2 illustrates the four categories of factors. The PEST analysis is helpful in converting desegregated information into systematic knowledge. This, in turn, can provide the foundation for defining goals and developing strategies.

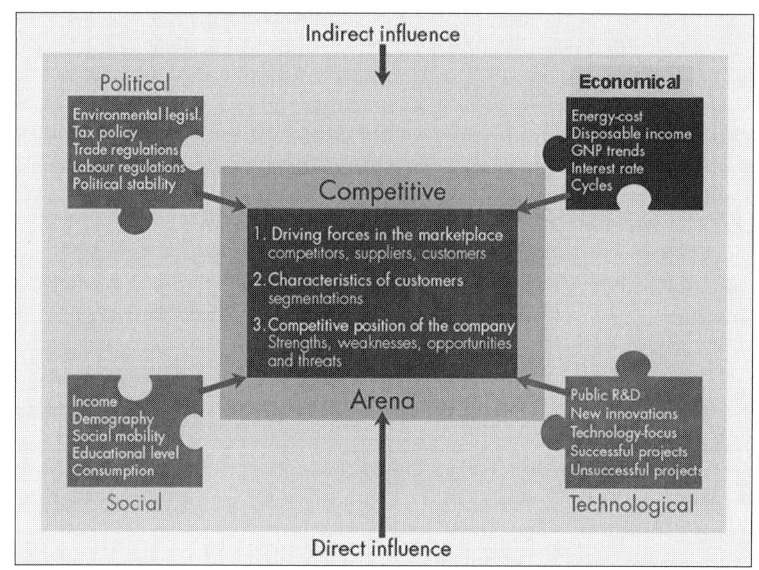

Source: Anliker AG, CH-6021 Emmenbrücke, www.anliker.ch

Figure 6.2.2 *Political, economic, social and technological (PEST) factors that influence the competitive arena*

SWOT analysis

SWOT analysis is used to understand a company's ability to seize opportunities and overcome threats, and provide the basis for strategic marketing decisions. It is performed using general knowledge about the market, experience within the company, and the product range specifications. The four elements of a SWOT analysis are:

1 *strengths*: positive attributes of the product and company (internal);
2 *weaknesses*: negative attributes of the product and company (internal);
3 *opportunities*: positive developments or attributes of the market, factors or actors which might affect business (external); and
4 *threats*: negative developments or attributes of the market, factors and actors which might affect the business (external).

6.3 A case study: Marketing new passive houses in Konstanz, Rothenburg, Switzerland

Edward Prendergast and Trond Haavik

6.3.1 Background

A private company chose to develop a housing type unfamiliar in the Swiss housing market. Their decision process resulted in a highly successful market introduction, which can be instructive for other firms with a similar goal.

At first the firm, Anliker AG, planned to build standard units for the housing estate in Konstanz. However, one of their architects, Arthur Sigg, convinced the company that a good business opportunity existed to develop sustainable (passive) housing at this location. The resulting success can be attributed to knowing the market and developing an appropriate sales approach. The main arguments for the company to take this risk, Arthur Sigg argued, were as follows:

- Building sustainable housing is consistent with the company philosophy of high quality construction with resulting low maintenance costs.
- The company can learn how to build and sell a new type of housing.
- This housing will improve the image of the company.
- A good profit can be made!

The company carefully considered the interaction between the buildings' infrastructure, green spaces and living space. To reduce the risk of building and selling such a new product, three different types of units were offered:

1 'villette': 3 blocks with 12 flats built to the Passivhaus standard (2 additional blocks were constructed afterwards);
2 'loft': 4 blocks with 32 units built to the less stringent Minergie standard;
3 'veranda': 6 blocks with 72 flats conventionally built.

The six-step process was applied in developing the marketing strategy, as follows.

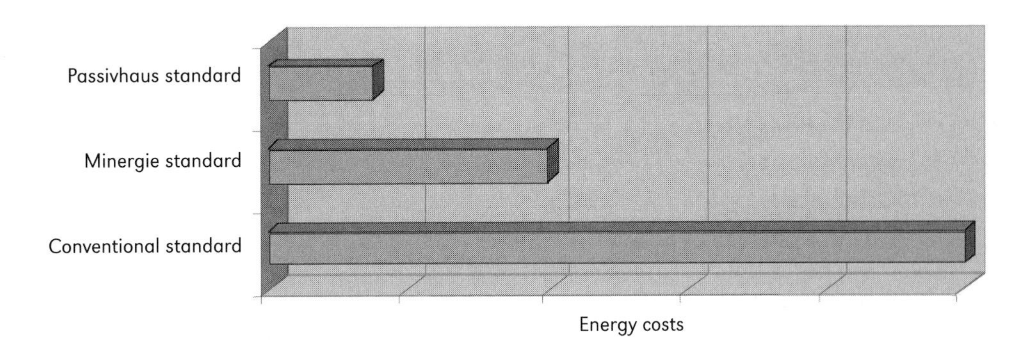

Source: Anliker AG, CH-6021 Emmenbrücke, www.anliker.ch

Figure 6.3.1 *Relative energy costs for housing: A marketing argument*

6.3.2 Information-gathering

Initial project phases drew on company knowledge of the market, competitors, suppliers and customers. The project team based their analyses on the following information:

- The Swiss typically rent their housing; only a small fraction of the population is homeowners.
- It is now possible to withdraw money from pension funds to finance the purchase of a home.
- The local cantonal bank offers two special loan types:
  - family loans with a fixed interest rate for five years – reduced by 0.50 per cent; 'environmental' loans that can be used as additional financing when buying eco-housing.
- The environment is in the national consciousness. Switzerland is a transit country for goods crossing Europe. This traffic causes pollution and has sensitized the populace.
- The Swiss energy demand is covered by hydropower (30 per cent), atomic power (30 per cent) and non-renewables, such as oil and gas (40 per cent).
- A low energy standard, Minergie, is widely recognized as a quality label with a market share of about 30 per cent of new housing construction.

Anliker collected information on the following:

- Who would be interested in this new housing type?
  - Who can be the suppliers and subcontractors?
  - Who are the competitors?
  - Which are the complementary industries/substitutes?
- What is the influence of each actor on the market?
- Who are the potential customers?
  - What are the characteristics of this market niche?
  - What are the needs of this market group?
- What is the right price for the product?
- How can Anliker achieve credibility for the product?
  - Are alliances possible with complementary industries?
  - Are alliances possible with potential competitors?

6.3.3 PEST analysis

This information was systematically evaluated in a PEST analysis to make decisions. A single supplier cannot change the PEST factors; but when they have been identified, it is possible to use the information to the advantage of the company or the product.

Product-cycle analysis

The classic product life cycle has five stages, illustrated in Figure 6.3.2. For each phase a different marketing strategy is needed.

The passive house as a product was situated between the introduction and growth phases. In the introduction phase, the market needs plentiful information from credible sources before a new product is accepted. Generally, in this phase a product earns little or no profit, but a company can make its name in the market. Later in the product life cycle the company has a significant advantage over its competitors and, accordingly, can earn a substantial profit.

During each phase different consumer types buy the product. It is possible to identify the market niches that 'match' each stage. In the introductory stage, these consumers (home buyers) either have a special interest in the product or they are so-called 'innovators', wanting to try something new. When these people have 'tested out' the product and given it credibility and a positive image, the 'followers' enter the market.

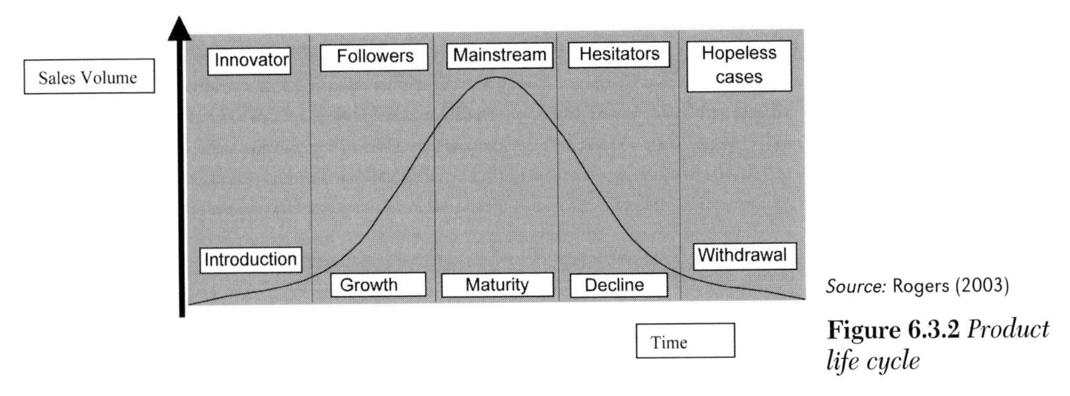

Source: Rogers (2003)

Figure 6.3.2 *Product life cycle*

Target-group analysis

Swiss residential planning is fragmented and strongly influenced by banks and private financing. There are no financial incentives from the government for sustainable housing, so market forces rule. Anliker singled out young families as a possible market niche for the passive houses. The questions, then, were: what drives this group of people? What kind of life situation are they in? What is important to them? What do they need?

By researching these questions, Anliker identified what is important for young families in their market niche. Young families and new house owners value:

- practical, economical flats adjusted to the needs of a young family with children;
- good architecture, design and environment; and
- a healthy environment for children.

This knowledge about the market and the product life cycle strongly influenced the planning of the project, how it was presented and, finally, how it was communicated to the targeted customers. Anliker identified the target group as being young families who are concerned about the environment and want to raise their children in a healthy environment.

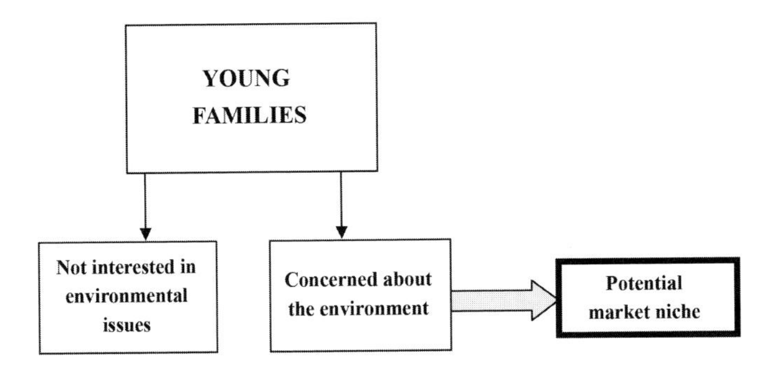

Source: Anliker AG, CH-6021 Emmenbrücke, www.anliker.ch

Figure 6.3.3 *Defining a potential market niche*

6.3.4 SWOT analysis

The SWOT factors for Anliker are summarized in Table 6.3.1, which is followed by an analysis of the different factors.

Table 6.3.1 *Anliker's strengths, weaknesses, opportunities and threats (SWOT) analysis of the passive house market*

Strengths	Weaknesses
1 Biggest company of its type within the Lucerne area 2 Appropriate business philosophy: high quality construction with low maintenance costs 3 Willingness and a drive to be innovative and take opportunities in the market 4 Good architecture, quality and design 5 Whole firm behind this project 6 Good reputation in the market; well known for service 7 Good public relations	1 7% higher construction costs 2 Lack of experience constructing passive housing 3 Lack of credibility in passive housing
Opportunities	**Threats**
1 An untapped market with significant growth potential 2 Increasing public awareness of environmental issues in the targeted market 3 Building to the Passivhaus standard is an investment in the future 4 Reduction of fossil fuels 5 The Cantonal Bank of Lucerne stimulates the growing interest in environmental thinking with a lower interest rate 6 The media has environmental concerns high on its agenda; free press coverage is possible	1 Private funding and banks strongly influence the market and are disinterested in developing the private housing market 2 House buyers still focus on initial costs and are less influenced by future operational costs (energy and maintenance)

Internal factors: The company's strengths

According to the company, they have a drive to innovate and take (or create) new opportunities in the market. Furthermore, their philosophy is to build good architecture of a high quality and with low maintenance costs. This means that the company philosophy mirrors the passive house product: innovation, new technology and know-how. Another simple factor also motivated the whole firm to build passive – all employees thought building passive houses was a good cause.

Anliker had several strengths that they could use in launching the product: the right business philosophy; competence in developing a new product; a good reputation in the market – the firm was solid and offered high quality products; and in the building sector they were also known for their good construction management, sensible pricing and ability to keep to schedule.

Internal factors: The company's weaknesses

Anliker has less than a 20 per cent share in the passive house market. Other companies in Switzerland already built similar houses.

It was important for Anliker to procure strategic alliances in order to gain credibility for this new product, of which they had little experience. Therefore, they asked the Technical College of Lucerne to certify their houses for the Passivhaus standard. This achieved instant credibility and recognition.

External factors: Opportunities

When analysing and discussing the market opportunities, Anliker also realized that there was another important reason for building the low energy housing: building to the Passivhaus standard and the

Minergie standard is a better investment. This is based on the fact that building requirements are getting higher and higher every year. By building better than required today, there is a smaller gap to bridge between the current requirements and those to come in the near future. This insight led Anliker to decide that the Minergie standard would be their minimum standard. Today, the company no longer builds conventional houses at all.

Regarding market shares, it is also easier for Anliker to take a certain market position in a new market niche right at the beginning than to try to penetrate an existing market. In this case, price is the only competitive factor. The main opportunity identified by Anliker AG was to develop a new market niche with interesting growth potential. By being among the first doing this, they were also able to generate free publicity.

External factors: Threats

The company managed to turn a threat into an opportunity regarding the financial incentives in the market. Knowing that, earlier, private funding and the banks strongly influenced the market, they saw the offer from the Cantonal Bank of Lucerne as the start of a new focus in the financial market. For this market, environmental awareness is also a developing market niche.

Lack of knowledge is always a problem when developing a new market. This concerns both the financial and the technical issues.

6.3.5 Setting goals

The primary goals for the project were to build and sell 3 passive house blocks with a total of 12 apartments during the year 2003. The units had to:

- be affordable, ecological and low energy for consumers;
- provide living space with good architecture;
- achieve the Passivhaus certificate; and
- be as profitable for Anliker as conventional housing.

6.3.6 Strategies

The four Ps

Product: being a private real estate developer, Anliker had to make the project a commercial success: 'Why build energy efficient houses if you can't sell them because they are too expensive or architecturally unattractive?' The answer was to do 'architecture' with unobtrusive energy features and include elements important to the target group of young, innovative families concerned about the environment.

To reduce risk, the company made a strategic marketing decision. Instead of constructing just one type of housing, three different types were planned:

1. veranda apartment blocks – conventional houses;
2. loft apartment blocks – to the Minergie standard; and
3. villette apartment blocks – passive houses to the Passivhaus standard.

Besides lowering the risk for Anliker, this product spread allowed for a greater freedom of choice for consumers.

The villette units are constructed in brick and concrete, and equipped with thick insulation, solar collectors, and ventilation with heat recovery to achieve the Passivhaus standard. The flats consist of generous rooms with corner windows over the full height to ensure bright living spaces. A very flexible floor plan can be easily adapted to fulfil individual buyers' needs. The ventilation system allows furnishing flexibility with the air outlets at the centre of the windows. To further address a concern of

Source: Anliker AG, CH-6021 Emmenbrücke, www.anliker.ch

Figure 6.3.4 *A passive house block as it was constructed*

the target audience, all electrical wiring is shielded to protect occupants from electro-smog in the bedrooms and the current can be switched off overnight. Figure 6.3.4 shows one of the passive house apartment blocks.

Price: the market in Switzerland is much more concerned about the initial costs than running costs. Therefore, it was important to Anliker to reduce costs substantially. They set the price at 7 per cent higher than for conventional houses to cover the expected 7 per cent higher construction costs.

Place: when introducing the project, Anliker invited interested home buyers to several information evenings at their head office. The architect, Arthur Sigg, presented the project.

Anliker had to present considerable information in order to create market credibility for the project. Building an information bridge to all potential project partners was of great importance. Potential buyers were registered with their names and contact addresses so that they could receive additional information with the aim of convincing them to buy an apartment.

Promotion: the challenge for Anliker was to promote ecological housing in the marketplace and to eliminate common prejudices. The new product needed a market 'breakthrough'. To accomplish this, Anliker invited the Technical College of Lucerne to test and certify the housing to the Passivhaus standard. The college is the licensed institution for this certification. By attaining Passivhaus certification, Anliker achieved instant recognition and credibility. The certification also generated much newspaper publicity, giving both the company and the product a high profile.

Bishof/Meier, a communications and advertising agency, was hired to create the marketing strategy and to carry out the marketing communications campaign. They designed:

- the logo;
- a sales folder;
- a 120 cm x 80 cm poster to put up on the building site; and
- a website (www.konstanzrothenburg.ch).

They also arranged the media coverage so that free publicity was generated at the start of the project and during the driving of the first stake. It was decided that the communications strategy was not about low energy, but about a way of living. The energy saving and payback time issues were therefore moved to the background and the core values of the target audience were addressed: comfortable and healthy living for the whole family.

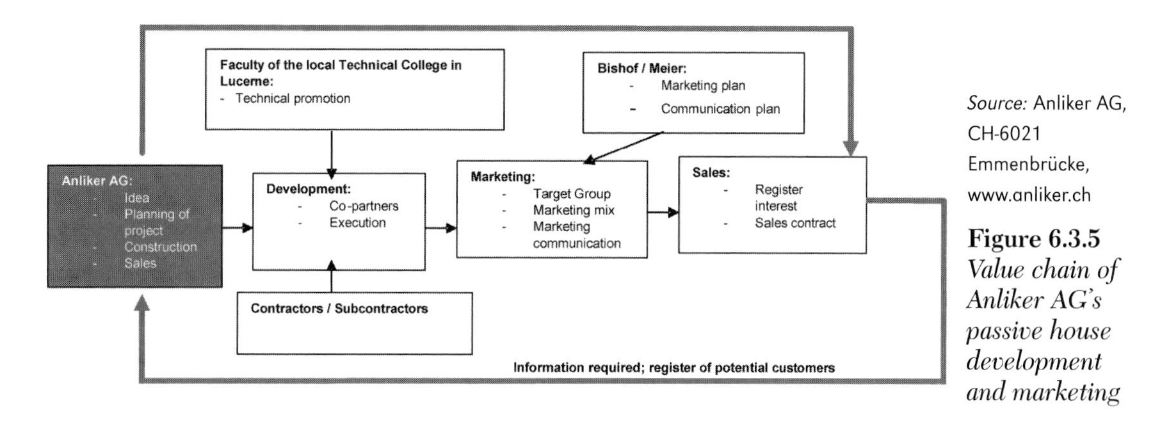

Source: Anliker AG, CH-6021 Emmenbrücke, www.anliker.ch

Figure 6.3.5 *Value chain of Anliker AG's passive house development and marketing*

6.3.7 Action plans

In order to act according to their strategies, Anliker chose to focus on the marketing activities. They had already decided that:

- Anliker should be in charge of the project, from start to finish;
- seven of the blocks should be to Minergie and Passivhaus standards; and
- the Konstanz development would consist of a total of 13 blocks, 6 being conventional (in order to keep a certain product range and freedom to choose for the customers, as well as for variation in the environment).

Taking the analysis into consideration, the essential focus was to get a market breakthrough and to create positive images around the project.

Build credibility

In order to build credibility in the market, Anliker had to invite other strategic partners to 'take part' in the project. In this way, they made it 'public' and emphasized the project rather than the developer. This underlined the authenticity of the project.

To get the Passivhaus certificate, the Technical College of Lucerne was invited to watch over the works and carry out measurements regarding humidity, temperature, etc. The certificate was awarded, and Anliker managed to build the necessary credibility both for the project and for the company.

Communication

To be successful, it was essential to find out what was important to their target group: what they preferred and what they were thinking. The general characteristics of the target group were:

- young people starting up a family (small children);
- concerned about the future (included ecological questions);
- following trends (innovative); and
- cost focused.

The paradox, however, is that when developing the marketing and communication plan, they chose to focus on other factors that they thought were even more important to this group and 'wrapped' the project in these positive images:

- a certain way of living;
- core family values;
- happy and healthy children;
- a trendy design;
- many green spaces;
- good architecture;
- a focus on:
  - clean air;
  - good indoor climate; and
- responsibility for the next generation, the future and the Earth:
  - low energy;
  - sustainability.

A good example of a positive image is the picture of a happy child playing on a swing in a summer meadow. This image was also used as the project's logo (see Figure 6.3.1) – although many would have been tempted to put the energy performance in the headlines!

One of the techniques used by Bishof/Meier to market the Konstanz estate was to create pleasant associations through images. It is worth noting how Bishof/Meier and Anliker targeted their market niche, using the critical success factors in this target group to form the communications strategy.

The communications concept for the target group occurred through four different main media channels, starting up with:

- the development of a flyer with information about the upcoming project – placed in post offices and various banks;
- newspaper advertisements;
- development of a website; and
- posters put up at the site, informing potential customers that Anliker was building a passive house.

Because of the rather small (but fast growing) interest in environmental building, Anliker put information in the headlines that they – at this stage without any similar project to refer to – thought would stimulate sales and reach the target market. They thought technical data referring to the Passivhaus standard would be less interesting information to start with, and that the main trigger points for the home buyers were:

- a nice location;
- high quality;
- price; and
- Passivhaus certification.

The concept of marketing through association is very effective and shows how important it is to promote positive images. It addresses people on an intuitive level.

Control and measurement

Using trend issues, both in developing the apartments and later when communicating with the customers, Anliker achieved an extra promotion/marketing effect in the market for their product. Several forces stimulated the market niche and finally resulted in good sales for the company.

In order to know how the results have been achieved, it is important to control and measure the project results. Knowing what caused both good and bad results, it is possible to have a better success rate on future decisions.

Anliker had to check on their goals, which were to build and sell houses based on the following criteria:

- affordable, ecological and with a minimum energy requirement;
- providing living space with good architecture;
- gaining profit equal to building conventional housing; and
- achieving Passivhaus certification.

Anliker reached these goals on time. The success of the Konstanz passive housing estate is demonstrated by the fast sales of 32 apartments in 8 blocks. The successes further led to the construction of five more blocks.

After evaluating a range of information, Anliker are certain that up to 95 per cent of the potential buyers received the necessary information from the advertisements. But they also think that information-sharing among potential home buyers (family and friends telling each other about the passive houses) also had a huge effect on the positive sales.

The project costs confirm the trend that passive housing is becoming an increasingly attractive way of building – one that meets the needs of forward-thinking home buyers.

As far as project accounting shows, the company managed to get a profit from the sales equal to that for the sales of conventional houses. Yet, comparable conventional flats (same size, design, etc.) are not built, and therefore the result is not entirely robust. Anliker should put effort into controlling the whole project in order to measure their goals, and this is a useful learning tool for the company.

6.4 Lessons learned from marketing stories

The analysis of marketing successes in Europe, North America and New Zealand carried out in the International Energy Agency (IEA) Task 28/38 framework led to the following recommendations.

Do:

- ride on the wave of increasing public awareness on environmental issues;
- join other players in the marketplace – commercial/interest groups/local, regional, national and international authorities – and develop win–win alliances; and
- think strategically:
  - accept the fact that you cannot sell to them all;
  - know your target group preferences;
  - clearly define how you differentiate yourself from others;
  - focus on added value.

Don't:

- have a one-sided focus on 'additional investment cost resulting in annual energy savings'.
- start immediately with communications without having done a strategic analysis.

6.4.1 Recommendations

Do not skip any step in the marketing process:

- Emphasize information-gathering.
- Focus on analysis:
  - internal: strengths and weaknesses of product, organization;
  - external: opportunities, threats, driving forces, life cycle, segmenting.
- Define goals – quantify:
  - people normally focus on what is measured – more success can be achieved by defining concrete and measurable objectives.

- Define strategies:
  - who you are going to sell to – define target group;
  - clear definition of all Ps: product, price, place and promotion.
- Make a plan of action:
  - all actions that need to be taken;
  - include the communications actions;
  - all costs (internal hours and direct external costs);
  - a time schedule.
- Control and measurement – modify:
  - to help you learn from what you have done and be able to adapt your strategies.

Take advantage of others' experiences:
- Why reinvent the wheel? There is a lot to learn from similar businesses in other countries and other types in your existing market.
- However, do not enter the trap of just copying concepts or ideas. Remember that there are differences between countries and companies. Therefore, a specific analysis must be executed in order to find out how to adapt an idea to your business environment.

Financial incentives are a potential blind spot:
- If starting a promotion with the message that 'if you buy this product you will receive xxx funding', the customer will perceive this as saying that there is a disadvantage to the product.
- Having initially strongly focused on financial incentives related to the product, it can later be difficult to sell it at a normal price.
- Very often those representing the 'early buyers' (in the introduction phase) are not very price sensitive.

Be market oriented instead of product oriented:
- Start with the question of 'who will I sell to?' Follow up with finding out what their real needs are. Then think 'wider' than just your own product range.
- Define the product that fulfils these needs. If you do not have the technology or skills to offer this on your own, you may do this through strategic partnerships with complementary businesses, suppliers or competitors.

Added value and branding
- Concept thinking through the use of images is more successful than a pure focus on technical aspects.
- Added values are elements beyond the purely physical (technical) aspects of the core product. Examples of value to the customer are 'non-energy benefits', such as better air quality, greater comfort, a sense of security, status and moral responsibility.
- Well-known brands (such as those of the World Wide Fund for Nature) may be used for a 'piggy-back ride' when entering the market.

Differentiation as a competitive instrument:
- Don't underbid your competitors' prices. Instead, focus your company and its products on specialities that draw attention away from those who are more concerned about other issues, such as short-term cost savings.
- Initial obstacles may be turned into opportunities, which may form the basis for differentiation.

Recognize the importance of key individuals:
- Entering the market with a new (sustainable) product requires highly motivated, skilled and respected individuals within the company to play a key role. They need to be competent about the subject – it is also crucial they have the ability to motivate other people in the organization.

If you can't beat them – join them:
- The experiences from the stories show clearly that strategic alliances with other actors have enabled the successful launches of new sustainable products and services.
- Alliances only work if all participants profit from the cooperation: make win–win alliances.

Conscious use of media:
- If a product or service is new to the region, the media will be quite keen to cover your story. Make use of this free publicity!
- Famous individuals attract attention and press coverage. Use them if you can!

Build credibility:
- A new product is not known or trusted to start with. It is necessary to establish a (good) name for your product.
- Build up credibility by working with established institutes
- Don't just act credible, be credible! Never 'sell' more than you can handle.

References

Note: a full version of this section is available at no cost on the website of the Solar Heating and Cooling Programme of the International Energy Agency at www.iea-shc.org/ under Task 28, outcomes or at www.moBiusconsult.nl/iea28.

ACEEE (2004) *Summer Study on Energy Efficiency in Buildings*, 2000 Sustainable Building Conference, Asilomar, CA, USA, August 23-27, www.aceee.org

Rogers, E. (2003) *Diffusion of Innovations*, fifth edition, Freepress, New York

Skumatz, L. A., Dickerson, C. A. and Coats, B. (2000) *Non-Energy Benefits in the Residential and Non-Residential Sectors: Innovative Measurements and Results for Participant Benefits*, ACEEE Summer Study on Energy Efficiency in Buildings, 8352-8364, Pacific Grove, California USA

Skumatz, V. and Stoecklein, A. (2004) *Using Non-Energy Benefits (NEBs) to Market Low Energy Homes New Zealand*, Sustainable Building Conference, Asilomar, CA, USA, August 23-27, www.aceee.org

Part II

SOLUTIONS

7

Solution Examples

Maria Wall

7.1 Introduction

7.1.1 Solution examples by climate, house type and strategy

This chapter presents example solutions for housing in cold, temperate and mild climates that achieve the targets of end energy and primary energy set by the International Energy Agency (IEA) Task 28/38. The example solutions are noteworthy in that they do not necessarily require extremes in either construction or equipment.

Two design approaches are differentiated:

1 conservation strategy (reducing losses); and
2 renewable energy supply (increasing gains).

The difference in approaches is in how far conservation or solar is taken. Both strategies, of course, involve conservation and the use of passive solar gains. Computer simulations were carried out to optimize the solutions. In addition, sensitivity analyses were carried out to quantify the importance of key design parameters.

Three climate regions were defined and reference climate data sets generated by the program Meteonorm (Meteotest, 2004):

1 cold (Stockholm);
2 temperate (Zurich); and
3 mild (Milan).

For the different building types in the cold, temperate and mild climates, different example solutions are shown based on strategy A (conservation) or strategy B (renewable energy). Each example solution was reached by applying different variations of envelope constructions and technical systems to reach the targets. Computed values for energy performance, summer comfort and carbon dioxide (CO_2) equivalent emissions are given for each solution. One example of life-cycle analyses for row houses in the temperate climate is also presented.

House types

Example solutions were developed for three reference housing types in the three climate regions. The reference house designs reflect typical houses built in the northern Mediterranean area, middle Europe and middle of the Nordic countries.

The single family house is in a one-and-a-half-storey home, with a floor area of 150 m^2 divided between the ground floor and the first floor, which has a smaller area. For this house type the energy conservation strategy is the more difficult approach because of the relatively large ratio of heat-losing envelope area to the floor area. On the other hand, a large envelope area offers ample space for solar systems.

The row house has 120 m^2 of floor area divided over two storeys. A row of six units is assumed in the analyses. The two end housing units have higher envelope heat losses than the four middle houses. The average energy demand for the row of six units should meet the targets. The resulting compact building geometry makes it easier to achieve the space heating target, compared to the single family detached house. Given the very small heating load, a sensible solution is to have one central heating system for the whole row. But energy efficient separate systems for each unit are also possible to design.

The four-storey apartment building has a total of 1600 m^2 floor area. Each apartment is assumed to be 100 m^2. This housing type is very compact and can therefore most easily achieve a very low space heating demand, even without using thick insulation. Very efficient ventilation heat recovery is still important.

7.2 Reference buildings based on national building codes, 2001

To start, common building geometries for a single family house, row house and apartment building were defined (see Appendix 1). For each country in the three climate regions, the allowable space heating demand and ventilation requirements set in the national building codes of the year 2001 were taken as the targets for the reference buildings. Working backward in an iterative process, the average envelope standards in each climate region were calculated to exactly meet the standards. In this way the regional reference buildings were defined.

Table 7.2.1 compares the average U-values for the different housing types in the three climate regions. The U-values for the cold region are relatively low due to a compensation for not using 50 per cent ventilation heat recovery, which otherwise should have been used for a Swedish reference building. As expected, the insulation standard is much higher in the north than in the south. However, in Figure 7.2.1 it can be seen that this is partially explained by the building codes in the south being less ambitious. The end result is that the space heating demands for all three housing types in the mild climate exceed the demands in the temperate and cold climates! Similarly, the space heating demand of the housing built to the regional codes in the temperate region exceeds the demand in the cold region. In all cases, logically, the space heating demand is lowest for the apartment building and highest for the single family detached house.

The space heating demand was calculated according to EN 832, using the Bilanz program (Heidt, 1999). All reference houses in the three climate regions are assumed to have ventilation losses equivalent to 0.6 air changes per hour (ach). No mechanical ventilation system or ventilation heat recovery is assumed.

The space heating demand is, in general, high for buildings constructed according to the actual building codes of 2001. This means that the potential to radically reduce the demand is high, and is extremely high for the mild climate region.

Table 7.2.1 *Mean regional U-values of the building envelope based on national building codes for the year 2001*

	Building envelope mean U-value (W/m²K)		
	Cold climate	**Temperate climate**	**Mild climate**
Single family house	0.29	0.47	0.74
Row house (six units)	0.33	0.55	0.86
Apartment building	0.35	0.60	0.94

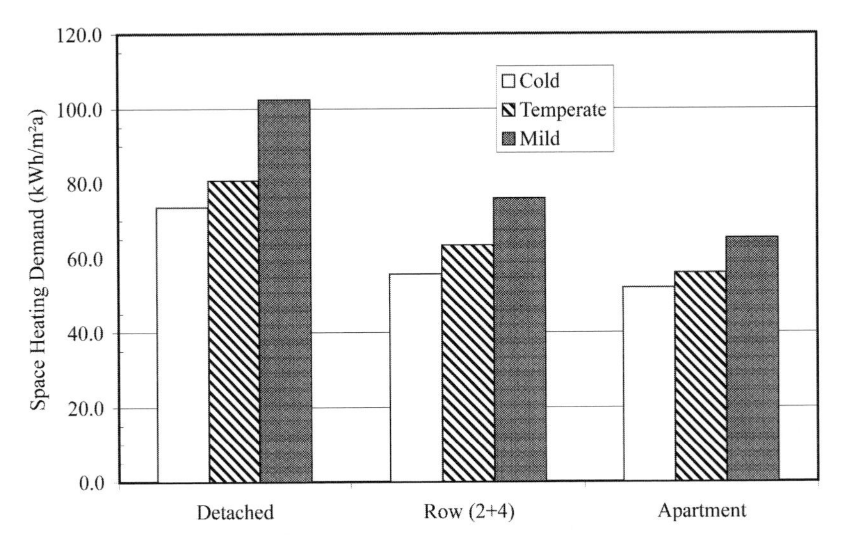

Source: Maria Wall

Figure 7.2.1 *Space heating demand for the regional reference buildings with standards based on building codes for the year 2001; the reference climates have been used*

7.2.1 Primary energy demand

In standard buildings today the energy use for heating and domestic hot water is mostly provided by fossil fuels (gas and oil), district heating and electricity. An important goal is to reduce the use of non-renewable energy. Therefore, it is also important to show results in terms of primary energy demand.

Since the focus is to reduce the non-renewable energy use, the primary energy target defined in this book only addresses the non-renewable part of the primary energy demand (see Appendix 2). To judge the different environmental impacts of buildings during operation, two indicators are used in this book:

1 The primary energy, which is the amount of energy use on site, plus losses that occur in the transformation, distribution and extraction of energy.
2 CO_2 emissions, which are related to the heat energy use, including the whole chain from extraction to transformation of the energy carrier to heat. Using the CO_2 equivalent values, not only CO_2 but all greenhouse gases are taken into account, weighted with their impact on global warming.

As a reference and as a base for finding a primary energy target, the non-renewable primary energy demand for the regional reference houses was calculated as follows:

- First, the end energy use, including space heating demand, DWH demand and system losses were summarized. Energy for electrical appliances such as pumps and fans was also summarized. The DHW demand was, in all cases, assumed to be 40 litres per person per day.
- Second, the DHW and space heating demand were multiplied with the primary energy factor 1.1 (representing a 'not unusual' system – for example, oil or gas). When electricity for fans and

pumps was assumed as part of the system, the electricity was multiplied with the primary energy factor 2.35 (EU mix + Switzerland + Norway).

Figure 7.2.2 shows an example of the primary energy use when no electricity for fans and pumps is assumed in the reference buildings. All examples are based on oil or gas for heating. The mean regional reference level of year 2001 were around 124 kWh/m^{2}a, varying between 110 and 156 kWh/m^{2}a.

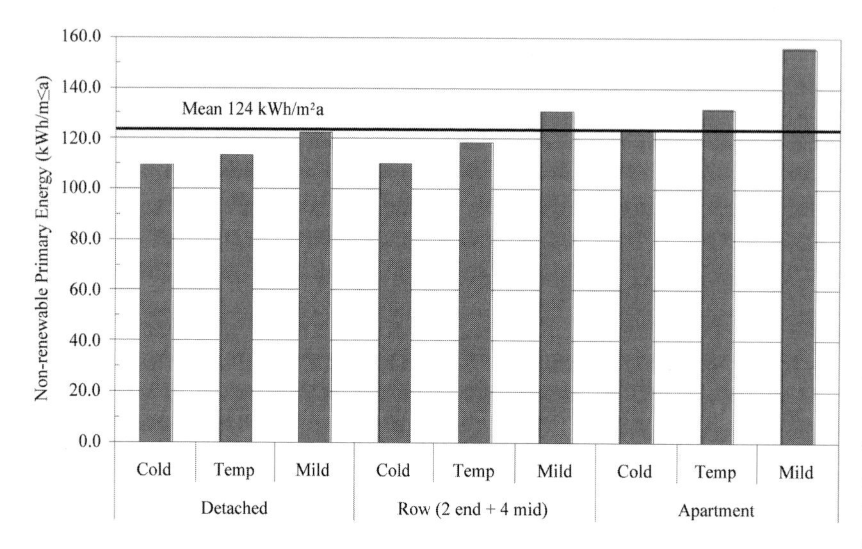

Note: This includes DHW and space heating demand. No electricity is assumed for fans and pumps.
Source: Maria Wall

Figure 7.2.2 *Non-renewable primary energy demand for the regional reference buildings*

7.3 Targets for space heating demand

A target was set for the space heating demand. This target ensures that the building itself will be designed with low transmission and ventilation losses. The building will not be influenced by external changes in the future, such as system changes and changes in supplied energy carriers. In addition, it is a more understandable target for occupants and building owners compared to primary energy, which is less readily understandable.

The defined space heating target has been set here to be stricter for the conservation strategy than for the renewable energy strategy to allow additional investments in renewable energy systems.

7.3.1 Strategy I: Energy conservation

The space heating target for the energy conservation strategy is 15 kWh/m^{2}a for the apartment building and the row house. For the single family detached house, the target is 20 kWh/m^{2}a. These levels are equivalent to a factor 4 compared with the space heating demand for the reference houses year 2001 (see Figure 7.3.1).

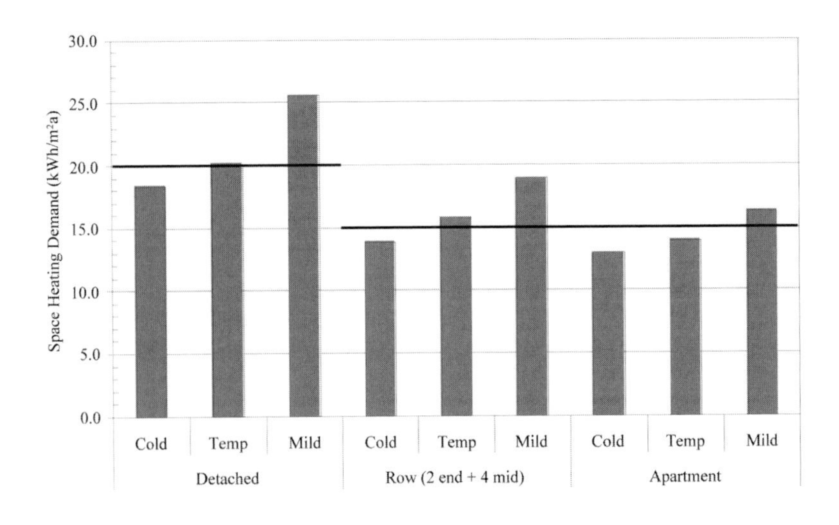

Note: Each bar shows the specific reference level (building code 2001 according to Figure 7.2.1) divided by 4. The lines show the chosen targets: 15 kWh/m^{2}a for the apartment building and the row house and 20 kWh/m^{2}a for the single family detached house.

Source: Maria Wall

Figure 7.3.1 *Factor 4 space heating target for the regional high-performance buildings conservation strategy*

7.3.2 Strategy II: Renewable energy

The space heating target for the renewable energy strategy is 20 kWh/m^{2}a for the apartment building and the row house. For the single family detached house, the target is 25 kWh/m^{2}a. These levels are equivalent to a factor 3 compared with the space heating demand for the reference houses year 2001 (see Figure 7.3.2).

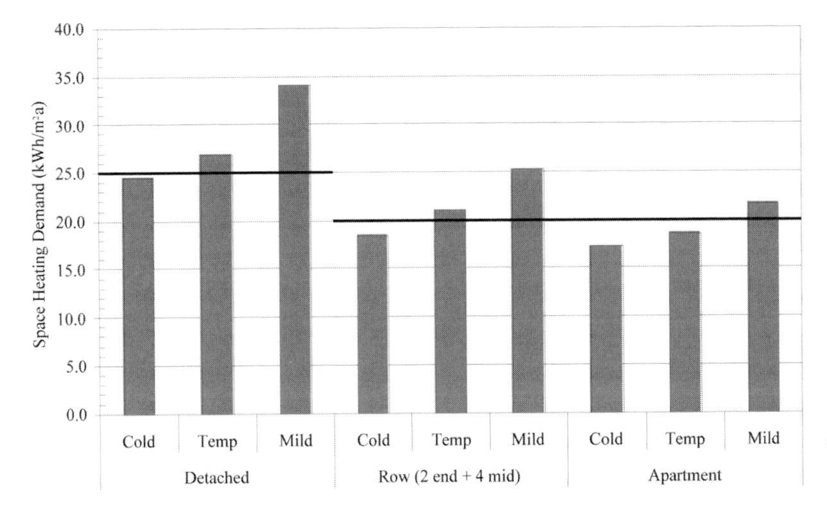

Note: Each bar shows the specific reference level (building code 2001 according to Figure 7.2.1) divided by 3. The lines show the chosen targets: 20 kWh/m^{2}a for the apartment building and the row house and 25 kWh/m^{2}a for the single family detached house.

Source: Maria Wall

Figure 7.3.2 *Factor 3 space heating target for the regional high-performance buildings renewable energy strategy*

7.4 Target for non-renewable primary energy demand

The target for non-renewable energy demand is 60 kWh/m^{2}a and includes DHW and space heating, system losses and electricity for fans and pumps. This target is the same for all climates and house types. Household electricity is not included in the target since this is very much influenced by the occupants. However, energy efficient household appliances are recommended and assumed in the simulations when estimating available internal gains.

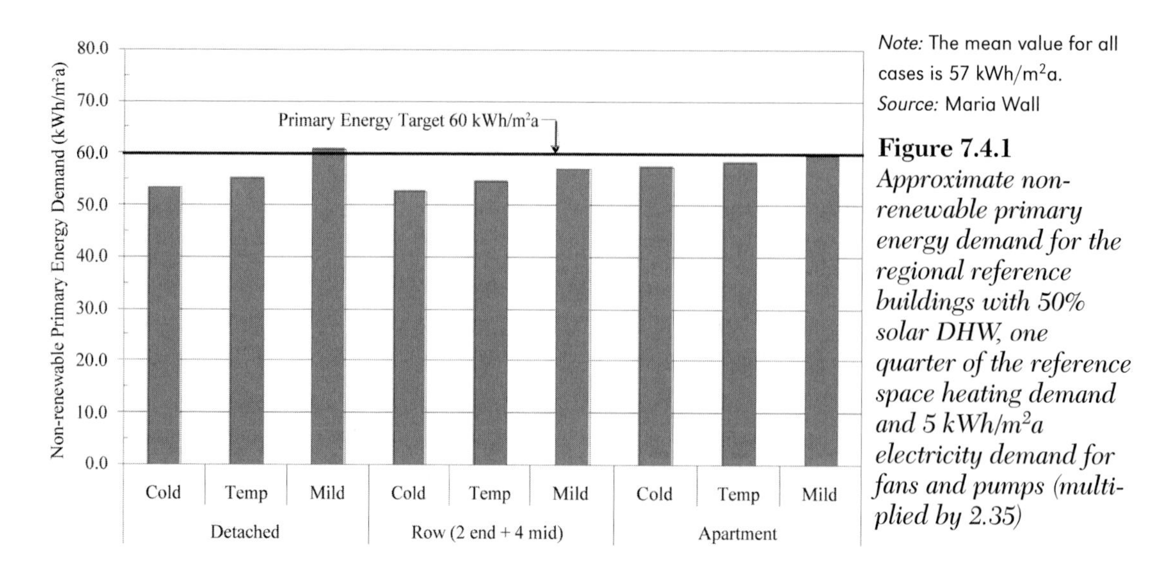

Note: The mean value for all cases is 57 kWh/m²a.
Source: Maria Wall

Figure 7.4.1 *Approximate non-renewable primary energy demand for the regional reference buildings with 50% solar DHW, one quarter of the reference space heating demand and 5 kWh/m²a electricity demand for fans and pumps (multiplied by 2.35)*

As a basis for deciding the target level, calculations were made for the regional reference buildings. The reference level of 124 kWh/m²a in non-renewable primary energy demand was reduced for each building type and climate by:

- reducing the space heating demand, reference level 2001, with a factor 4 before calculating the primary energy demand;
- reducing delivered DHW heating by 50 per cent compared to the reference level, assuming that 50 per cent of the demand is supplied by solar collectors (the solar energy supply does not give rise to any non-renewable primary energy demand); and
- adding electricity use for fans and pumps of 5 kWh/m²a since the high-performance solutions use mechanical ventilation and other electrical appliances.

The remaining energy demand is then multiplied with the primary energy factors 1.1 (for example, oil or gas) and 2.35 (electricity), respectively. As a result, the non-renewable primary energy demand for each building type and climate was given (see Figure 7.4.1). The level of non-renewable primary energy demand roughly represents half of the demand for the reference buildings, year 2001 (see Figure 7.2.1). Thus, the target of 60 kWh/m²a represents a factor 2 compared to the standard of new buildings in many European countries today.

The target of 60 kWh/m²a could be achieved in many ways. The target is chosen as realistic both when applying a conservation strategy and a renewable energy strategy. The non-renewable primary energy demand is highly influenced by the choice of energy carrier. If a high fraction of renewables is used for the solution, the non-renewable primary energy demand could be much lower than the 60 kWh/m²a, which can be seen in the following chapters.

7.4.1 Computer tools

Different computer tools have been used in order to find solutions for the building envelope design and the system design. The programmes TRNSYS (TRNSYS, 2005), DEROB-LTH (Kvist, 2005), SCIAQ Pro (ProgramByggerne, 2004) and Polysun (Polysun, 2002) have been used. A special version of the program Polysun, the 'Larsen edition', can read a heat demand output file from the building simulation program DEROB-LTH. The whole building can thus be simulated as accurately as possible by having the interface coupling between the two programs. TRNSYS includes simulations both for the building and the systems.

The goal of this chapter is to show the sensitivity of energy use to variations in selected design features. No detailed analyses of thermal bridges were carried out; these losses were reflected in setting the average U-values of the building envelope. When a real design of a building is made, details must be conceived so as to avoid thermal bridges.

References

Heidt, F. D. (1999) *Bilanz Berechnungswerkzeug, NESA-Datenbank*, Fachgebiet Bauphysik und Solarenergie, Universität-GH Siegen, D-57068 Siegen, Germany

Kvist, H. (2005) *DEROB-LTH for MS Windows, User Manual Version 1.0–20050813*, Energy and Building Design, Department of Architecture and Built Environment, Lund University, Lund, Sweden

Meteotest (2004) *Meteonorm 5.0 – Global Meteorological Database for Solar Energy and Applied Meteorology*, Bern, Switzerland, www.meteotest.ch

Polysun (2002) *Polysun 3.3, Thermal Solar System Design: User's Manual*, SPF, Institut für Solartechnik, Rapperswil, Switzerland, www.solarenergy.ch

ProgramByggerne (2004) *ProgramByggerne ANS, 'SCIAQ Pro 2.0 – Simulation of Climate and Indoor Air Quality': A Multizone Dynamic Building Simulation Program*, www.programbyggerne.no

TRNSYS (2005) *A Transient System Simulation Program*, Solar Energy Laboratory, University of Wisconsin, Madison, WI

8

Cold Climates

8.1 Cold climate design

Johan Smeds

The example solutions for conservation and renewables in this section of Chapter 8 are compared to reference buildings, which fulfil the building codes from the year 2001 in Sweden, Norway and Finland (see Appendix 1). The examples solutions have been designed to fulfil the energy targets of this book while achieving superior comfort.

8.1.1 Cold climate characteristics

The climate of Stockholm was used to represent the cold climate region. Stockholm, located at latitude 59.2° N has an average yearly temperature of 6.7°C (Meteotest, 2004). Similar climate conditions can be found in Oslo, while Helsinki is somewhat colder. A comparison of the heating degree days for these three cities and Zurich and Milan, representing the temperate and mild climate regions, is illustrated in Figure 8.1.1.

In this cold climate, meeting the energy targets set for the high-performance examples requires that the building envelope, including the windows, has to have a very low U-value. There can be very little infiltration, so good air quality depends on mechanical ventilation and heat recovery is essential.

The combined effects of the northern latitude and cold climate create special limitations on supplying energy from solar gains. Calculated monthly average temperatures and solar radiation are shown in Figure 8.1.2. The amount of annual solar radiation statistics seems encouraging: approximately 1000 kWh/m^2 global horizontal. The average amount of sunshine hours with beam radiation exceeding 120 W/m^2 is 1600 hours per year (SMHI, 2005). However, the sun angles are very low, ranging from 9° to 55° at noon on 21 December and 21 June. In addition, most of the solar radiation occurs in summer when the days are very long.

The implications for design are that DHW, as an all-year demand, is a good end use for solar energy. The very weak solar energy in mid winter, coming at a low angle a long distance through the atmosphere, is a very limited resource. The low power of the radiation makes space heating (with a target temperature of 20°C) a better application for the solar radiation than domestic water heating (with a target temperature of 60°C). The low sun angles also mean that façades intercept the sun at a relatively direct angle. In the winter, the window heat losses from the cold temperatures cannot offset the weak gains from the low-angle short-duration sun. In summer, the still low angles relative to more southern latitudes mean that overheating due to windows has to be addressed.

In the cold climate, the advantage of the very low area to volume ratio (A/V) of the apartment building is especially pronounced compared to the single family house.

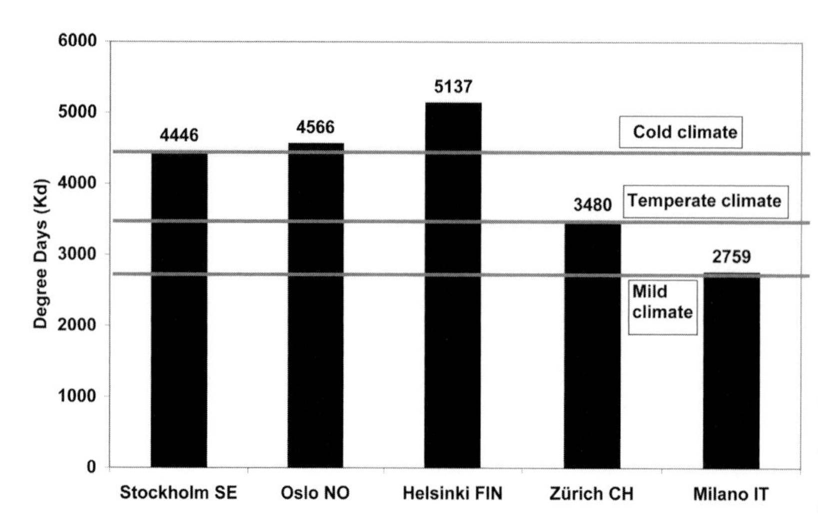

Source: Johan Smeds

Figure 8.1.1 *Degree days (20/12) in cold, temperate and mild climate cities*

How much CO_2 equivalent emissions are reduced in the example solutions relative to the reference buildings depends very much on the source of the heat production. In Scandinavian cities, district heating is very common and is therefore assumed for some of the example solutions. If the district heat is produced with a high share of renewable energy, the reduction in primary energy for the high-performance house will not be so large. In case of a reference building using electric resistance heating, a larger reduction is possible.

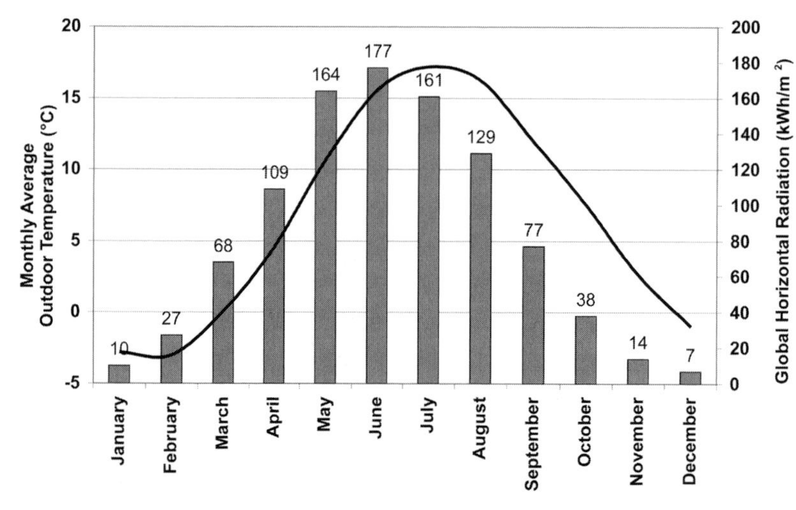

Source: Johan Smeds

Figure 8.1.2 *Monthly average outdoor temperature and solar radiation (global horizontal) for Stockholm*

8.1.2 Single family house

Single family house reference design

Building envelope: the exterior walls and roof are in wooden lightweight frame construction with mineral wool as insulating material. The windows have a frame ratio of 30 per cent, double glazing, one low-e coating and are filled with air. As is typical for a single family house, the floor is an externally insulated concrete slab on the ground without a cellar. The U-values of the building components are shown in Table 8.1.1. The U-values are relatively low due to a compensation for not using 50 per cent ventilation heat recovery, which otherwise should have been used for a Swedish reference building.

Table 8.1.1 *Building component U-values for the single family house*

Component	U-value (W/m^2K)
Walls	0.20
Roof	0.19
Floor (excluding ground)	0.20
Windows (frame + glass)	1.81
Window frame	1.70
Window glass	1.85
Whole building envelope	0.29

Mechanical systems: a mechanical exhaust air ventilation system without heat recovery extracts room air through ducts by an electrical fan. The ventilation rate is 0.5 air changes per hour (ach) and the infiltration rate is 0.1 ach. An electrical boiler for space heating and domestic hot water is used and the heat is distributed by a hot water radiant heating system.

Household electricity: the household electricity use is assumed to be 29 kWh/m^2a for two adults and two children. Household electricity is not included as part of the energy targets.

Space heating demand: the monthly space heating demands of the house, calculated with DEROB-LTH (Kvist, 2005), are shown in Figure 8.1.3. For the heating season from October to April, the annual space heating demand is approximately 69 kWh/m^2a. This assumes that the heating set point is 20°C and the maximum room temperature is 23°C during winter and 26°C during summer. In order to reduce excess temperatures, shading devices and window ventilation or forced ventilation are used.

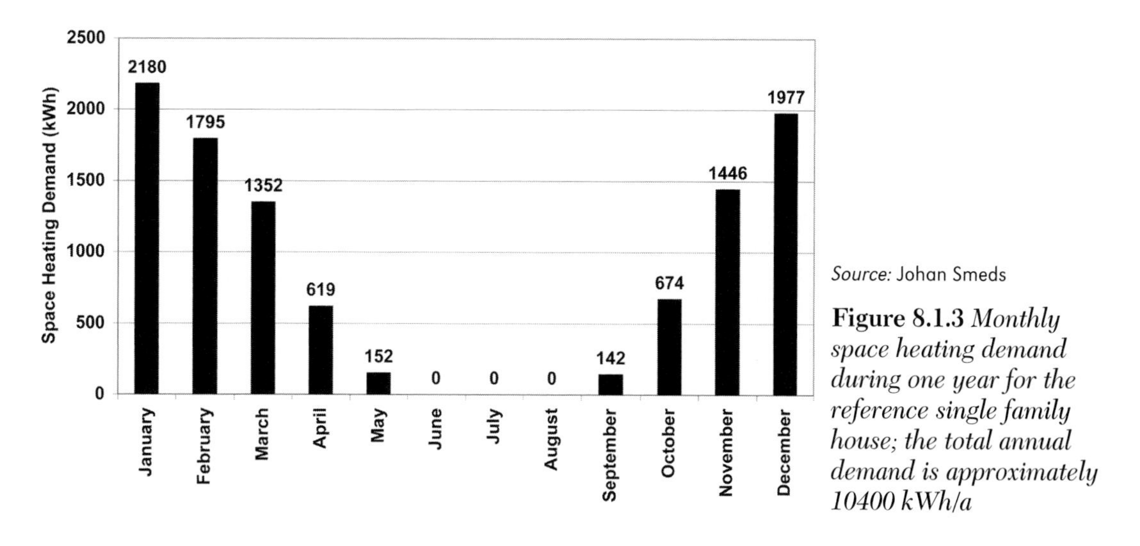

Source: Johan Smeds

Figure 8.1.3 *Monthly space heating demand during one year for the reference single family house; the total annual demand is approximately 10400 kWh/a*

Balance point temperature and peak load: the balance point temperature is the outdoor temperature at which a comfortable indoor temperature can be kept without running the heating system. Transmission and ventilation losses are in balance with heat gains from household appliances, body heat and from passive solar gains. The balance point temperature therefore indicates the energy performance of a building.

Simulations in DEROB-LTH of the required heat load of a shaded single family house built according to the reference construction given above result in a balance point temperature of approx-

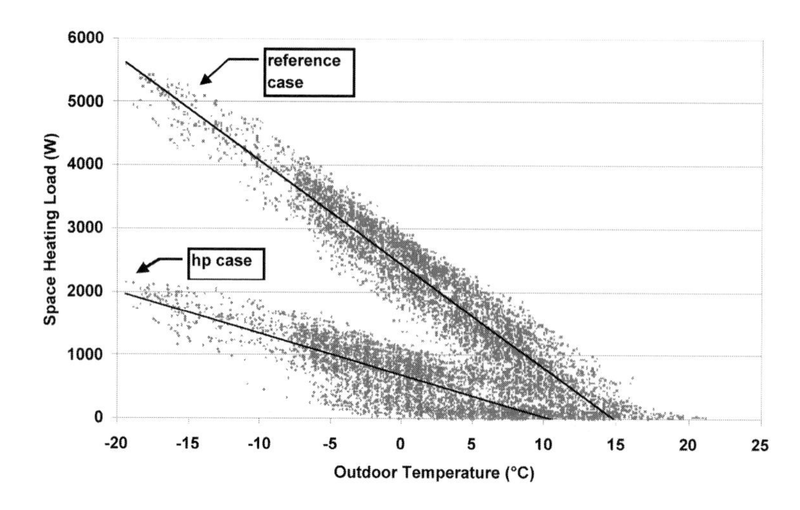

Source: Johan Smeds

Figure 8.1.4 *Space heating load and balance point temperature of a single family house, a high-performance (hp) case (20 kWh/m²a) and a reference case (69 kWh/m²a)*

imately 15°C. In comparison, a high-performance house has an approximate balance point temperature of only 10°C. The peak load of the heating system of the reference building is 5500 W, while the peak load of the high-performance house is only 2200 W. The space heating demand for the high-performance house is 20 kWh/m^{2}a – thus, 29 per cent of the space heating demand for the reference house. The results of the simulations are illustrated in Figure 8.1.4.

Non-renewable primary energy demand and CO_2 emissions: the total energy demand for space heating, DHW, system losses and mechanical systems, delivered energy, primary energy and CO_2 equivalent emissions for the reference house are shown in Table 8.1.2. Household electricity is not included. The demand for non-renewable primary energy is 231 kWh/m^{2}a and the CO_2 equivalent emissions sum up to 42 kg/m^{2}a.

Table 8.1.2 *Total energy demand, non-renewable primary energy demand and CO_2 equivalent emissions for the reference single family house*

Net Energy (kWh/m²a)		Total energy use (kWh/m² a)				Delivered energy (kWh/m²a)		Non-renewable primary energy		CO_2 equivalent emissions	
		Energy use		Energy source				factor (-)	(kWh/m²a)	factor (kg/kWh)	(kg/m²a)
Mechanical systems	5.0	Mechanical systems	5.0	Electricity	5.0	Electricity	5.0	2.35	11.8	0.43	2.2
Space heating	68.9	Space heating	68.9	Electricity	93.4	Electricity	93.4	2.35	219.5	0.43	40.2
DHW	21.0	DHW	21.0								
		Circulation losses	3.5								
		Conversion losses	0.0								
Total	94.9		98.4		98.4		98.4		231.3		42.4

Single family house example solutions

A problem, even in this cold climate, is to find an inexpensive heating system, given that the annual energy consumption will be very low with 3000 kWh/a for space heating and 3150 kWh/a for DHW, equalling 6150 kWh/a in total. Correspondingly, the required heating capacity is small, with a peak load for space heating of only 2200 W.

Different example solutions for the single family house in the cold climate are described in sections 8.2 and 8.3. The examples are as follows.

CONSERVATION: SOLUTION 1A

U-value of the whole building:	0.12 W/m^2K
Space heating demand:	11.5 kWh/m^2a
Heating distribution:	supply air
Heating system:	direct electric resistance heating
DHW heating system:	Solar collectors with electrical backup

CONSERVATION: SOLUTION 1B

U-value of the whole building:	0.17 W/m^2K
Space heating demand:	20 kWh/m^2a
Heating distribution:	hot water radiant heating
DHW and space heating system:	heat pump (outdoor air to water) or a borehole heat pump for a group of houses (local district heating)

RENEWABLE ENERGY: SOLUTION 2A AND 2B

U-value of the whole building:	0.21 W/m^2K
Space heating demand:	25 kWh/m^2a
Heating distribution:	hot water radiant heating
DHW and space heating system:	solar combi-system with biomass boiler (solution 2a) or condensing gas boiler (solution 2b)

In Figure 8.1.5, the total energy demand, delivered energy and non-renewable primary energy demand are shown for the different solutions. Compared to the reference building, all solutions have a drastically reduced total energy use, and with the considered system solutions the non-renewable primary energy use is cut by 76 per cent or more. Solution 1b needs the least delivered energy, 20 kWh/m^2a, while solution 2a needs the least non-renewable primary energy, 17 kWh/m^2a. Also, the CO_2 equivalent emissions illustrated in Figure 8.1.6 are lowest in solution 2a, with only 3.8 kg/m^2a, which is 9 per cent of the emissions of the reference building. This means that the choice of a heating system using renewable energy sources is of major importance; but a reduction of emissions and energy use can also be achieved by strict energy conservation measures.

A borehole heat pump supplying heat to a group of buildings via a local district heating system requires co-operative ownership of the heating system. This is already common in Scandinavian countries. It would be hard to justify the investment cost of a borehole heat pump for only one high-performance detached house. Both the borehole system and the outdoor air to water heat pump depend totally on a reliable and inexpensive supply of electricity.

Increasing the south-facing window area results in a minor increase in heating demand, but may be justified for daylighting or the view. The heating demand increases more when increasing the window areas facing east, west and north.

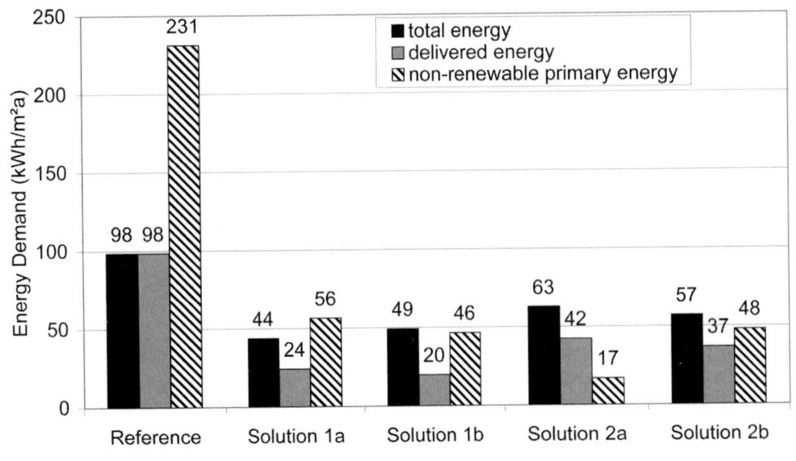

Source: Johan Smeds

Figure 8.1.5 *Overview of the total energy use, the delivered energy and the non-renewable primary energy demand for the single family houses; the reference building has electric resistance heating*

Summer comfort is achieved by using shading devices for windows, a bypass of the ventilation heat recovery and rational window ventilation.

An air-tight construction of the building envelope is most critical in the cold climate. The assumed air tightness here is 0.6 ach at 50 Pa, resulting in approximately 0.05 ach infiltration rate under normal conditions. Were only common practice construction and air tightness provided (i.e. 0.20 ach), the space heating demand would increase by 50% from 20 kWh/m^{2}a to almost 30 kWh/m^{2}a.

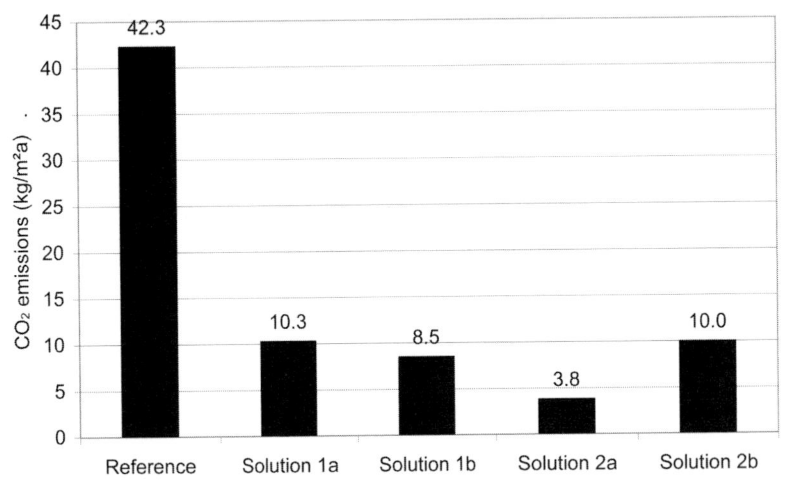

Source: Johan Smeds

Figure 8.1.6 *Overview of the CO_2 equivalent emissions for the single family houses; the reference building has electric resistance heating*

8.1.3 Row houses

Row house reference design

The exterior walls and roof are of wooden lightweight frame construction with mineral wool as insulating material. The façade has a brick veneer. The windows have a frame ratio of 30 per cent, triple glazing (4-30-(D4-12)) and are filled with air. The floor is an externally insulated concrete slab on the ground without basement. The U-values of the building components are shown in Table 8.1.3. The U-values are relatively low due to a compensation for not using 50% ventilation heat recovery, which otherwise should have been used for a Swedish reference building.

Table 8.1.3 *Building component U-values for the row house*

Component	U-value (W/m²K)
Walls	0.20
Roof	0.20
Floor (excluding ground)	0.20
Windows (frame + glass)	1.74
Window frame	1.70
Window glass	1.75
Whole building envelope	0.33

Mechanical systems: a mechanical exhaust air ventilation system without heat recovery provides 0.5 ach. The infiltration rate is 0.1 ach. District heating is used for space heating and DHW. The heat is distributed in the building by a hot water radiant heating system.

Household electricity: the household electricity use is assumed to be 36 kWh/m²a for two adults and two children. Household electricity is not included as part of the energy targets.

Space heating demand: the average annual space heating demand for the row house with six units is 58 kWh/m²a. The use of shading devices and window ventilation or forced ventilation for temperature regulation during the summer months is assumed.

Non-renewable primary energy demand and CO_2 emissions: the total energy demand for space heating, DHW, system losses and mechanical systems, delivered energy, primary energy and CO_2 equivalent emissions for the reference house are shown in Table 8.1.4. The non-renewable primary energy demand is 79 kWh/m²a and the CO_2 equivalent emissions sum up to 23 kg/m²a. The total energy demand for space heating, DHW, system losses and mechanical systems is 93 kWh/m²a or 11100 kWh/a for one unit.

Table 8.1.4 *Total energy demand, non-renewable primary energy and CO_2 equivalent emissions for the reference row house*

Net energy (kWh/m²a)		Total energy use (kWh/m²a)				Delivered energy (kWh/m²a)		Non-renewable primary energy		CO_2 equivalent emissions	
		Energy use		Energy source				factor (-)	(kWh/m²a)	factor (kg/kWh)	(kg/m²a)
Mechanical systems	5.0	Mechanical systems	5.0	Electricity	5.0	Electricity	5.0	2.35	11.8	0.43	2.2
Space heating	58.0	Space heating	58.0								
DHW	25.5	DHW	25.5	District heating	87.9	District heating	87.9	0.77	67.7	0.24	21.2
		Circulation losses	4.4								
		Conversion losses	0.0								
Total	88.5		92.9		92.9		92.9		79.5		23.4

Row houses example solutions

A general design problem with row houses is the small window area in mid units. Therefore, deep and narrow units should be avoided since this is likely to cause problems with insufficient daylighting. Increasing the area of south-facing windows can be done without dramatically increasing the space heating demand if high-performance windows are used. The additional solar gains can, to some

Conservation: Solution 1

U-value of the whole building:	0.21 W/m²K
Space heating demand:	15 kWh/m²a
Heating distribution:	hot water radiant heating
DHW and space heating system:	district heating

Renewable energy: Solution 2

U-value of the whole building:	0.28 W/m²K
Space heating demand:	20 kWh/m²a
Heating distribution:	hot water floor or wall radiant heating, supply air heating
DHW and space heating system:	solar combi-system and condensing gas boiler

extent, compensate for higher thermal losses. But if a north-facing window area has to be increased, it is essential to use a window with the lowest U-value available in order to limit the thermal losses.

The use of shading devices is highly recommended for providing a good summer comfort. During summer, the use of increased night ventilation with a bypass of the heat exchanger will help to avoid excessive temperatures.

Since the geometry of the row houses requires a rather large hot water grid, it is important that

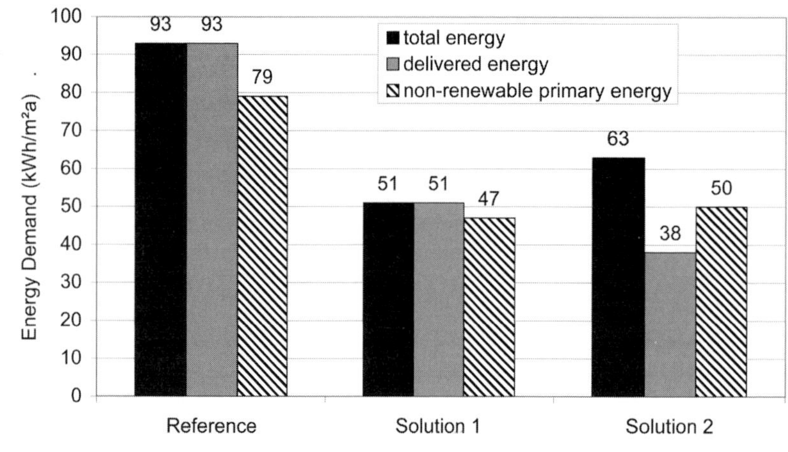

Source: Johan Smeds

Figure 8.1.7 *Overview of the total energy use, the delivered energy and the use of non-renewable primary energy for the row houses; the reference house is connected to district heating*

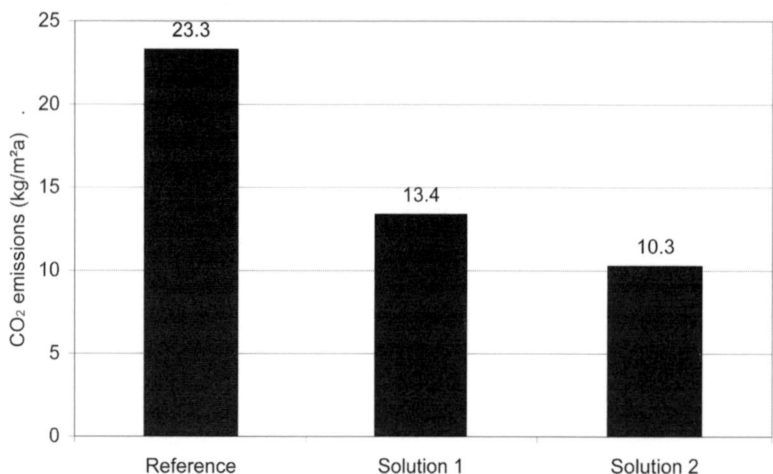

Source: Johan Smeds

Figure 8.1.8 *Overview of the CO_2 equivalent emissions for row houses; the reference house is connected to district heating*

the hot water pipes for domestic hot water and heating are well insulated.

Different example solutions for the row house in the cold climate are described in sections 8.4 and 8.5. The examples are as follows.

As can be seen in Figures 8.1.7 and 8.1.8, the total energy demand is lowest in solution 1 due to the highly insulated building envelope. The use of delivered energy is lowest in solution 2 where renewable energy sources are used. The CO_2 equivalent emissions are also lowest in solution 2. Compared to the reference building, the CO_2 equivalent emissions are reduced by 42 per cent in solution 1 and by 56 per cent in solution 2. The non-renewable primary energy use is reduced by 42 per cent in solution 1 and by 37 per cent in solution 2.

8.1.4 Apartment buildings

Apartment building reference design

Building envelope: the apartment building has reinforced concrete structure and wooden frame façades with mineral wool. The windows have a frame ratio of 30 per cent, triple glazing (4-30-(D4-12)) and are filled with air. The U-values of the building components are shown in Table 8.1.5. The U-values are relatively low due to a compensation for not using 50 per cent ventilation heat recovery, which otherwise should have been used for a Swedish reference building.

Table 8.1.5 *Building component U-values for the apartment building*

Building component	U-value (W/m^2K)
Walls	0.18
Roof	0.10
Floor (excluding ground)	0.20
Windows (frame + glass)	1.74
Window frame	1.70
Window glass	1.75
Whole building envelope	0.35

Mechanical systems: an exhaust air ventilation system without heat recovery provides 0.5 ach. The infiltration rate is 0.1 ach. District heating, which is very common in urban areas of the cold climate region, is used for space heating and domestic hot water. The heat is distributed in the building by a hot water radiant heating system.

Household electricity: the annual use of household electricity in one apartment is approximately 3800 kWh (38 kWh/m^2) for two adults and one child. Household electricity is not included as part of the energy targets.

Space heating demand: the space heating demand is 44 kWh/m^2a. The monthly space heating demand is shown in Figure 8.1.9. Hourly loads of the heating system are calculated with DEROB-LTH without direct solar radiation in order to simulate a totally shaded building. The annual peak load is 46 kW or 29 W/m^2 and occurs in January.

Non-renewable primary energy demand and CO_2 emissions: the total energy demand for space heating, DHW, system losses and mechanical systems is 81 kWh/m^2a. According to Table 8.1.6, the use of district heating and electricity results in a non-renewable primary energy demand of 70 kWh/m^2a (approximately 112,600 kWh/a) and CO_2 equivalent emissions of 20 kg/m^2a for the whole building in one year.

The system losses consist mainly of circulation losses in the DHW distribution system, which are set to 100 W per apartment (8.8 kWh/m^2a). The circulation losses are taken into account as internal

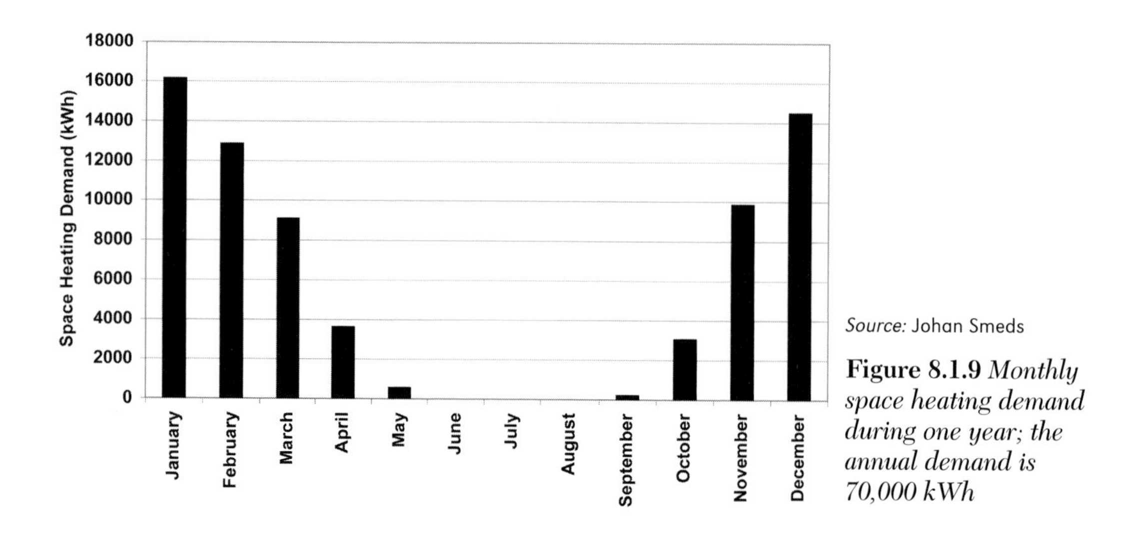

Source: Johan Smeds

Figure 8.1.9 *Monthly space heating demand during one year; the annual demand is 70,000 kWh*

Table 8.1.6 *Total energy demand, non-renewable primary energy demand and CO_2 equivalent emissions for the reference apartment building*

Net energy (kWh/m^2a)		Total energy use (kWh/m^2a)				Delivered energy (kWh/m^2a)		Non-renewable primary energy		CO_2 equivalent emissions	
		Energy use		Energy source				factor (-)	(kWh/m^2a)	factor (kg/kWh)	(kg/m^2a)
Mechanical systems	5.0	Mechanical systems	5.0	Electricity	5.0	Electricity	5.0	2.35	11.8	0.43	2.2
Space heating	43.8	Space heating	43.8	District heating	76.2	District heating	76.2	0.77	58.7	0.24	18.4
DHW	23.6	DHW	23.6								
		Circulation losses	8.8								
		Conversion losses	0.0								
Total	72.4		81.2		81.2		81.2		70.5		20.6

gains in the simulations. The losses of the heat exchanger connected to the district heating system are set to zero.

Apartment building example solutions

Different example solutions for the apartment building in the cold climate are described in sections 8.6 to 8.8. The examples are as follows.

As can be seen in Figures 8.1.10 and 8.1.11, the total energy demand and the delivered energy are lowest in solution 1a due to the highly insulated building envelope. However, the CO_2 emissions and the non-renewable primary energy demand are lowest in solution 2 due to the extensive use of renewable energy sources. Compared to the reference building (connected to district heating), the CO_2 equivalent emissions are reduced by 82 per cent and the non-renewable primary energy use is reduced by 76 per cent in solution 2.

CONSERVATION: SOLUTION 1A

U-value of the whole building:	0.21 W/m^{2}K
Space heating demand:	6.5 kWh/m^{2}a
Heating distribution:	supply air heating
Heating system:	direct electric resistance heating
DHW heating:	solar system with electrical boiler

CONSERVATION: SOLUTION 1B

U-value of the whole building:	0.34 W/m^{2}K
Space heating demand:	15 kWh/m^{2}a
Heating distribution:	hot water radiant heating
DHW and space heating system:	district heating

RENEWABLE ENERGY: SOLUTION 2

U-value of the whole building:	0.41 W/m^{2}K
Space heating demand:	20 kWh/m^{2}a
Heating distribution:	hot water radiant heating
DHW and space heating system:	solar combi-system with biomass burner

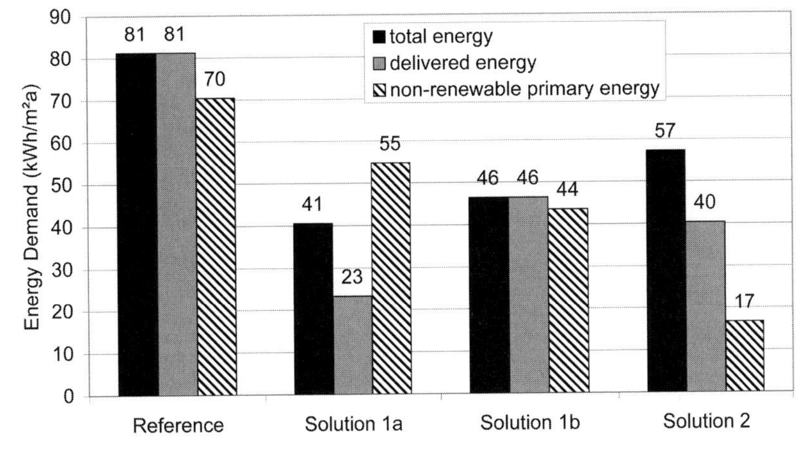

Source: Johan Smeds

Figure 8.1.10 *Overview of the total energy demand, the delivered energy and the non-renewable primary energy demand for the apartment buildings; the reference building is connected to district heating*

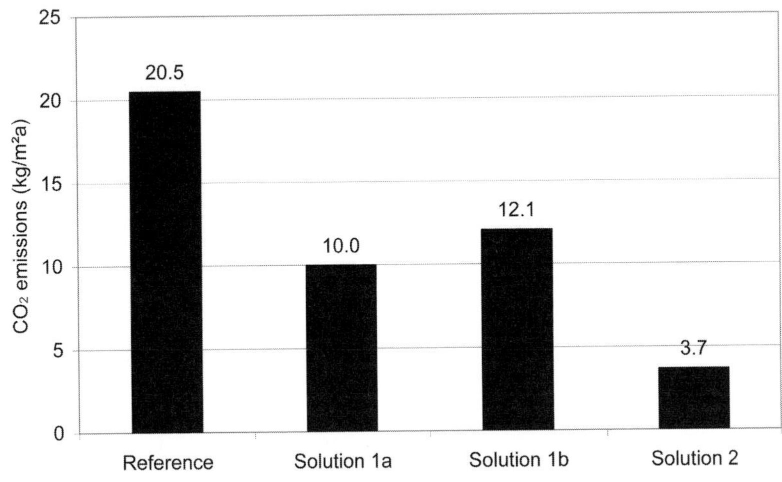

Source: Johan Smeds

Figure 8.1.11 *Overview of the CO_2 emissions for the apartment buildings; the reference building is connected to district heating*

8.1.5 Design advice

A solar DHW system could be a suitable system solution for a high-performance house in all climates since it is difficult to cover much more than the DHW demand by solar energy for houses with such short heating seasons. The choice of heating system in the building is important for the design of a solar thermal system.

In cold climates, where the heating season of high-performance houses is slightly longer compared to warmer climates, a solar combi-system with a common storage tank for DHW and space heating can be a suitable solution if a radiator heating system is needed in the building. The solar energy received during the heating season will still not cover much more than the DHW demand (if dimensioned for the summer demand). However, when a common storage tank is used for DHW and space heating, the withdrawal of space heat from the tank lowers the collector operation temperature and thus improves the efficiency. This effect can be enhanced by designing the system with a possibility of delivering solar energy directly to a low temperature heating system without passing the tank. This lowers the overall heat losses of the system and increases the efficiency of the collector.

If the apartment building is situated in an urban area where district heating systems use large quantities of renewable fuels (as, for example, in Scandinavia, where 80 per cent of the fuel used in district heating systems is renewable), solution 1b is preferable. In rural areas where district heating is not available, solution 1a, with a solar domestic hot water system and electrical auxiliary heating, is a good alternative, especially as the solar system can cover the energy demand in summer. Of course, the generation method of electricity is an important issue that significantly affects the resulting primary energy demand of a building using electricity (as, for example, in solution 1a). If the opportunity of choosing an energy source for electricity generation is given, the environmentally friendly choice has a positive impact on the primary energy use.

References

Kvist, H. (2005) *DEROB-LTH for MS Windows, User Manual Version 1.0–20050813*, Energy and Building Design, Lund University, Lund, Sweden, www.derob.se

Meteotest (2004) *Meteonorm 5.0 – Global Meteorological Database for Solar Energy and Applied Meteorology*, Bern, Switzerland, www.meteotest.ch

SMHI (2005) www.smhi.se

TRNSYS (2005) *A Transient System Simulation Program*, Solar Energy Laboratory, University of Wisconsin, Madison, WI

8.2 Single family house in the Cold Climate Conservation Strategy

Johan Smeds

This section presents two solutions for the single family house in the cold climate. As a reference for the cold climate, the city of Stockholm is used. The solutions are based on energy conservation minimizing the heat losses of the building. A balanced mechanical ventilation system with heat recovery is used to reduce the ventilation losses.

Table 8.2.1 *Targets for the single family house in the Cold Climate Conservation Strategy*

	Targets
Space heating	20 kWh/m^{2}a
Non-renewable primary energy: (space heating + water heating + electricity for mechanical systems)	60 kWh/m^{2}a

8.2.1 Solution 1a: Conservation with electric resistance heating and solar DHW

Building envelope

In this solution the envelope and ventilation system were designed so that the peak space heating load is limited to approximately 10 W/m^2. This allows a central heating element and heat distribution by the supply air, which is more economical than individual room heaters. This solution requires a well-insulated and air-tight building envelope.

The opaque part of the building envelope is a wooden lightweight frame with mineral wool insulation. The windows have a frame ratio of 30 per cent, triple glazing, two low-e coatings and are filled with krypton gas. The window frame consists of a sandwich construction with wood and insulation material. The U-values of the construction are shown in Table 8.2.2. The detailed layers of the construction are shown in Table 8.2.8.

Table 8.2.2 *Building component U-values for solution 1a with supply air heating*

Component	U-value (W/m^{2}K)
Walls	0.08
Roof	0.08
Floor (excluding ground)	0.10
Windows (frame + glass)	0.60
Window frame	0.83
Window glass	0.50
Whole building envelope	0.12

Mechanical systems

A balanced mechanical ventilation system with 80 per cent heat recovery and a bypass for summer ventilation is used. Heat for space heating is supplied by electric resistant heating of the supply air. Energy for domestic hot water is supplied by a solar domestic hot water system in combination with electrical backup.

Energy performance

Space heating demand: simulation results from DEROB-LTH give the monthly space heating demand of the building. A comparison of this ultra high-performance case and a reference standard building according to current standards is shown in Figure 8.2.1. The heating season for the proposed solution extends from November to March and the annual space heating demand is approximately 1700 kWh (11.5 kWh/m^{2}a). Other assumptions made for the simulations in DEROB-LTH are as follows:

- heating set point: 20°C;
- maximum room temperatures: 23°C during winter, 26°C during summer (assumes use of shading devices and window ventilation or enforced ventilation through the bypass);
- ventilation rate: 0.45 ach through the heat exchanger from September to March and through a bypass from April to August; 1.5 ach for cooling at night from June to August;
- infiltration rate: 0.05 ach; and
- heat recovery: 80 per cent efficiency, in use from September to March.

Peak load for space heating: hourly loads of the heating system are calculated with results from DEROB simulations without direct solar radiation in order to simulate a totally shaded building. Figure 8.2.2 illustrates the peak load for each month of the high-performance case and the reference building. The annual peak load of the high-performance building is approximately 1600 W and occurs in January. Near peak demands also occur in February and December. Outside of these three months, the peaks fall off very sharply.

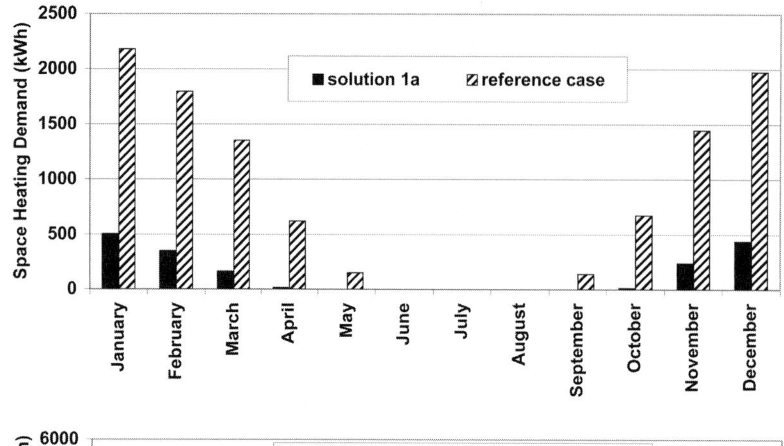

Source: Johan Smeds

Figure 8.2.1 *Space heating demand for the high-performance solution (annual total 1700 kWh/a) and the reference house (annual total 10,400 kWh/a)*

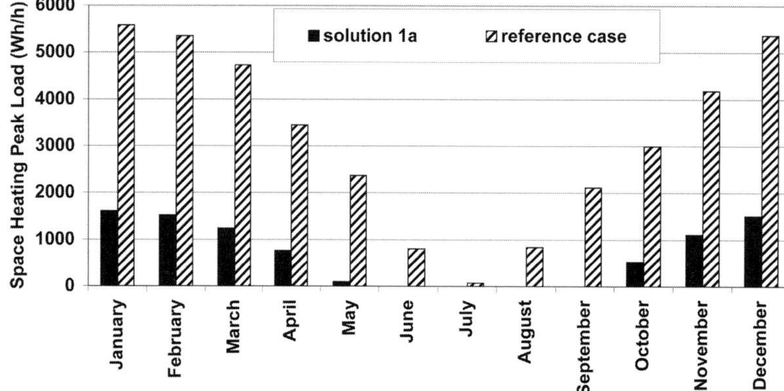

Source: Johan Smeds

Figure 8.2.2 *Space heating peak load for the high-performance solution and the reference building*

Domestic hot water demand: the net DHW heat demand for two adults and two children is approximately 3150 kWh/a or 21 kWh/m^{2}a. The DHW demand is larger than the space heating demand. The system losses consist mainly of losses from the hot water storage tank, but also from pipe losses in the distribution system.

Household electricity: the household electricity use is assumed to be 2500 kWh/a (16.6 kWh/m^{2}a) for two adults and two children. The household electricity use is not taken into account in the primary energy target since the use of household electricity can vary considerably depending on the occupants' behaviour.

Total energy use: the energy use for space heating, DHW and system losses is 5790 kWh/a and the use of electricity for household appliances and mechanical systems is approximately 3250 kWh/a (see Table 8.2.3). The total energy use would, in this case, amount to 9040 kWh/a.

Table 8.2.3 *Total energy use for solution 1a*

Total energy use	**kWh/m²a**	**kWh/a**
Space heating	11.50	1725
DHW heating	21.00	3150
System losses	6.10	915
Electricity, mechanical systems	5.00	750
Household electricity	16.60	2500

Table 8.2.4 *Total energy demand, non-renewable primary energy demand and CO_2 equivalent emissions for the solution with supply air heating and solar DHW heating*

Net energy (kWh/m²a)		Total energy use (kWh/m²a)				Delivered energy (kWh/m²a)		Non-renewable primary energy		CO_2 equivalent emissions	
		Energy use		Energy source				factor (-)	(kWh/m²a)	factor (kg/kWh)	(kg/m²a)
Mechanical systems	5.0	Mechanical systems	5.0	Electricity	5.0	Electricity	5.0	2.35	11.8	0.43	2.2
Space heating	11.5	Space heating	11.5	Electricity	19.0	Electricity	19.0	2.35	44.7	0.43	8.2
DHW	21.0	DHW	21.0								
		Tank and circulation losses	6.1	Solar	19.6						
		Conversion losses	0.0								
Total	37.5		43.6		43.6		24.0		56.4		10.3

Non-renewable primary energy demand and CO_2 emissions: the total energy demand for DHW and space heating, mechanical systems and system losses is 43.6 kWh/m²a (see Table 8.2.4). The non-renewable primary energy demand is 56.4 kWh/m²a. The CO_2 equivalent emissions sum up to 10.3 kg/m²a.

8.2.2 Solution 1b: Conservation with outdoor air to water heat pump

Building envelope

The opaque part of the building envelope is a wooden lightweight frame with mineral wool as insulating material. The windows have a frame ratio of 30 per cent, triple glazing, one low-e coating and are filled with krypton gas. The U-values of the construction components are shown in Table 8.2.5. The detailed layers of the construction are shown in Table 8.2.9.

Table 8.2.5 *Building component U-values for solution 1b*

Component	U-value (W/m²K)
Walls	0.11
Roof	0.11
Floor (excluding ground)	0.17
Windows (frame + glass)	0.92
Window frame	1.20
Window glass	0.80
Whole building envelope	0.17

Mechanical systems

A balanced mechanical ventilation system with 80 per cent heat recovery and a bypass for summer ventilation is considered. Heat is supplied by an outdoor air to water heat pump for space heating and domestic hot water, and the heat is distributed by hot water radiant heating. An alternative to the exhaust air heat pump is a ground-coupled heat pump system serving a group of houses.

Energy performance

Space heating demand: simulation results from DEROB-LTH give the monthly space heating demand of the building, as shown in Figure 8.2.3. The heating season extends from November to March and the annual space heating demand is approximately 2950 kWh or 19.6 kWh/m^{2}a. Other assumptions made for the simulations are the same as for solution 1a.

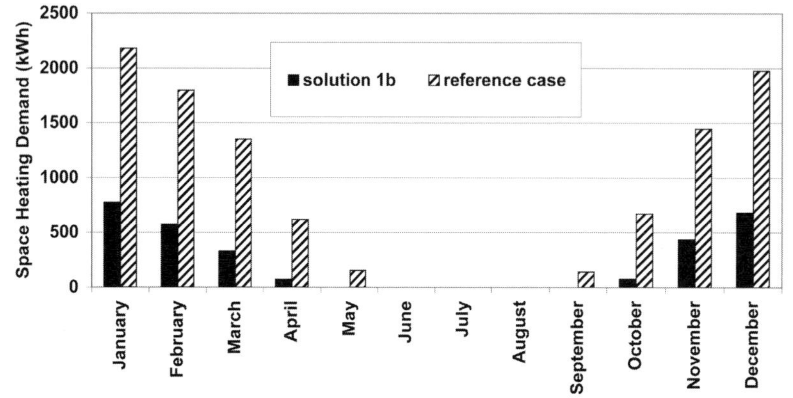

Source: Johan Smeds

Figure 8.2.3 *Space heating demand for the high-performance solution (annual total 2950 kWh/a) and the reference building (annual total 10,400 kWh/a)*

Peak load for space heating: Figure 8.2.4 illustrates the peak load for each month of this high-performance case and the reference building. The annual peak load is approximately 2200 W and occurs in January. While the peak occurs in January, near peak demands also occur in February and December. Outside of these three months, the peaks fall off very sharply.

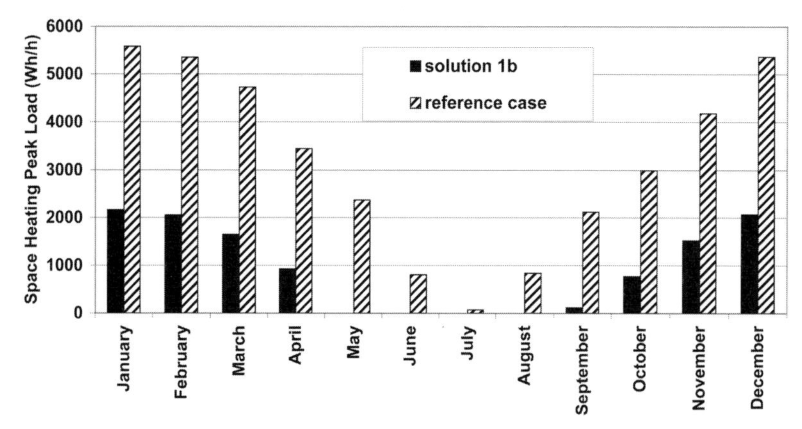

Source: Johan Smeds

Figure 8.2.4 *Space heating peak load for the high-performance solution and the reference building*

Domestic hot water demand: the net DHW heat demand is approximately 3150 kWh/a (21 kWh/m^{2}a). The system losses consist mainly of losses from the hot water storage tank, but also from pipe losses in the distribution system.

Household electricity: the household electricity use is assumed to be 2500 kWh/a or 16.6 kWh/m^{2}a for two adults and two children, as for solution 1a.

Total energy use: the energy use for space heating, DHW and system losses is 6675 kWh/a and the use of electricity for household appliances and mechanical systems is approximately 3190 kWh/a (see Table 8.2.6).

Table 8.2.6 *Total energy use for solution 1b*

Total energy use	kWh/m²a	kWh/a
Space heating	20.00	3000
DHW heating	21.00	3150
System losses	3.50	525
Electricity, mechanical systems	4.60	690
Household electricity	16.60	2500

Non-renewable primary energy demand and CO_2 emissions: the total energy demand for DHW, space heating, mechanical systems and system losses is 49.1 kWh/m²a (see Table 8.2.7). The non-renewable primary energy demand is 46 kWh/m²a. The CO_2 equivalent emissions sum up to 8.5 kg/m²a.

Table 8.2.7 *Total energy demand, non-renewable primary energy demand and CO_2 equivalent emissions for the solution with outdoor air to water heat pump*

Net Energy (kWh/m²a)		Total Energy Use (kWh/m²a)				Delivered energy (kWh/m²a)		Non renewable primary energy		CO_2 equivalent emissions	
		Energy use		Energy source				factor (-)	(kWh/m²a)	factor (kg/kWh)	(kg/m²a)
Mechanical systems	5.0	Mechanical systems	5.0	Electricity	5.0	Electricity	5.0	2.35	11.8	0.43	2.2
Space heating	19.6	Space heating	19.6	Electricity heat pump COP=3	14.7	Electricity	14.7	2.35	34.5	0.43	6.3
DHW	21.0	DHW	21.0								
		Circulation losses	3.5	Outdoor air	29.4						
		Conversion losses	0.0								
Total	45.6		49.1		49.1		19.7		46.3		8.5

8.2.3 Summer comfort

The following analysis on summer comfort is made for solution 1b: conservation with outdoor air to water heat pump.

Two ventilation strategies are tested by simulations in DEROB-LTH. The first ventilation strategy uses a bypass of the heat recovery with enforced ventilation at 1.5 ach between 7 pm and 6 am from June to August. The second ventilation strategy for achieving a good summer comfort is to bypass the heat recovery totally from April to August and to enforce night ventilation by opening windows or increasing the ventilation rate of the mechanical ventilation system to 2.5 ach between 7 pm and 6 am from June to August. For both ventilation strategies, external shading for windows facing west, south and east with 50 per cent transmittance and 10 per cent absorption is considered.

As shown in Figure 8.2.5, the use of ventilation heat recovery during April and May according to ventilation strategy 1 still causes overheating of the building. Consequently, a bypass of the heat exchanger even during spring can be a very important means to reduce excessive indoor temperatures. One way of characterizing the summer comfort is by the sum of degree hours for temperatures exceeding 26°C. The amount of overheating in this case is 727 Kelvin hours per year. The maximum indoor temperature is 31.9°C in late May.

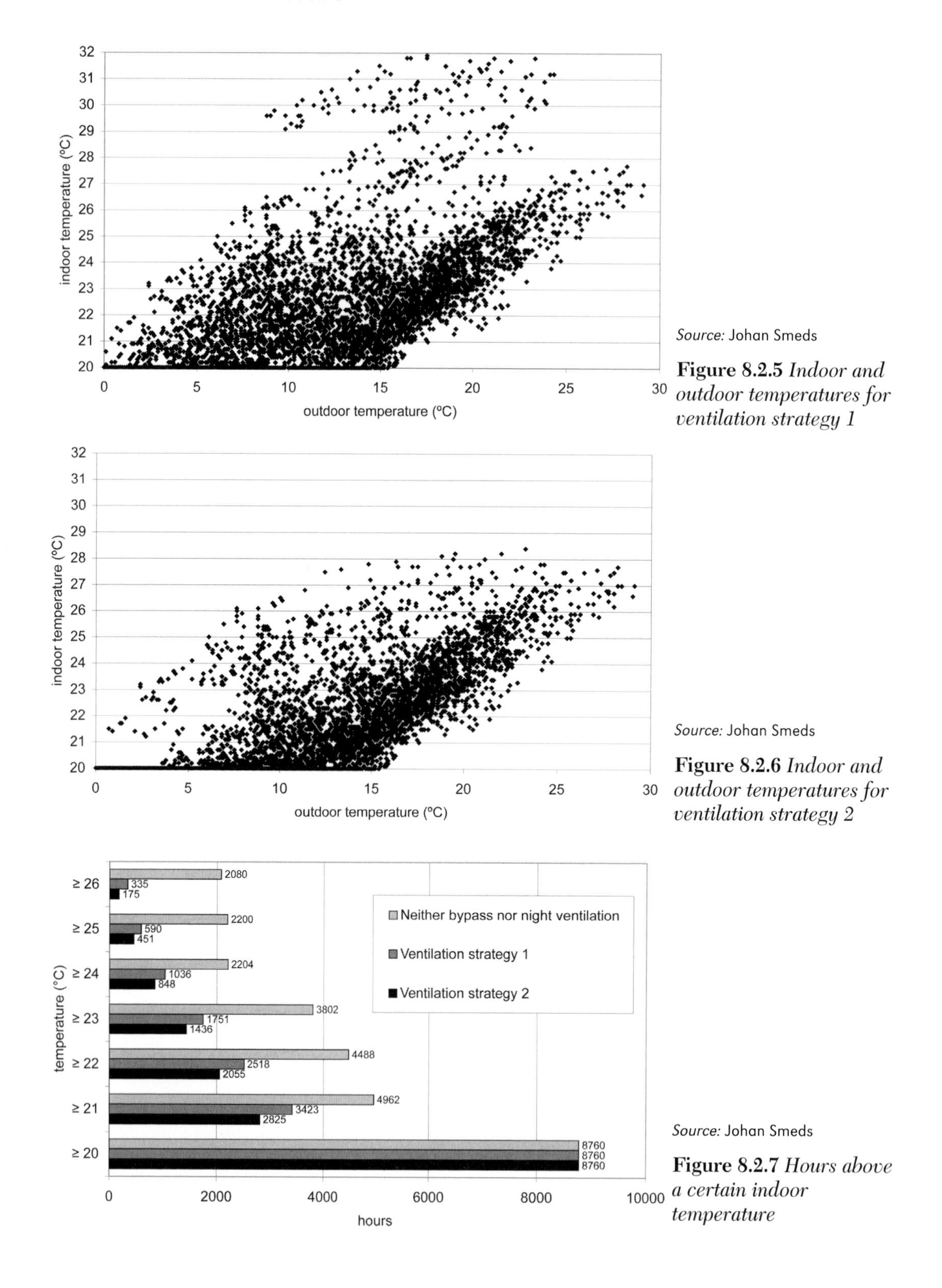

Source: Johan Smeds

Figure 8.2.5 *Indoor and outdoor temperatures for ventilation strategy 1*

Source: Johan Smeds

Figure 8.2.6 *Indoor and outdoor temperatures for ventilation strategy 2*

Source: Johan Smeds

Figure 8.2.7 *Hours above a certain indoor temperature*

For the second ventilation strategy, the amount of overheating is 129 Kh/a and the maximum indoor temperature of 28.4°C occurs in spring. The temperature variations of ventilation strategy 2 are illustrated in Figure 8.2.6.

The number of overheating hours is illustrated in Figure 8.2.7. In addition to ventilation strategies 1 and 2, this figure also shows the effects of using the heat recovery all year round and not using any shading devices. For ventilation strategy 2, the amount of hours over 26°C is limited to 175. With the heat recovery in use the whole year and without using shading devices, the amount of hours with temperatures over 26°C increases to 2080. Simulations in DEROB-LTH show that the heating demand would increase by 6 kWh/m^{2}a, assuming that the windows would be shaded continuously from April to August and that ventilation strategy 2 applies. Thus, it is important to use operable shading devices that do not reduce useable solar gains.

8.2.4 Sensitivity analysis

The following sensitivity analysis is made for solution 1b: conservation with outdoor air to water heat pump.

Window size and orientation

The effect of changing the orientation of the building and increasing the glazed area of the originally south-facing window is shown in Figure 8.2.8.

Increasing the glazed area of the south-facing windows from 6.3 m^2 to 15 m^2 leads to an increase of the annual space heating demand of 6 per cent from approximately 2950 kWh to 3120 kWh. Rotating the original building by 180 degrees, letting the originally south-facing façade now face north, and increasing the glazed area of the now north-facing façade from 6.3 m^2 to 15 m^2 will increase the annual space heating demand by 20 per cent from approximately 3360 kWh to 4030 kWh.

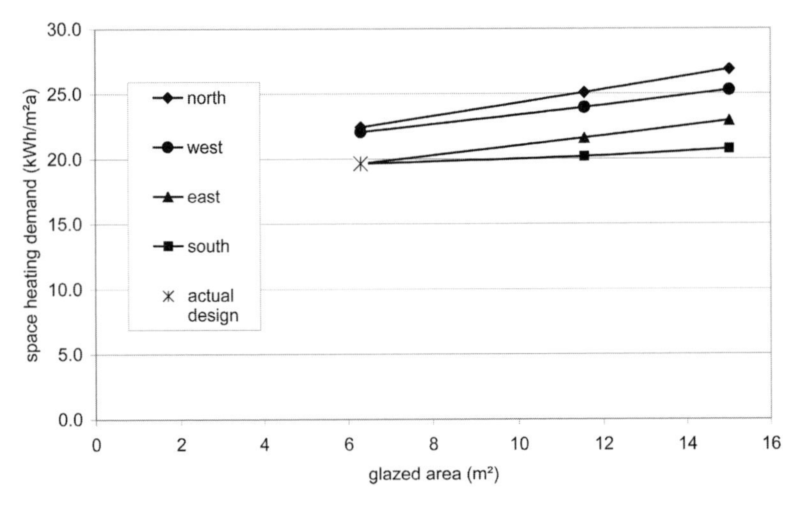

Source: Johan Smeds

Figure 8.2.8 *Effect of window size and orientation on space heating demand; the star shows the actual design of the building according to solution 1b*

Air tightness of building envelope

Air leakage through the building envelope drastically increases the space heating demand, as shown for solution 1b in Figure 8.2.9. The high-performance detached house has a very air-tight building envelope with an infiltration rate of 0.05 ach. Doubling the infiltration rate to 0.10 ach causes an increase in space heating demand by 16 per cent from 19.6 kWh/m^{2}a to 22.8 kWh/m^{2}a.

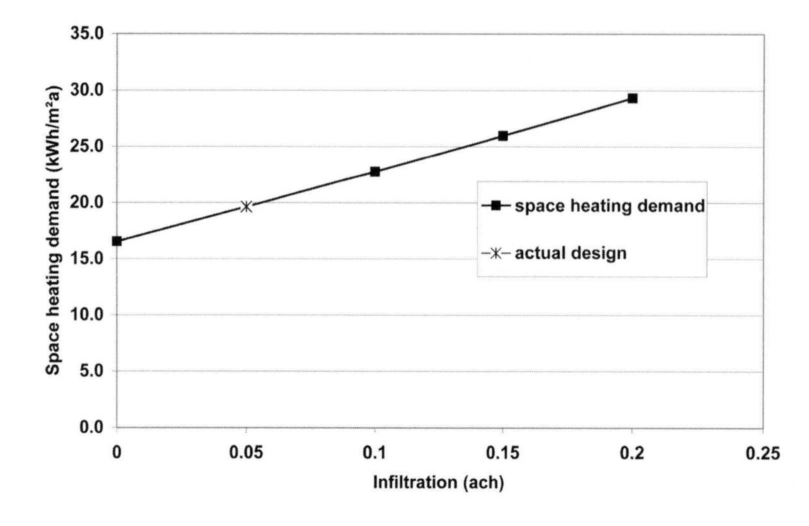

Source: Johan Smeds

Figure 8.2.9 *Air tightness of the building envelope; the star shows the actual design for solution 1b*

8.2.5 Design advice

Because, even in this cold climate, the annual space and domestic water heating demand is very low (4875 kWh/a for solution 1a and 6150 kWh/a for solution 1b), the capital investment in the heating system must be kept small. Two example solutions have been presented. The first solution presented (solution 1a) is based on direct electric resistant heating in an extremely well-insulated house. The second solution (solution 1b) is based on an outdoor air to water heat pump for each house. This is already common in Scandinavian countries. Another way can also be a borehole heat pump system with a local district heating system serving a group of houses.

Table 8.2.8 *Solution 1a: Conservation with electric resistance heating and solar DHW – building envelope construction*

	Material	Thickness	Conductivity	Per cent	Studs	Studs	Resistance without Rsi, Rse	Resistance with Rsi, Rse	U-value
		m	λ (W/mK)	%	λ (W/mK)	%	(m²K/W)	(m²K/W)	(W/m²K)
wall	exterior surface							0.04	
	wooden panel	0.045		100%					
	air gap	0.025		100%					
	mineral wool hd	0.100	0.030	100%			3.33		
	mineral wool	0.440	0.036	85%	0.14	15%	8.53		
	plastic foil			100%					
	mineral wool	0.045	0.036	85%	0.14	15%	0.87		
	plaster board	0.013	0.220	100%			0.06		
	interior surface							0.13	
		0.668					**12.79**	**12.96**	**0.077**
roof	exterior surface							0.04	
	roof tiles	0.050		100%					
	roof felt	0.002		100%					
	wooden panel	0.022		100%					
	air gap	0.025		100%					
	mineral wool	0.470	0.036	95%	0.14	5%	11.41		
	plastic foil			100%					
	mineral wool	0.045	0.036	85%	0.14	15%	0.87		
	plaster board	0.013	0.220	100%			0.06		
	interior surface							0.10	
		0.627					**12.34**	**12.48**	**0.080**
floor									
	mineral wool	0.350	0.036	100%			9.72		
	concrete	0.100	1.700	100%			0.06		
	interior surface							0.17	
		0.450					**9.78**	**9.95**	**0.100**
window				emissivity					
	pane	0.004	low emissivity	5%	reversed				
	gas	0.012	krypton						
	pane	0.004	clear	83.7%					
	gas	0.012	krypton						
	pane	0.004	low emissivity	5%					**0.500**
		0.036							
	frame	0.093			0.09		1.03	1.20	**0.831**

Summer comfort will be achieved by a combination of measures, including shading of windows, bypassing the ventilation heat recovery and increasing the ventilation rate at night by opening windows or by enforced mechanical ventilation.

Increasing the size of south-facing windows of such highly insulated houses does not save energy; but it also does not significantly increase the heating demand either if windows with low U-values are chosen. This is not the case for window areas facing other directions, where increasing the glass area does increase the heating demand.

Air-tight construction of the building envelope is essential in the cold climate. The assumed tightness here is 0.6 ach by a pressurization of 50 Pa, resulting in approximately 0.05 ach infiltration rate under normal conditions. With common practice construction (i.e. 0.20 ach at normal conditions), the space heating demand would increase by 50 per cent from 20 kWh/m²a to almost 30 kWh/m²a.

Table 8.2.9 *Solution 1b: Conservation with outdoor air to water heat pump – building envelope construction*

	Material	Thickness	Conductivity	Per cent	Studs	Studs	Resistance without Rsi, Rse	Resistance with Rsi, Rse	U-value
		m	λ (W/mK)	%	λ (W/mK)	%	(m²K/W)	(m²K/W)	(W/m²K)
wall	exterior surface							0.04	
	wooden panel	0.045		100%					
	air gap	0.025		100%					
	mineral wool hd	0.100	0.030	100%			3.33		
	mineral wool	0.240	0.036	85%	0.14	15%	4.65		
	plastic foil			100%					
	mineral wool	0.045	0.036	85%	0.14	15%	0.87		
	plaster board	0.013	0.220	100%			0.06		
	interior surface							0.13	
		0.468					**8.92**	**9.09**	**0.110**
roof	exterior surface							0.04	
	roof tiles	0.050		100%					
	roof felt	0.002		100%					
	wooden panel	0.022		100%					
	air gap	0.025		100%					
	mineral wool	0.330	0.036	95%	0.14	5%	8.01		
	plastic foil			100%					
	mineral wool	0.045	0.036	85%	0.14	15%	0.87		
	plaster board	0.013	0.220	100%			0.06		
	interior surface							0.10	
		0.487					**8.94**	**9.08**	**0.110**
floor									
	mineral wool	0.210	0.036	100%			5.83		
	concrete	0.100	1.700	100%			0.06		
	interior surface							0.17	
		0.310					**5.89**	**6.06**	**0.165**
window				emissivity					
	pane	0.004	low emissivity	5%	reversed				
	gas	0.012	krypton						
	pane	0.004	clear	83.70%					
	gas	0.012	krypton						
	pane	0.004	clear	83.70%					
		0.036							**0.800**
	frame	0.093	wood		0.14		0.66	0.83	**1.20**

References

GEMIS (2004) *Gemis: Global Emission Model for Integrated Systems*, Öko-Institut, Darmstadt, Germany

Kvist, H. (2005) *DEROB-LTH for MS Windows, User Manual Version 1.0–20050813*, Energy and Building Design, Department of Architecture and Built Environment, Lund University, Lund, Sweden

Meteotest (2004) *Meteonorm 5.0 – Global Meteorological Database for Solar Energy and Applied Meteorology*, Bern, Switzerland, www.meteotest.ch

8.3 Single family house in the Cold Climate Renewable Energy Strategy

Tobias Boström and Johan Smeds

Table 8.3.1 *Targets for the single family house in the Cold Climate Renewable Energy Strategy*

	Targets
Space heating	25 kWh/m^{2}a
Non-renewable primary energy: (space heating + water heating + electricity for mechanical systems)	60 kWh/m^{2}a

This section presents a solution for the single family house in the cold climate. As a reference for the cold climate, the city of Stockholm is considered. The solution is focused on the use of renewable energy sources.

8.3.1 Solution 2: Renewable energy with solar combi-system and biomass or condensing gas boiler

Building envelope

The opaque part of the building envelope is a wooden lightweight frame with mineral wool as insulating material. The windows have a frame ratio of 30 per cent, triple glazing, one low-e coating and are filled with krypton gas. The U-values of the construction components are shown in Table 8.3.2. The detailed layers of the construction are shown in Table 8.3.8.

Table 8.3.2 *Building component U-values for solution 2*

Component	U-value (W/m^{2}K)
Walls	0.14
Roof	0.15
Floor	0.20
Windows (frame + glass)	0.92
Window frame	1.20
Window glass	0.80
Whole building envelope	0.21

Mechanical systems

A balanced (mechanical) ventilation system with 80 per cent heat recovery and a bypass for summer ventilation is considered.

Heat is supplied by a solar combi-system with a biomass burner and 6 m^2 of direct-flow vacuum tube collectors or 7.5 m^2 of flat-plate collectors. The solar collectors are mounted at an angle of 40° and a storage tank of 0.5 m^3 is assumed. The solar collector area is optimized for achieving 100% coverage of the heat demand during summer. The heat is distributed in the building by hot water radiant heating. An alternative to using the biomass boiler is a solar combi-system with a condensing gas boiler.

A question that could be posed is whether or not the solar collector system should be of a DHW or combi-type. The better a house is insulated, the less advantage is achieved from a combi-system since the heating season is greatly reduced. By using a DHW solar collector system instead of a combi-system, the overall performance of the solar system for this typical example solution will decrease by 2 per cent to 3 per cent. The combi-system does not achieve a higher overall efficiency and it costs more. On the other hand, a space heating distribution system is needed in any case. If this is achieved by a water-based system, the added cost for a combi-system is mainly just the additional collector area and the upsizing of the storage tank.

Energy performance

Space heating demand: the space heating demand was computed DEROB-LTH (Kvist, 2005). Figure 8.3.1 compares this high-performance building and a reference building according to current building standards. The heating season extends from November to March and the annual space heating demand is 3700 kWh. Other assumptions made for the DEROB simulations are:

- heating set point: 20°C;
- maximum room temperatures: 23°C during winter and 26°C during summer (assumes use of shading devices and window ventilation);
- ventilation rate: 0.45 ach;
- infiltration rate: 0.05 ach; and
- heat recovery: 80% efficiency.

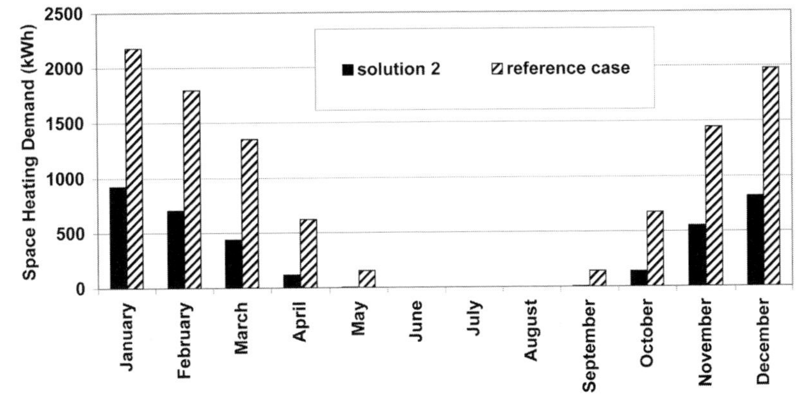

Source: Tobias Boström and Johan Smeds

Figure 8.3.1 *Space heating demand (annual total 3701 kWh/a)*

Peak load of space heating system: the hourly heat loads of the heating system are calculated with DEROB-LTH. The simulation is performed without direct solar radiation in order to simulate a totally shaded building. A comparison of the peak loads for each month for this high-performance building and a reference building according to current building standards is shown in Figure 8.3.2. The peak load for the most extreme hour of the year is 2515 W and occurs in January. While the peak occurs in January, near peak loads also occur during some hours in February, March and December. Outside of these months, the peaks fall off very sharply.

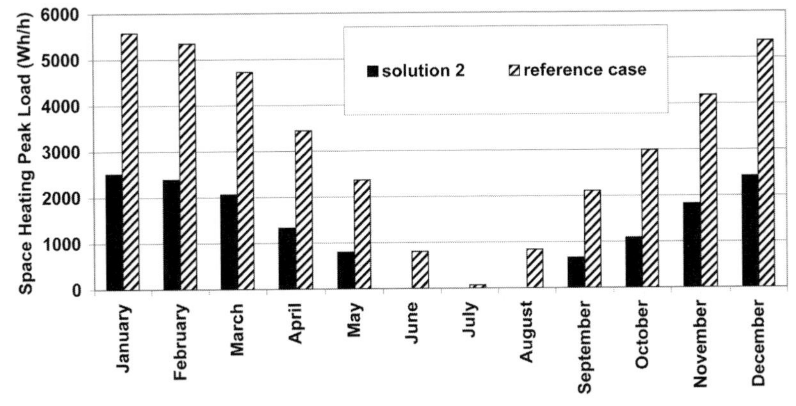

Source: Tobias Boström and Johan Smeds

Figure 8.3.2 *Space heating peak load*

Domestic hot water demand: the net DHW heat demand is approximately 3150 kWh/a or 21 kWh/m^{2}a for two adults and two children who occupy a typical single family detached house. The DHW temperature is set to 55°C and the average DHW consumption per person is 40 litres per person per day. Consequently, the model single family house consumes 160 litres of domestic hot water per day. The average temperature over the year of cold tap water in Stockholm is 8.5°C. The temperature of the hot water was set to 50°C at the faucet. The on/off temperature set points for the thermostat in the tank were set to 55/57°C.

System losses: the system losses consist mainly of losses from the hot water storage tank, but also from pipe losses in the distribution system and conversion losses in the boiler. The system losses are dependent on several parameters and how the losses actually are defined. The losses provided by the used solar collector simulation programme are the tank losses, which include the total heat losses through the tank wall, base and cover, and the connection losses. The tank losses become larger with an increase in tank size and/or increase of solar collector area. The tank losses for a solar collector system with 7.5 m^2 of collectors and a 600 litre tank are about 950 kWh per year or 6.3 kWh/m^{2}a (per living area). The coefficient of performance (COP) for the biomass boiler is 85 per cent, resulting in conversion losses of 5.6 kWh/m^{2}a. Due to the assumed COP of 100 per cent for a condensing gas boiler, the conversion losses are set to zero.

Household electricity: the amount of household electricity used by two adults and two children is approximately 2500 kWh or 16.6 kWh/m^{2}a. The primary energy calculations shown in Tables 8.3.4 and 8.3.5 do not include household electricity since this factor can vary considerably, depending on the occupant's behaviour.

Total energy use: the total energy use of heat for DHW, space heating and system losses is 7800 kWh/a and the end energy use of electricity for household electricity and mechanical systems is approximately 3240 kWh/a (see Table 8.3.3).

Table 8.3.3 *Total energy use for solution 2*

Total energy use	kWh/m^{2}a	kWh/a
Space heating	24.7	3700
DHW heating	21.0	3150
System losses	6.3	950
Electricity, mechanical systems	5.0	750
Household electricity	16.6	2490

Non-renewable primary energy demand and CO_2 equivalent emissions: the conversion factors for the furnace are set to 0.85 for the pellet burner and 1.0 for the condensing gas burner. The factors for primary energy and CO_2 emissions are taken from GEMIS (GEMIS, 2004). The amount of remaining energy for space heating, DHW, tank and pipe losses after taking solar gains into account is calculated in Polysun. The thermal solar combi-system consists of 7.5 m^2 of flat-plate collectors tilted 40° and a 600 litre storage tank.

According to Tables 8.3.4 and 8.3.5, the use of non-renewable primary energy is 17 kWh/m^{2}a for the solar combi-system with a biomass boiler and 47.9 kWh/m^{2}a if a solar combi-system with condensing gas boiler is used. The CO_2 equivalent emissions sum up to 3.8 kg/m^{2}a for the solar combi-system with a biomass boiler and 10 kg/m^{2}a if a solar combi-system with condensing gas boiler is used.

Table 8.3.4 *Primary energy demand and CO_2 emissions for solar combi-system with biomass boiler*

Net Energy (kWh/m²a)		Total Energy Use (kWh/m²a)				Delivered energy (kWh/m²a)		Non renewable primary energy		CO_2 equivalent emissions	
		Energy use		Energy source				factor (-)	(kWh/m²a)	factor (kg/kWh)	(kg/m²a)
Mechanical systems	5.0	Mechanical systems	5.0	Electricity	5.0	Electricity	5.0	2.35	11.8	0.43	2.2
Space heating	24.7	Space heating	24.7	Wood pellets	37.3	Wood pellets	37.3	0.14	5.2	0.04	1.6
DHW	21.0	DHW	21.0								
		Tank and circulation losses	6.3	Solar	20.3						
		Conversion losses	5.6								
Total	50.7		62.6		62.6		42.3		17.0		3.8

Table 8.3.5 *Primary energy demand and CO_2 emissions for solar combi-system with condensing gas boiler*

Net Energy (kWh/m²a)		Total Energy Use (kWh/m²a)				Delivered energy (kWh/m²a)		Non renewable primary energy		CO_2 equivalent emissions	
		Energy use		Energy source				factor (-)	(kWh/m²a)	factor (kg/kWh)	(kg/m²a)
Mechanical systems	5.0	Mechanical systems	5.0	Electricity	5.0	Electricity	5.0	2.35	11.8	0.43	2.2
Space heating	24.7	Space heating	24.7	Gas	31.7	Gas	31.7	1.14	36.1	0.25	7.8
DHW	21.0	DHW	21.0								
		Tank and circulation losses	6.3	Solar	20.3						
		Conversion losses	0.0								
Total	50.7		57.0		57.0		36.7		47.9		10.0

8.3.2 Sensitivity analysis to key solar active parameters

The solar collector system has been studied and simulated extensively using the Swiss simulation programme Polysun (see Solartechnik Prüfung Forschnung, www.spf.ch). Pipe losses and storage losses in Polysun are calculated and taken into account as internal gains. The most important *fixed* parameters for the simulations are:

- auxiliary boiler power: 4 kW;
- Stockholm climate: Meteonorm generated (Meteotest, 2004);
- insulation levels: 150 mm for the tank and 25 mm for the piping;
- piping lengths: 3 m outdoors and 12 m indoors; and
- collector types: either a high-performing flat plate or an evacuated tube.

In the following sections we have chosen to show the solar collector system's sensitivity to a few key *optimization* parameters, such as:

- azimuth effect;
- absorber area;
- tank volume;
- tilt angle 40° or 90° (roof or wall placement);
- collector type, evacuated tube or flat plate;
- hot water set temperature; and
- combi- or DHW system.

Rather than presenting the energy conversion efficiency or solar fraction for the solar thermal collector systems, most figures show the remaining auxiliary demand needed to cover the DHW and space heating demands.

Collector orientation

The first question is: what happens if the collector is not facing directly towards south? The simulations show that the direction is not critical. The efficiency of the collector will be more than 95 per cent of maximum output as long as the orientation or azimuth direction is within +/–30° from south.

Collector area

Figure 8.3.3 shows how the auxiliary demand for a flat-plate solar combi-system varies during the summer months for collector areas between 5 m^2 and 10 m^2. When designing a conventional solar thermal collector system, one should try to obtain 100 per cent coverage during June to August. The demand during the summer only consists of DHW, which amounts to 262 kWh/month (excluding storage tank losses). It can be seen that 5 m^2 is too small in area in order to get complete coverage during the summer months. If the size is increased to 10 m^2, the auxiliary demand becomes almost non-existant from May to August and it is likely that the solar system will suffer from overheating problems during the summer.

Table 8.3.6 shows what happens with the solar collector output and the solar fraction when the area is increased from 5 m^2 to 10 m^2. The *useful* collector output drastically decreases when the area is increased. This is mainly due to the fact that the 5 m^2 collector almost already covers the whole demand during the summer. A larger collector will produce non-useful waste heat during the summer, which can be harmful for the solar collector. The absorber itself can sustain periods with tempera-

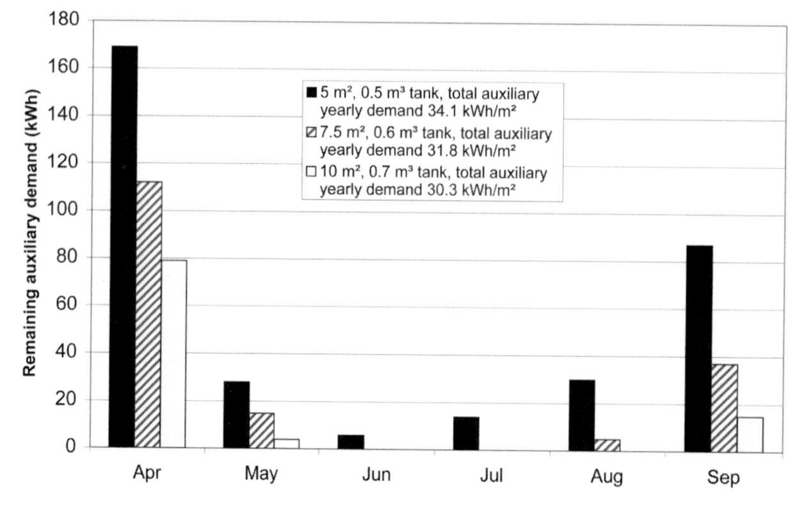

Source: Tobias Boström and Johan Smeds

Figure 8.3.3 *The auxiliary demand's dependence upon collector area during the summer months and the total auxiliary annual demand in kWh/m^2 (living area)*

tures above 200°C; but the water/glycol mixture will start to disintegrate above about 140°C. The disintegration results in the fact that glycol lumps are formed inside the tubes, which eventually blocks, and the absorber will be unusable.

A collector of about 7.5 m^2 is, however, needed if one wants to be completely independent of an auxiliary system during the summer. A 7.5 m^2 system will, during one year, only get about 6 collector temperature spikes over 140°C, while 10 m^2 gets about 40 spikes over 140°C. About 10 spikes over the limit are acceptable, but not more.

Table 8.3.6 *Collector area effect on various system parameters*

Collector area (m^2)	Useful output per m^2 collector (kWh/m^2a)	Solar fraction (%)	Total efficiency increase (%)	Remaining aux demand per living area (kWh/m^2a)
5	420	29	–	34.1
7.5	320	34	17	31.8
10	260	37	28	30.3

Storage tank size

Another important factor is the size of the tank. Figure 8.3.4 shows the influence on the yearly auxiliary demand for a 7.5 m^2 flat-plate solar collector combi-system with increasing tank sizes. The utilization of the solar gains from the solar collector increases with the size of the tank; but, on the other hand, the heat losses also become larger as the size of the tank is increased. The maximum allowed tank temperature is 95°C and the night cool-off temperature was set to 80°C. According to the simulation program, the tank size does not play a central role. An optimum can still be seen at 0.5 m^3. Below this threshold the solar gains cannot be fully utilized, and above, the heat loss increase exceeds the increase in solar gains. However, the auxiliary demand does not vary more than 1 per cent as long as the tank size is within a reasonable volume span (0.4–1.0 m^3). When the size exceeds this span, the heat losses become quite substantial and, hence, the auxiliary demand increases a great deal. What the figure does not show is that the tank size should not be less than 400 litres in order to be able to cover the DHW and space heating demand. Too small a tank cannot store sufficient enough heat in order to supply warm water when the power requirements are high (i.e. during the winter or when the shower is used for a long time).

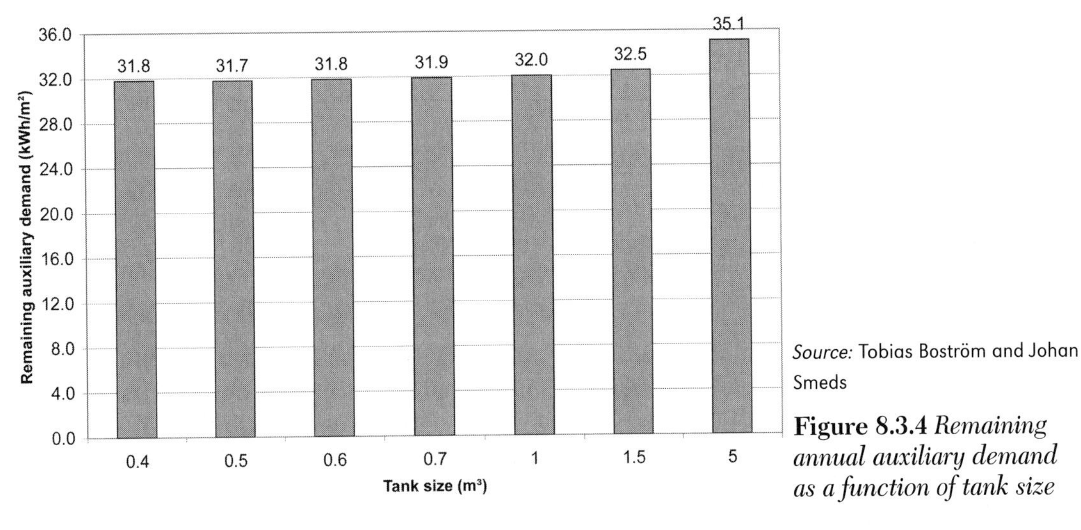

Source: Tobias Boström and Johan Smeds

Figure 8.3.4 *Remaining annual auxiliary demand as a function of tank size*

Tilt effect

To increase the solar collector output over the year, the output during spring and autumn must be increased while the output during the summer is suppressed. One way of achieving this is to use concentrating solar collector systems with various acceptance angles; but this is not studied here.

In order to increase the solar fraction during the winter for non-concentrating systems, the tilt angle of the collector has to be raised. The low-standing winter sun can be more efficiently utilized by placing the collector vertically. By having vertical standing collectors, the high-standing summer sun is also suppressed, making it possible to have larger collectors without creating an overheating problem (Boström el al, 2003).

Figure 8.3.5 shows how the annual auxiliary demand diminishes as the collector area is increased. Small systems tilted 40° are more effective than the vertical equivalent. However, the 40° tilted system creates a large amount of unusable heat during the summer when the collector area is larger than 10 m^2, which results in the vertical system becoming more effective for large areas. Nevertheless, it is not economically justified to double the collector area just to get a 13 per cent decrease in auxiliary demand. It might also be a problem to find a large enough non-shaded south-facing façade for the solar collector. Advantages and disadvantages with either mounting position can be found in the Table 8.3.7.

Table 8.3.7 *Pros and cons with a roof- or wall-mounted collector for cold climates*

40º tilt		90º tilt (vertical)	
Pros	**Cons**	**Pros**	**Cons**
Possible to shut off the auxiliary system during the summer	Susceptible to overheating during the summer	Acquires the highest solar fraction (large systems)	Needs larger areas
Small collector areas give a higher solar fraction	Becomes covered by snow in the winter	Cheaper to install	More easily shaded
Generally the more economical choice		Boosted solar radiation through snow reflections during the winter	Smaller available surface area to install the collectors on

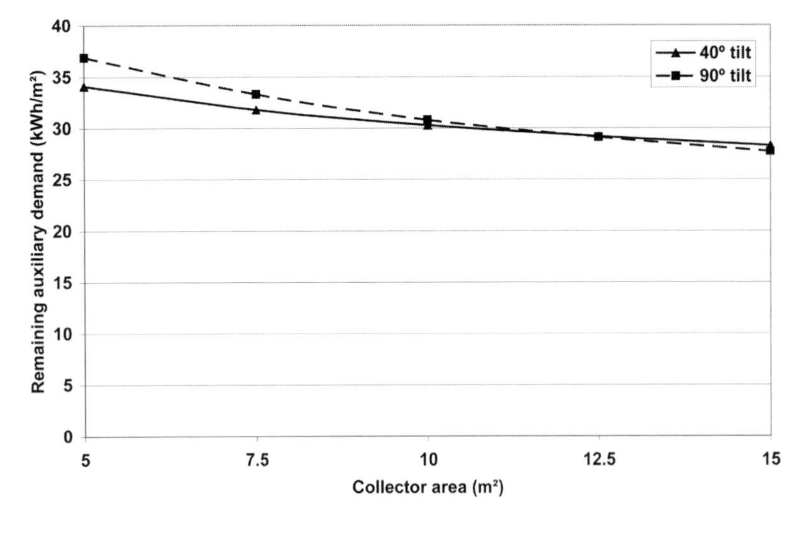

Source: Tobias Boström and Johan Smeds

Figure 8.3.5 *Solar collector tilt effect on the remaining auxiliary annual demand for flat-plate systems*

Evacuated tube versus flat-plate collector

An alternative to using flat-plate collectors is evacuated tubular collectors. Vacuum systems have a higher efficiency compared to flat-plate systems but are, on the other hand, about twice as expensive per m^2. To compare, the optimal flat-plate combi-system with 7.5 m^2 tilted 40° which needs an auxiliary demand of 31.8 kWh/m^2a is matched with the vacuum system that needs the same amount of auxiliary energy. Simulations show that 5 m^2 of evacuated collectors tilted 40° are sufficient to achieve the same efficiency. In other words, the evacuated collector is about 50 per ent more effective per area unit than the flat-plate collector; but since the evacuated collector is twice as expensive, it does not become an economically justified choice today. These findings correspond well with an in situ measurement (Kovacs and Pettersson, 2002), which showed that evacuated collectors were between 45 per ent and 60 per cent more effective than flat plates per m^2, depending on the load applied. Both the flat plate and the evacuated tube simulated in this chapter are modern high-performance collectors.

Hot water set temperature

The discussion has so far concerned the solar collectors and the storage tank; but the auxiliary system is also important. One very important parameter is the domestic hot water set temperature, T_{DHW}, which is the actual temperature at the faucet. The temperature at the top of the tank is usually a few degrees higher in order to cope with the temperature drop from the tank to the faucet. T_{DHW} has been set to 50°C in the simulations above, which is an adequate temperature for household purposes. Many solar collector systems have a much too high T_{DHW}, quite often up to 70°C or more, which results in the fact that the needed auxiliary energy drastically increases. This is often due to the fear of Legionella disease. However, a temperature of 50°C is sufficient to prevent growth of Legionella. Simulations show that the auxiliary energy demand for a 7.5 m^2 flat-plate combi-system tilted 40° decreases with 5% if T_{DHW} is decreased from 60 to 50°C.

DHW versus combi-system

Both systems have 7.5 m^2 of flat-plate collectors tilted 40° and a 500 litre tank. The remaining auxiliary demand for the combi-system was 31.7 kWh/m^2a (living area); the corresponding figure for the DHW system was 32.2 kWh/m^2a. Highly insulated buildings will consequently only benefit to a very small degree from having a conventional solar combi-system; a DHW system will provide about the same amount of useful heat since it is mainly the spring, summer and autumn DHW load that is covered by both systems. A roof-mounted collector will produce very little useful heat during the short heating season of November to March.

There are a number of approaches in order to increase the efficiency of a solar combi-system. As mentioned earlier, the collector can be tilted vertically and thus achieve a higher efficiency for the low-standing winter sun. Alternatively, one might use concentrating systems that can boost the collector performance for specific solar angles. When choosing your heating system, you should also opt for a low temperature heating alternative that allows the collector to work more efficiently (i.e. floor heating).

8.3.4 Design advice

The environmental benefits of using a biomass boiler instead of a condensing gas boiler are obvious according to results shown in this chapter. A system with a biomass boiler is to be preferred. On a yearly basis, the difference in emissions of CO_2 equivalents between the two heating systems totals 930 kg. The emissions of CO_2 equivalents for the solar combi-system with condensing gas burner have the same magnitude as the electrically heated building according to the conservation strategy (10.3 kg/m^2a).

A less-insulated building can have the same or even less environmental effect in comparison to the detached single family house according to the conservation strategy, even though more energy is used. It must be emphasized that the construction of the building envelope according to the solar and renewable strategy is also better insulated than that required according to current building standards. The importance of air tightness and efficient ventilation heat recovery is also crucial for achieving low energy demands.

Solar active system recommendation

Since a hot water radiant heating system is required because of a high peak power demand, there is no extra work or cost in having a combi-system. Consequently, the extra 1 per cent given by the combi-system might as well be used and the recommendation is:

- 7.5 m^2 flat-plate combi-system tilted 40° with a 0.6 m^3 tank, which results in a remaining auxiliary demand of 31.8 kWh/m^2a.

The corresponding total solar fraction is about 34 per cent. The corresponding solar fraction for the DHW coverage is about 68 per cent.

Summary of advice for solar systems

- Evacuated collectors are not economically justified today.
- The size of the tank is not crucial; thus, a smaller tank is preferable (less expensive).
- The tank design is of great importance (insulation, connections, heat exchangers, etc.).

Table 8.3.8 *Solution 2: Renewable energy with the solar combi-system and biomass or condensing gas boiler*

	Material	Thickness	Conductivity	Per cent	Studs	Studs	Resistance without Rsi, Rse	Resistance with Rsi, Rse	U-value
		m	λ (W/mK)	%	λ (W/mK)	%	(m^2K/W)	(m^2K/W)	(W/m^2K)
wall	exterior surface							0.04	
	wooden panel	0.045		100%					
	air gap	0.025		100%					
	mineral wool hd	0.070	0.030	100%			2.33		
	mineral wool	0.185	0.036	85%	0.14	15%	3.59		
	plastic foil			100%					
	mineral wool	0.045	0.036	85%	0.14	15%	0.87		
	plaster board	0.013	0.220	100%			0.06		
	interior surface							0.13	
		0.383					**6.85**	**7.02**	**0.142**
roof	exterior surface							0.04	
	roof tiles	0.050		100%					
	roof felt	0.002		100%					
	wooden panel	0.022		100%					
	air gap	0.025		100%					
	mineral wool	0.230	0.036	95%	0.14	5%	5.58		
	plastic foil			100%					
	mineral wool	0.045	0.036	85%	0.14	15%	0.87		
	plaster board	0.013	0.220	100%			0.06		
	interior surface							0.10	
		0.387					**6.51**	**6.65**	**0.150**
floor									
	mineral wool	0.170	0.036	100%			4.72		
	concrete	0.100	1.700	100%			0.06		
	interior surface							0.17	
		0.270					**4.78**	**4.95**	**0.202**
window				emissivity					
	pane	0.004	low emissivity	5%	reversed				
	gas	0.012	krypton						
	pane	0.004	clear	83.70%					
	gas	0.012	krypton						
	pane	0.004	clear	83.70%					
		0.036							**0.800**
	frame	0.093	wood		0.14		0.66	0.83	**1.20**

- The collector area should be chosen in order to achieve a 100 per cent solar coverage during the summer months.
- The azimuth angle is not crucial; +/–30° is satisfactory.
- The collectors should preferably be placed on the roof, not the wall, unless one wants a high solar fraction (not economical).
- Lowering the DHW set temperature from 60°C to 50°C gives a 5 per cent lower auxiliary demand. But one has to look into the legislation for each country and check what temperature levels are allowed.
- Using a stratifying heat exchanger unit, instead of coil, gives approximately up to 10 per cent lower auxiliary demand.
- The benefits of a combi-system are drastically reduced for highly insulated buildings.

References

Boström, T., Wäckelgård, E. and Karlsson, B. (2003) *Design of a Thermal Solar System with High Solar Fraction in an Extremely Well Insulated House*, Proceedings of ISES 2003, Göteborg, Sweden

GEMIS (2004) *Gemis: Global Emission Model for Integrated Systems*, Öko-Institut, Darmstadt, Germany

Kovacs, P. and Pettersson, U. (2002) *Solar Combisystem: A Comparison between Vacuum Tube and Flat Plate Collectors using Measurements and Simulations*, SP Swedish National Testing and Research Institute, Borås, Sweden

Kvist, H. (2005) *DEROB-LTH for MS Windows, User Manual Version 1.0–20050813*, Department of Construction and Architecture, Lund Institute of Technology, Lund University, Lund, Sweden

Meteotest (2004) *Meteonorm 5.0 – Global Meteorological Database for Solar Energy and Applied Meteorology*, Bern, Switzerland, www.meteotest.ch/en

Solartechnik Prüfung Forschnung (2005) *Polysun Program*, Switzerland, www.spf.ch

8.4 Row house in the Cold Climate Conservation Strategy

Udo Gieseler

Table 8.4.1 *Targets for row houses in the Cold Climate Conservation Strategy*

	Targets
Space heating	15 kWh/m²a
Non-renewable primary energy: (space heating + water heating + electricity for mechanical systems)	60 kWh/m²a

This section presents a solution for the row houses in the cold climate. As a reference for the cold climate, the city of Stockholm is used. The solution is based on energy conservation minimizing the heat losses of the building. A balanced mechanical ventilation system with heat recovery is used to reduce the ventilation losses.

8.4.1 Solution 1: Conservation with district heating

Building envelope and space heating demand

In the cold climate, the standard reference row house has a remaining space heating demand of about 58 kWh/m²a. To reach the energy demand target of 15 kWh/m²a, the building losses must be significantly reduced. Performance indicators for the reference house as well as for the high-performance house (solution 1) are given in Table 8.4.2. The strategy to achieve the target includes three energy saving measures, as follows.

Insulation: the walls and the roof are highly insulated. The east–west walls get the thickest insulation, leading to a U-value of U = 0.10 W/m^{2}K. The massive construction of the reference house is not changed and thus these walls become quite thick, approximately 50 cm. Therefore, this high insulation standard is only used for the east–west walls, which have a small window area. The massive construction improves comfort in the summer and increases the usability of solar gains in winter. The insulation between the ground floor and the ground is not increased since it is usually more expensive and saves less energy (Gieseler et al, 2004).

Windows: for the south-facing façade, high-performance windows with U = 0.7 W/m^{2}K for the glazing are used. Because these windows are still quite expensive, the window size was somewhat reduced. A side benefit is the reduction of overheating in summer. Comfort is also improved in winter because the surface temperature of the glazing is closer to that of the walls. These effects are not so significant for the small window areas of the north–east–west sides of the building. To save costs, windows with U = 1.1 W/m^{2}K for the glazing are used here.

Ventilation: the improved air tightness of the high-performance house with 1 ach at 50 Pa pressure difference leads to a very small infiltration of 0.05 ach. The necessary fresh air is provided by a central ventilation system with a heat recovery efficiency of 75 per cent. An efficiency of 75 per cent is necessary to reach the target for the remaining energy demand. Note that the energy demand for defrosting the heat exchanger increases with increasing efficiency. In cold climates, a more efficient heat exchanger (in the range of 65 per cent to 90 per cent) does not substantially reduce the primary energy demand. The primary energy demand is almost independent of the efficiency in the range of 65 per cent to 90 per cent if electric defrosting is used (Gieseler et al, 2002). Since the costs today for the heat recovery unit increase with the efficiency, but there is no significant reduction in primary energy use, a very high efficiency unit is not chosen.

These measures lead to a remaining space heating demand of 13.0 kWh/m^{2}a for the mid row house unit, and 19.1 kWh/m^{2}a for the end units. The average space heating demand for a row with four mid and two end units is 15 kWh/m^{2}a, which meets the target.

For a more detailed comparison of the reference and the high-performance case, the five contributions of the energy balance during the heating period are given in Table 8.4.3. The average values for the row with six units are also given and shown in Figure 8.4.1 for the reference and the high-performance houses.

The aim of the conservation strategy was to reduce the heat losses, and the means of achieving this is primarily by reducing ventilation losses by 66 per cent. This is due to the mechanical ventilation system with heat recovery, as discussed above. Because the U-value of 0.2 W/m^{2}K for the opaque envelope of the reference case is already quite low, the savings for the transmission losses are not that high, but are still significant. Altogether, the losses of the high-performance house were cut to half those of the reference case. These losses are partially offset by internal gains; but the high-performance solution has less gains than the reference house. The assumed higher efficiency of the appliances produce less free heat, and smaller windows with a lower g-value deliver less passive solar heat (see Table 8.4.3). However, the reduced gains are easily compensated for by the significant reduction in the ventilation and transmission heat losses. The remaining space heating demand for the high-performance house is only 15 kWh/m^{2}a.

Table 8.4.2 *Comparison of key numbers for the construction and energy performance of the row house (areas are per unit)*

	Reference building	Conservation Strategy
Walls		
Area north and south (m^2)	39.40	41.40
U-value north/south (W/m^2K)	0.20	0.18
Area east or west (m^2)	57.00	57.00
U-value east/west (W/m^2K)	0.20	0.10
Roof (area: 60 m^2)		
U-value (W/m^2K)	0.20	0.14
Floor (area: 60 m^2)		
U-value (W/m^2K)	0.20	0.20
Windows		
South		
Area (m^2)	14.00	12.00
U-value glazing (W/m^2K), 70%	1.75	0.70
U-value frame (W/m^2K), 30%	2.30	0.70
g-value	0.68	0.58
North		
Area (m^2)	3.00	3.00
U-value glazing (W/m^2K), 60%	1.75	1.10
U-value frame (W/m^2K), 40%	2.30	1.80
g-value	0.68	0.59
East/west		
Area (m^2)	3.00	3.00
U-value glazing (W/m^2K), 60%	1.75	1.10
U-value frame (W/m^2K), 40%	2.30	1.80
g-value	0.68	0.59
Air change rate (air volume: 275 m^3)		
Infiltration (ach)	0.60	0.05
Ventilation (ach)	0.00	0.45
Heat recovery (–)	0.00	0.75
Space heating demand		
(simulation for 1 January–31 December)		
Mid unit (kWh/m^2a)	53.4	13.0
End unit (kWh/m^2a)	65.8	19.1
Row of four mid and two end units (kWh/m^2a)	57.5	15.0

Table 8.4.3 *Simulation results for the energy balance during the heating period*

Energy balance Simulation period: 1 October–30 April		**Gains**		**Losses**	
	Remaining space heating demand	**Solar**	**Internal**	**Transmission**	**Ventilation**
			(kWh/m^2a)		
Reference					
Mid unit	52.2	16.7	22.5	46.0	45.4
End unit	63.3	18.9	22.5	59.4	45.2
Row house (four mid + two end units)	55.9	17.4	22.5	50.5	45.3
Conservation strategy					
Mid unit	13.0	10.4	19.2	27.0	15.7
End unit	19.1	11.5	19.2	35.3	14.5
Row house (four mid + two end units)	15.0	10.8	19.2	29.8	15.3

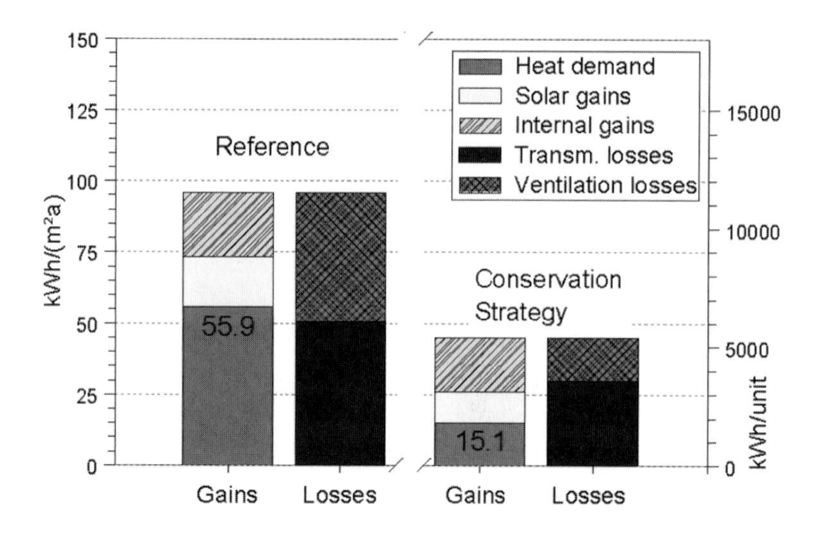

Source: Udo Gieseler

Figure 8.4.1 *Simulation results for the energy balance of the row houses (six units) according to Table 8.4.3*

Mechanical systems

Space heating and DHW system: district heating is used as the heat source for DHW and space heating. For the actual sizing of the heating system, the maximum peak load is important. Figure 8.4.2 shows simulation results for the hourly heat load of an end unit. To model the worst case of weather conditions, only diffuse solar radiation is taken into account (overcast sky). The heat load is calculated for an occupied building (i.e. with internal gains). The resulting maximum heat load is 1730 W. In the simulation described here, heating system capacity was given as 1800 W, which, in this case, could cover the peak load. The monthly space heating demand of an average row house unit and its distribution over the year is shown in Figure 8.4.3.

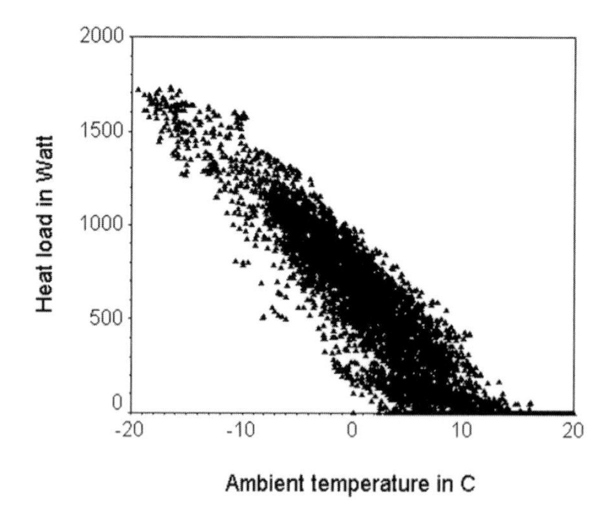

Source: Udo Gieseler

Figure 8.4.2 *Simulation results for the hourly heat load without direct solar radiation; the maximum heat load for an end unit is 1730 W*

Domestic hot water system: a domestic hot water demand of 160 litres/day at a temperature of 55° C for each row house was assumed. Each unit has its own 300 litre storage tank. The tank losses are calculated on the basis of a tank height of 1.6 m and thermal insulation of U = 0.28 W/m^{2}K. Freshwater enters the system at a temperature of 8.5°C. The heating is supplied by district heating. The efficiency of the DHW heating is assumed as 85 per cent.

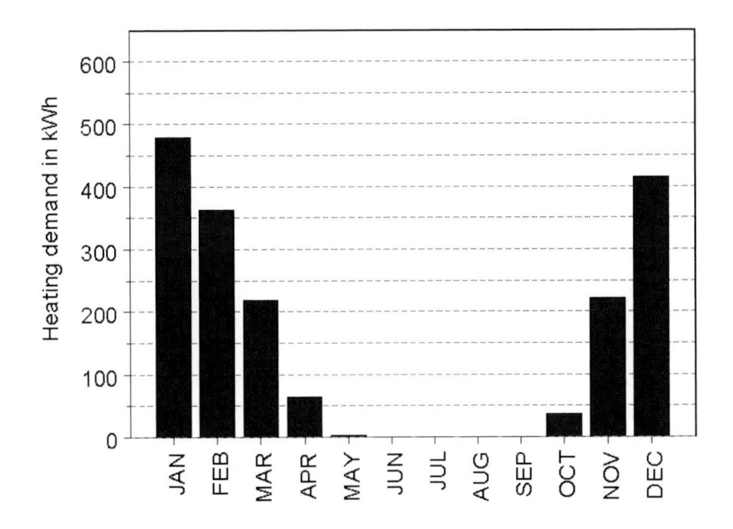

Source: Udo Gieseler

Figure 8.4.3 *Monthly heating demand for a row house unit (average over four mid and two end units)*

Electricity for mechanical systems: the electricity demand for pumps, fans and controls is estimated as 5 kWh/m²a.

Energy performance

Non-renewable primary energy demand and CO_2 emissions: the energy demands of the high-performance row house, as well as the corresponding non-renewable primary energy demand and CO_2 emissions, are shown in Table 8.4.4. The suggested high-performance house has a non-renewable primary energy demand of 47 kWh/m²a, which fulfils the target. Note that the primary energy factor of a district heating system can vary considerably. For an oil-based district heating without combined heat and power (CHP), the PEF could be up to 1.5, whereas district heating with higher CHP fraction and/or renewable energy sources could have a PEF of well below 1. The CO_2 equivalent emissions amount to 13 kg/m²a.

Table 8.4.4 *Total energy demand, non-renewable primary energy demand and CO_2 equivalent emissions for the solution with district heating; all numbers are related to the heated floor area (120 m²)*

Net Energy (kWh/m²a)		Total Energy Use (kWh/m²a)				Delivered energy (kWh/m²a)		Non renewable primary energy factor (-)	Non renewable primary energy (kWh/m²a)	CO_2 factor (kg/kWh)	CO_2 equivalent emissions (kg/m²a)
		Energy use		Energy source							
Mechanical systems	5.0	Mechanical systems	5.0	Electricity	5.0	Electricity	5.0	2.35	11.8	0.43	2.2
Space heating	15.0	Space heating	15.0	District heating	46.4	District heating	46.4	0.77	35.7	0.24	11.2
DHW	26.2	DHW	26.2								
		Tank losses	0.6								
		Conversion losses	4.6								
Total	46.2		51.4		51.4		51.4		47.5		13.4

8.4.2 Summer comfort

To evaluate summer comfort, separate simulations were performed. From 1 May to 30 September the ventilation is increased by 1 ach at night between 7pm and 6am. The heat recovery unit is not used. Additional shading reduces the direct and diffuse sunlight on windows by 50 per cent during the summer period. The indoor temperature is not allowed to fall below 20°C at any time in order not to overestimate the cooling effect. These assumptions represent a reasonably good passive cooling strategy.

Figures 8.4.4 and 8.4.5 show the number of hours with the average indoor temperature exceeding certain limits. The two shown cases in each figure are the overheating hours with increased night ventilation only (dark columns) and the overheating hours with increased night ventilation and shading (light columns), as described above.

Figure 8.4.4 shows that in the row house end unit, the average indoor temperature is above 22°C during 2840 hours (77 per cent of the summer time) if only increased night ventilation is used. With additional shading (of 50 per cent), the average indoor temperature is above 22°C during 1120 hours (30% of the time). With this cooling strategy, the indoor temperature does not exceed 26°C, which could be considered as comfortable in most cases.

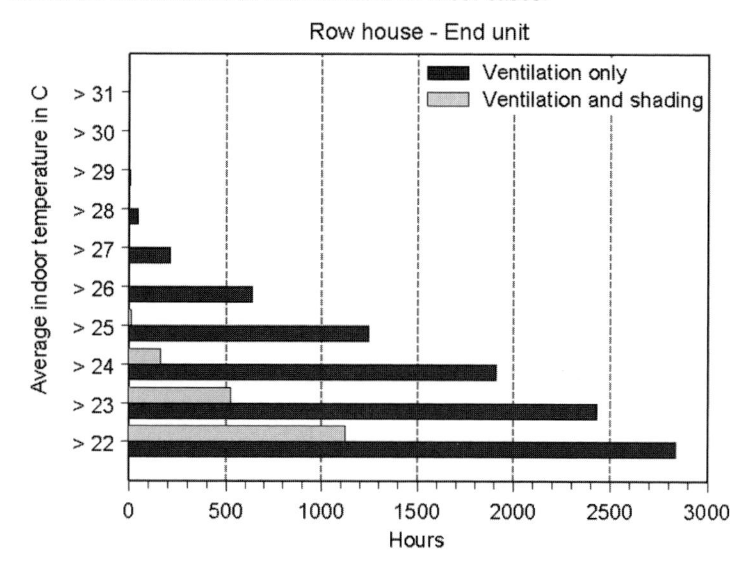

Source: Udo Gieseler

Figure 8.4.4 *Number of hours with average indoor temperature exceeding certain limits for an end unit; the simulation period is 1 May to 30 September*

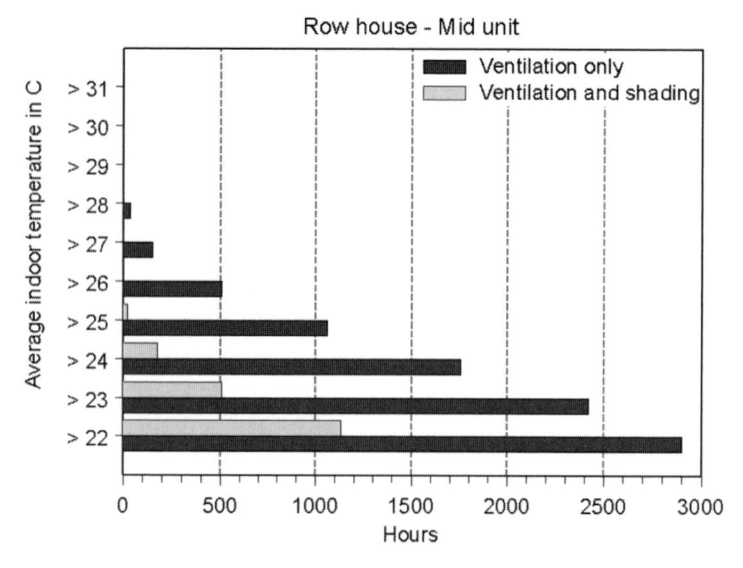

Source: Udo Gieseler

Figure 8.4.5 *Number of hours with average indoor temperature exceeding certain limits for a mid unit; the simulation period is 1 May to 30 September*

8.4.3 Sensitivity analysis

This section shows the importance of window type and window area on the space heating demand. Figure 8.4.6 shows the space heating demand with varying south window area. The window area includes 30 per cent frame area. The windows have an overall U-value of 0.7 W/m^{2}K. The results show that for these windows the solar gains are almost exactly balanced by the transmission losses through the windows. The window size for the south façade is 12 m^2 per unit (i.e. 6 m^2) per unit and floor in the suggested high-performance solution. A larger window area would increase costs and overheating hours; but the heating demand could not be reduced.

In Figure 8.4.7, simulation results for the variation of the window on the north-facing façade are presented. Results for two types of windows are shown. Both window types include 40 per cent frame area. The house has 3 m^2 of standard type windows for each unit (11 per cent window fraction). With high-performance windows (U = 0.7 W/m^{2}K for glass and frame), the energy target can be met with a window fraction of 35 per cent. This corresponds to 9.5 m^2 for each unit. However, the costs for such windows are today at least twice as high as for standard windows with U = 1.1 W/m^{2}K for the glass and U = 1.8 W/m^{2}K for the frame. If a larger window area to the north is preferred, high-performance windows should be chosen in order not to increase the space heating demand.

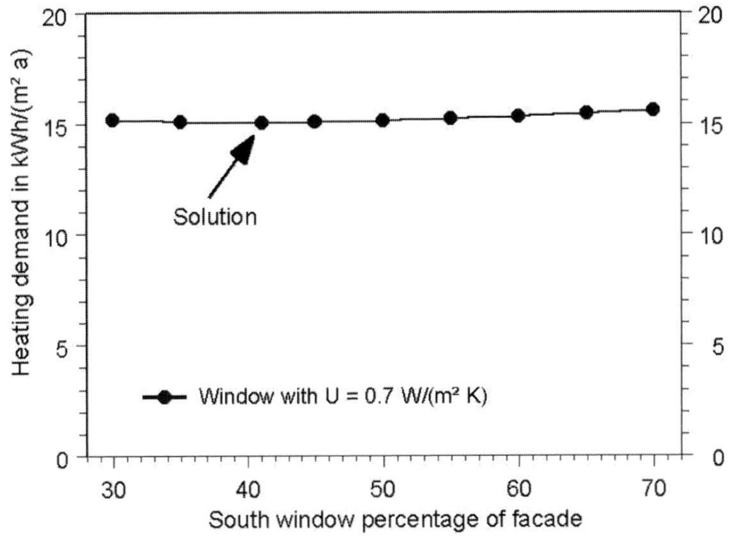

Source: Udo Gieseler

Figure 8.4.6 *Space heating demand for the south window variation in the row house (average unit); U-values shown are for the glazing*

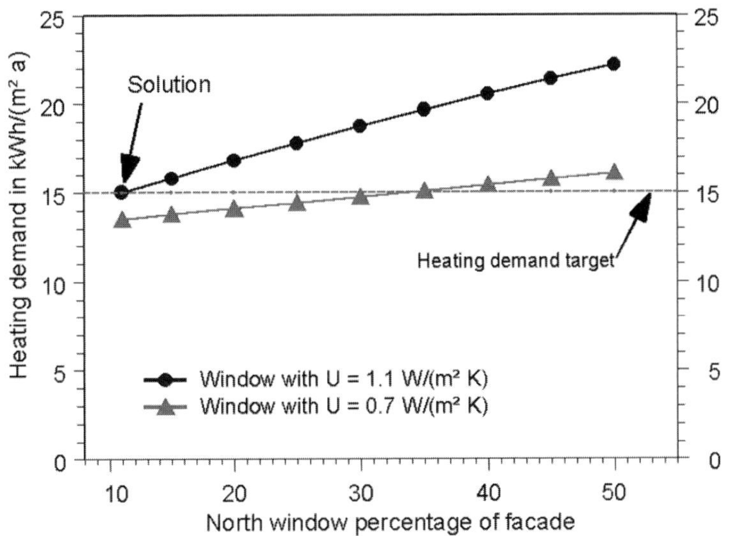

Source: Udo Gieseler

Figure 8.4.7 *Space heating demand for the north window variation in the row house; U-values shown are for the glazing*

The effect of shading is presented in Figure 8.4.8 for the south-facing façade. The space heating demand for the row house (all six units) is plotted for different shading coefficients. Shading is applied to direct sunlight on the windows in the south façade. Diffuse radiation is not changed. This is a rough model for shading from buildings, trees or other objects. The heating demand can increase to more than 20 kWh/m^{2}a if no direct solar radiation reaches the south-facing façade. In such a case, the construction has to be improved to meet the space heating target.

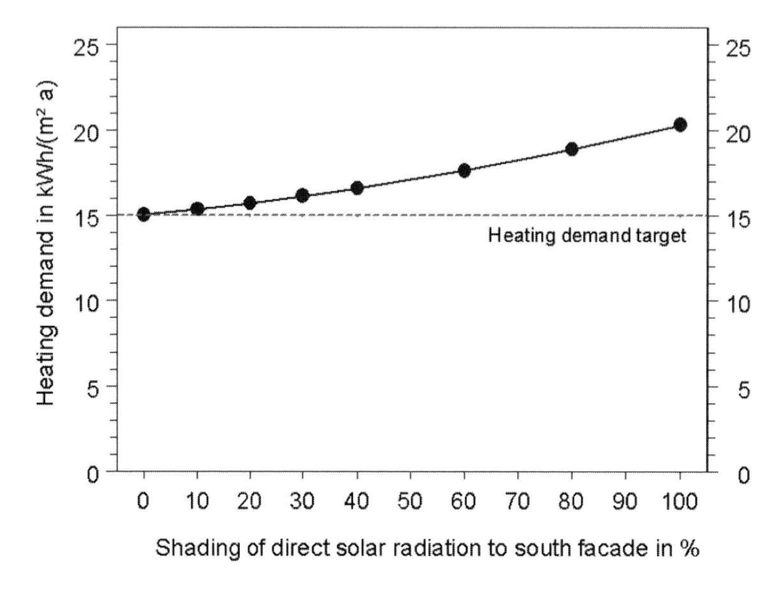

Source: Udo Gieseler

Figure 8.4.8 *Space heating demand for the row house (all six units) for different shading coefficients*

8.4.4 Conclusions and design advice

In this section, the conservation strategy for a row house in a cold climate has been presented. The conservation strategy is based on a ventilation system with heat recovery, a high insulation level of the building envelope and reduced window area.

The suggested conservation strategy fulfils the quite ambitious energy target of a space heating demand of 15 kWh/m^{2}a and provides high comfort in summer, as well as in winter, at reasonable construction costs. The space heating demand is only 25 per cent of the demand for a row house built according to normal standards in the cold climates.

Care should be taken not to increase the north window area. However, the window area to the south could be increased without increasing the space heating demand. Note that an increased window area may easily increase the number of overheating hours.

Furthermore, if the building is shaded during wintertime, the passive solar gains will be reduced, giving rise to a higher space heating demand not fulfilling the target. Thus, if the houses are planned to be built in, for example, a dense area or close to a forest, this should be taken into account in the planning.

A district heating system was used in this example. Other systems are also possible – for example, an outdoor air to water heat pump or a borehole heat pump system serving a group of houses. Solar collectors for domestic hot water are a good solution to further reduce the non-renewable energy use.

Table 8.4.5 *Details of the construction of row houses in the Cold Climate Conservation Strategy (layers are listed from inside to outside)*

Element	Layer	Thickness (m)	Conductivity (W/mK)	Resistance (m^2K/W)	U-value (W/m^2K)
Wall south/north	Plaster	0.015	0.700	0.021	
	Lightweight concrete	0.170	0.120	1.417	
	Polystyrol	0.140	0.035	4.000	
	Plaster	0.020	0.869	0.023	
	Surface resistances	-	-	0.170	
	Σ	0.345	-	5.631	0.18
Wall east/west	Plaster	0.015	0.700	0.021	
	Lightweight concrete	0.170	0.120	1.417	
	Polystyrol	0.300	0.035	8.571	
	Plaster	0.020	0.869	0.023	
	Surface resistances	-	-	0.170	
	Σ	0.505	-	10.202	0.10
Roof 90%	Plaster board	0.013	0.211	0.062	
	Mineral wool	0.370	0.040	9.250	
	Pantile	0.020	-	-	
	Surface resistances	-	-	0.170	
	Σ	0.403	-	9.482	0.11
Roof 10%	Plaster board	0.013	0.211	0.062	
	Timber	0.370	0.131	2.824	
	Pantile	0.020	-	-	
	Surface resistances	-	-	0.170	
	Σ	0.403	-	3.056	0.33
Floor	Parquet	0.020	0.200	0.100	
	Anhydrite	0.060	1.200	0.050	
	Polystyrol	0.162	0.035	4.629	
	Concrete	0.120	2.100	0.057	
	Surface resistances	-	-	0.170	
	Σ	0.362	-	5.006	0.20
Window glazing north/east/ west	Glass (low-e)	0.004	-	-	
	Argon	0.016	-	-	
	Glass	0.004	-	-	
	Σ	0.024	-	-	1.10
Window glazing south	Glass (low-e)	0.004	-	-	
	Argon	0.016	-	-	
	Glass	0.004	-	-	
	Argon	0.016	-	-	
	Glass (low-e)	0.004	-	-	
	Σ	0.044	-	-	0.70

References

Gieseler, U. D. J., Bier, W. and Heidt, F. D. (2002) *Cost Efficiency of Ventilation Systems for Low-Energy Buildings with Earth-to-Air Heat Exchange and Heat Recovery*, Proceedings of the 19th International Conference on Passive and Low Energy Architecture (PLEA), Toulouse, France, pp577–583

Gieseler, U. D. J., Heidt, F. D. and Bier, W. (2004) 'Evaluation of the cost efficiency of an energy efficient building', *Renewable Energy Journal*, vol 29, pp369–376

TRNSYS (2005) *A Transient System Simulation Program*, Solar Energy Laboratory, University of Wisconsin, Madison, WI

8.5 Row house in the Cold Climate Renewable Energy Strategy

Joachim Morhenne

Table 8.5.1 *Row house targets in the Cold Climate Renewable Energy Strategy*

	Targets
Space heating	20 kWh/m²a
Non-renewable primary energy: (space heating + water heating + electricity for mechanical systems)	60 kWh/m²a

This section presents a renewable energy solution for the row houses in the cold climate. As a reference for the cold climate, the city of Stockholm is used.

8.5.1 Solution 2: Solar domestic hot water and solar-assisted heating

The use of a solar combi-system and efficient mechanical ventilation with heat recovery are essential to achieve the space heating target of 20 kWh/m²a (2400 kWh/a per unit) in this climate for row houses. Applying these two measures then allows some freedom as to how much transmission losses must be reduced and still meet the target. These two measures, together with the good A/V ratio of row houses, mean that the target can be met with an envelope construction only slightly better than current building code requirements. Consequently, this strategy is also applicable to building retrofit where improving the building envelope may be difficult.

To achieve the primary energy target, a solar domestic water system with a 60 per cent solar fraction is adequate. A larger solar combi-system will therefore reduce the primary energy demand well below the target of 60 kWh/m²a.

Why follow this strategy?

Solar gains make it possible to reach the target without having to apply excessive conservation measures. Given that most high-performance houses today have a solar domestic hot water system, this strategy proposes to increase the solar system to also provide some space heating. As a result, the target can be met without, for example, using expensive high-performance windows.

Because this solution uses surface and not ventilation air to deliver the needed heat, the heat delivery capacity is not limited by the ventilation rate. Furthermore, surface heating provides superior comfort. Finally, this strategy can compensate for low passive gains if the building is not optimally orientated or is shaded.

Building envelope

Opaque construction: massive or lightweight walls with exterior insulation. Thermal mass increases summer comfort and slightly increases the usefulness of solar gains. In case of lightweight walls, floors and ceilings should at least be massive.

Windows: frame ratio of 30 per cent, double low-e coated glass with argon gas.

Table 8.5.2 *Building envelope U-values*

Component	U-Value (W/m²K)
Floor	0.21
Walls	0.20
Walls east/west (end houses)	0.16
Roof	0.16
Window glass	1.2
Window frame	1.7

Mechanical systems

Ventilation: mechanical ventilation with 80 per cent heat recovery. Ventilation rate: 0.45 ach. Infiltration rate: 0.05 ach. Electric consumption: 0.3 W/m^{2}h.

Heat supply: central condensing gas furnace, biomass boiler or connection to local grid. Four-pipe heating grid for each row of houses.

Solar system: central solar combi-system or individual solar combi-systems for each unit.

Heat distribution: hot water floor or wall radiant heating, supply air heating.

Figure 8.5.1 explains the solar heating system. The solar system is a typical solar combi-system with six houses each with 10 m^2 collector area. Important questions are whether the solar system is private or common property and how investment and maintenance costs are shared. In contrast to collective systems, individual systems need more space for the storage, investment costs are higher and surplus heat when one family is away cannot be shared. However, there are no grid heat losses.

Solar heat is used to raise the temperature of the return flow from the floor heating (see section 8.5.3). For further information about solar combi-systems, see the final report of the IEA-SHC Task 26 (Weiss, 2003).

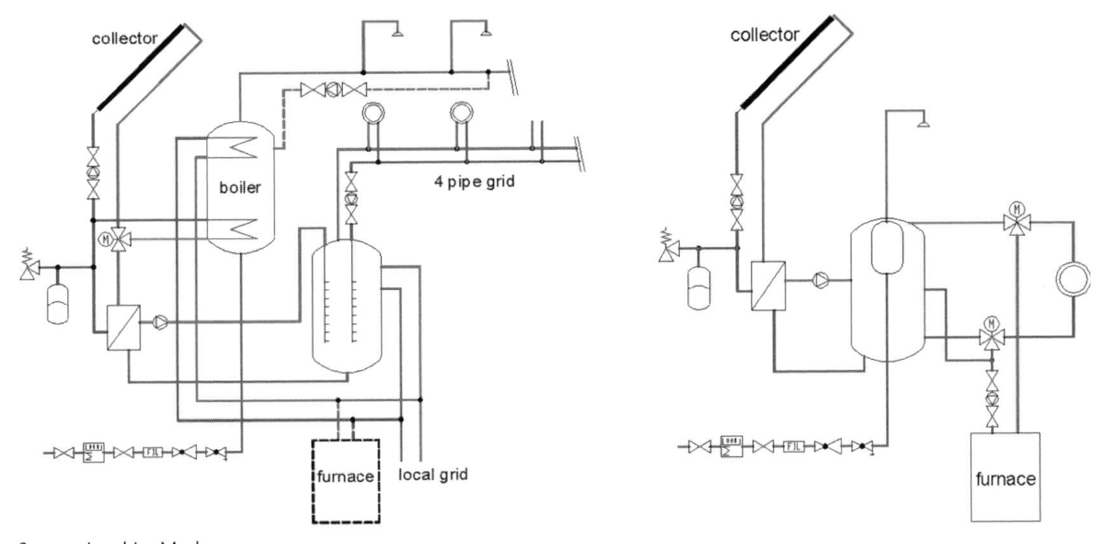

Source: Joachim Morhenne

Figure 8.5.1 *Scheme of the solar assisted heating system: (a) central system for a row of houses and (b) individual solution*

Energy performance

Space heating demand: the monthly space heating demand of the row houses was computed with TRNSYS and is shown in Figure 8.5.2. The row consists of two end houses and four mid houses. Results in Table 8.5.3 are mean values for the row of houses. The heating season extends from 1 October to 31 May.

Without the solar system the net energy demand for heating the building is 21.7 kWh/m^{2}a. In reality, the heating demand totals 26.6 kWh/m^{2}a because:

- the indoor temperature often exceeds 20°C; and
- the grid also loses heat before reaching the house.

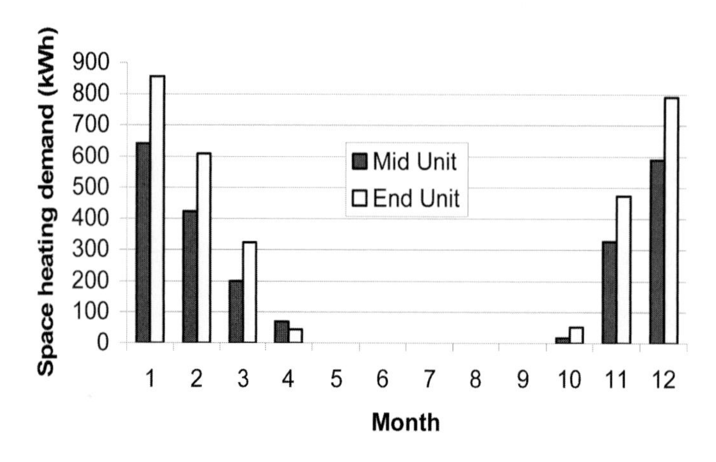

Source: Joachim Morhenne

Figure 8.5.2 *Space heating demand*

Table 8.5.3 shows the performance of the building, including the savings of the solar system. The delivered energy to cover heat demand includes losses of the control system, storage and pipes, as well as from the collector and its circuit. The system losses include the combustion losses of a condensing gas furnace for heating and hot water preparation.

Table 8.5.3 *Performance of the building, including the system*

Delivered energy for space heating (mean)	19.3 kWh/m^{2}a (2300 kWh/a)
System losses	1.5 kWh/m^{2}a
Solar contribution to space heating demand	16%
Heating set point	20°C

Peak load for space heating: the peak load is 2350 W for the end units and 1800 W for the mid units. While the peak occurs in January by an ambient temperature of –18.9°C, near peak demands also occur in February and December. Outside of these three months, the peaks fall off very sharply.

Delivered energy for DHW: the energy demand for DHW is 1360 kWh/a or 11.3 kWh/m^{2}a. The solar contribution to the delivered energy for DHW is 62 per cent.

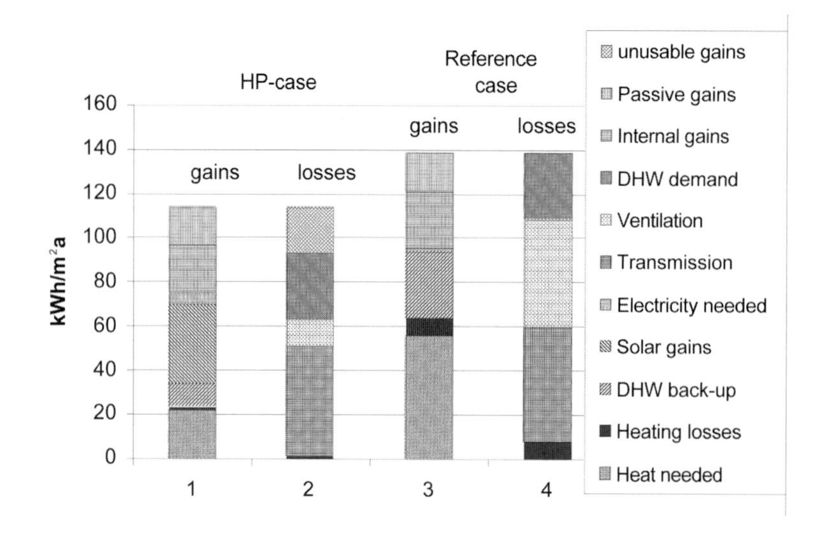

Source: Joachim Morhenne

Figure 8.5.3 *Energy balance of the reference and solar base case (columns 1 and 3 are gains, columns 2 and 4 are losses)*

Delivered energy use: the total end energy use for DHW and space heating is 3850 kWh/a. The electric consumption by fans, pumps and controls is 670 kWh/a. Figure 8.5.3 explains the energy balance for the reference case and the high-performance case.

Primary energy demand and CO_2 emissions: the primary energy demand and the CO_2 emissions are shown in Table 8.5.4. Factors are taken from GEMIS (2004).

The primary energy factors are:

- gas: 1.14;
- biomass: 0.06; and
- electricity: 2.35.

The CO_2 emission factors are:

- gas: 0.247 kg/kWh;
- biomass: 0.035 kg/kWh; and
- electricity: 0.430 kg/kWh.

The primary energy demand is 49.8 kWh/m^{2}a and the corresponding CO_2 emissions are 10.3 kg/m^{2}a when using gas as a heating source for the remaining energy. If, instead, biomass is used as fuel the primary energy demand is only 15 kWh/m^{2}a and the CO_2 emissions are 3.6 kg/m^{2}a.

Table 8.5.4 *Total energy use, non-renewable primary energy demand and CO_2 emissions*

Net Energy (kWh/m^{2}a)		Total Energy Use (kWh/m^{2}a): Energy use		Total Energy Use (kWh/m^{2}a): Energy source		Delivered energy (kWh/m^{2}a)		Non renewable primary energy factor (-)	Non renewable primary energy (kWh/m^{2}a)	CO_2 factor (kg/kWh)	CO_2 equivalent emissions (kg/m^{2}a)
Mechanical systems	5.6	Mechanical systems	5.6	Electricity	5.6	Electricity	5.6	2.4	13.2	0.43	2.4
Space heating	21.7	Space heating	21.7	Gas	32.1	Gas	32.1	1.1	36.6	0.25	7.9
DHW	30.0	DHW	30.0								
		Tank, circulation and conversion losses	5.9	Solar	25.5						
Total	57.3		63.2		63.2		37.7		49.8		10.3

8.5.2 Summer comfort

The heating system is not active in summer; therefore, only internal and passive solar gains and the ventilation air contribute to overheating. The indoor temperature never exceeds 26°C during the simulated year (see Figure 8.5.4). Due to the chosen shading and ventilation strategy, summer comfort is achieved. Better shading devices and increased night ventilation could further improve comfort.

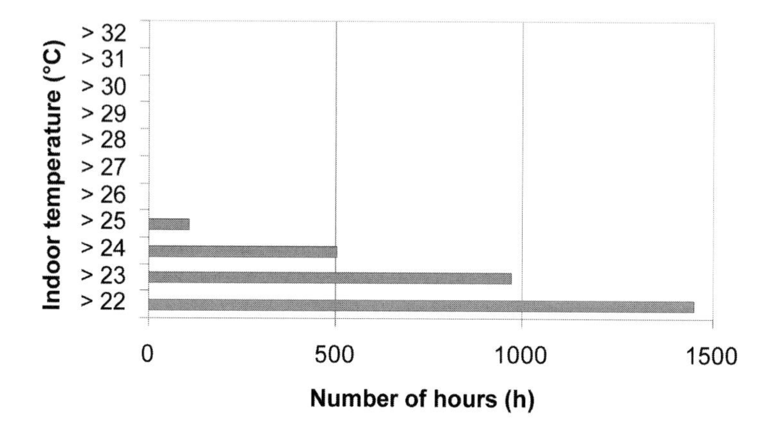

Source: Joachim Morhenne

Figure 8.5.4 *Number of hours of the indoor temperature distribution*

To reduce electric consumption, night cooling with ambient air by opening windows could be used instead of a mechanical ventilation system. If the ventilation system is used during the summer, an automatic bypass of the heat exchanger is recommended. Note that in these very air-tight houses, it is important to ventilate since the infiltration rate is extremely low.

8.5.3 Sensitivity analysis

System design

The evaluated system is shown in Figure 8.5.5. Due to the cold climate, the performance of the solar system in winter is very sensitive to the return flow temperature of the heating system (see Figure 8.5.7). Therefore a four-pipe grid was chosen to be able to operate the grid with different tempera-

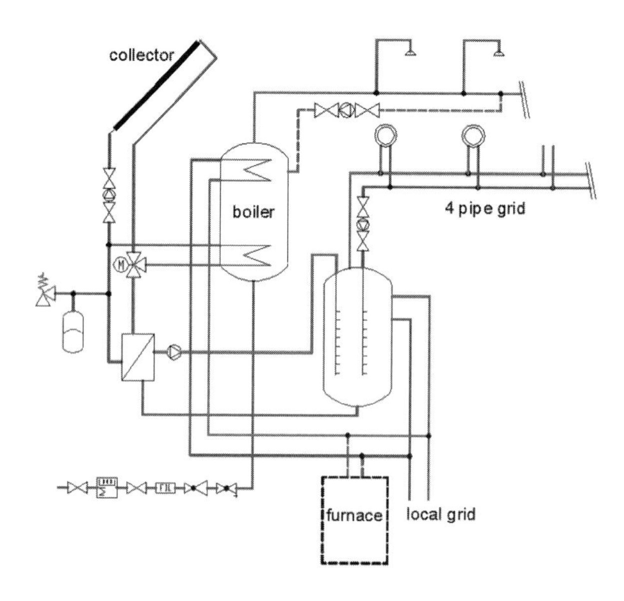

Source: Joachim Morhenne

Figure 8.5.5 *Scheme of the central system*

tures for space heating and DHW. The space heating grid has, in this case, a lower supply and return temperature then the hot water grid. This results in a better use of solar gains. Compared to two-pipe grids, they have higher grid losses; but the grid losses can partly be recovered if the grid is installed inside the building envelope.

The most important parameters of the heating system are shown in Table 8.5.5.

Table 8.5.5 *Important system parameters*

	Temperate climate; solar strategy
Design temperature	35/30°C
Heated surface/heating power	2550 W
Collector area:	
base high-performance retrofit	10 m^2
Collector type	Flat plate
Collector slope, south	54°
Flow rate	12 l/m^2
Control	Maximum efficiency
Heat exchanger	92%
Storage	45 l/m^2
Main façade	South
Shading coefficient	0.5
Construction	Heavyweight

Collector area

The influence of collector slope and azimuth angle can be taken from standard tables (see Duffie and Beckman, 1991). The optimum values (azimuth south, slope 54°) are used here. Figure 8.5.6 shows the influence of the collector area on the usable collector gains. The usable collector gains are 255 kWh/m^2a (per m^2 collector) for the collector area of 10 m^2. The specific collector gain for heating is about 44 kWh/m^2a and cannot be increased much (see Figure 8.5.7). A further reduction of the total energy demand by increased solar gains is only possible by increasing the solar fraction for DHW.

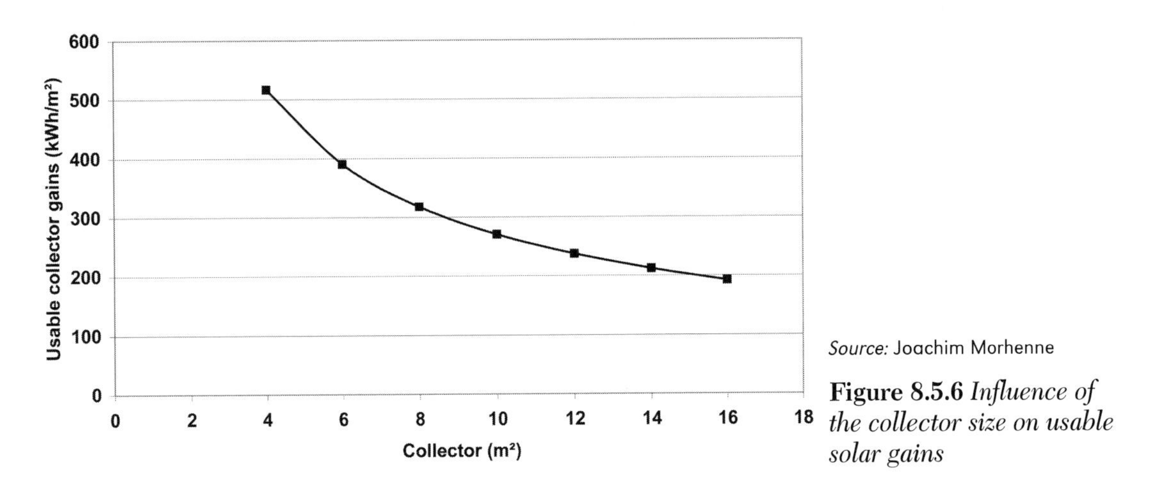

Source: Joachim Morhenne

Figure 8.5.6 *Influence of the collector size on usable solar gains*

The design temperature for the heating system is very important for the operation of the solar combi-system. The useful solar contribution drops drastically as the return flow temperature increases. For this reason, only low temperature floor or wall heating systems are recommended.

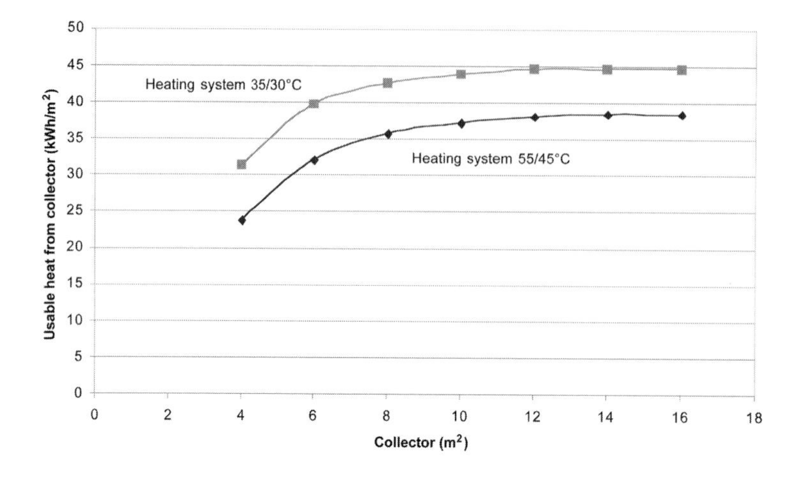

Source: Joachim Morhenne

Figure 8.5.7 *Influence of the collector size on usable solar gains for a reduction in heating demand depending upon the supply temperature of the heating system*

Primary energy demand depending upon collector area: Figure 8.5.8 shows the primary energy demand depending on the collector area when natural gas supplies the remaining needed energy.

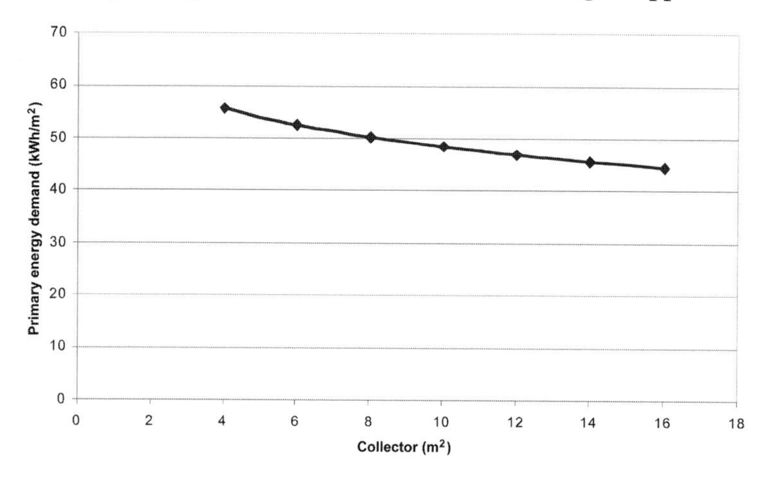

Source: Joachim Morhenne

Figure 8.5.8 *Primary energy demand depending upon collector area*

Influence of the supply temperature of the heating system: the temperature of the supply and return flows of the heating system has a high influence since the solar system is always connected to the return flow of the system. The influence of supply temperatures is approximately 8 per cent to 10 per cent, as can be seen in Figure 8.5.7.

Influence of the storage size: the storage volume of the solar system does not strongly affect system performance unless the storage size exceeds a critical value. Within the range of 42 to 57 l/m^2 of collector the influence of the storage is less than 3 per cent.

Influence of the flow rate in the collector circuit: since the solar collector heat is transferred to the return water flow for space heating, logically, the collector outlet temperature must exceed the return flow temperature for heat to be transferred. Therefore, the flow rate has to be reduced until the necessary outlet of the collector is reached independently of the collector efficiency. A dynamic control of the flow rate of the collector loop optimizes system performance. In the simulations, 12 l/m^2h have been used. Other parameters, such as the thermal mass of the building, have been analysed for the temperate climate.

8.5.4 Design advice

The largest energy savings come from solar hot water heating. Therefore, the storage volume for DHW is most important if DHW and space heating tanks are separate. The most important design parameters are the supply temperature of the heating system and the flow rate of the collector. It is not recommended to exceed the temperatures shown here.

The hot water distribution grid of many built projects often experiences high losses. Generous insulation for the pipes is therefore essential. To avoid heat losses of the heating grid outside the heating season, it is recommended to use a manual switch to stop the circulation. A temperature-based switch will activate the grid each time the ambient temperature falls below a fixed value, even if there is no heating demand.

For further information about solar combi-systems, see the results from IEA SHC Task 26 (Weiss, 2003).

Table 8.5.6 *Construction according to the space heating target of 20 kWh/m²a*

	Material	Thickness		Conductivity	Per cent	Studs	Studs	Resistance		U-value	
		m		λ (W/mK)	%	λ (W/mK)	%	(m²K/W)		(W/m²K)	
		cold ref	cold basis					cold ref	cold basis	cold ref	cold basis
wall	exterior surface							0.04	0.04		
	plaster	0.02	0.02	0.9	100%			0.02	0.02		
	polystyrol	0.16	0.16	0.035	100%			4.57	4.57		
	limestone	0.175	0.175	0.560	100%			0.31	0.31		
	plaster	0.015	0.015	0.7	100%			0.02	0.02		
								0.13	0.13		
		0.37	**0.37**					**5.10**	**5.10**	**0.20**	**0.20**
wall	exterior surface							0.04	0.04		
E/W	plaster	0.02	0.02	0.9	100%			0.02	0.02		
	polystyrol	0.16	0.2	0.035	100%			4.57	5.71		
	limestone	0.175	0.175	0.560	100%			0.31	0.31		
	plaster	0.015	0.015	0.7	100%			0.02	0.02		
								0.13	0.13		
		0.37	**0.41**					**5.10**	**6.24**	**0.20**	**0.16**
roof	exterior surface							0.04	0.04		
	roof tiles	0.050	0.050		100%						
	air gap	0.045	0.045		100%						
	protection foil	0.0025	0.0025		100%						
	mineral wool	0.200	0.300	0.039	85%			4.59	6.88		
	wood	0.220	0.320			0.13	15%	1.69	2.46		
	PE- foil				100%						
	gypsum board	0.013	0.013	0.210	100%			0.06	0.06		
	interior surface							0.10	0.10		
		0.5305	**0.7305**					**4.36**	**6.42**	**0.23**	**0.16**
floor											
	concrete	0.012	0.012	2.1	100%			0.01	0.01		
	mineral wool	0.150	0.150	0.035	100%			4.29	4.29		
	pavement	0.060	0.060	0.800	100%			0.08	0.08		
	wood	0.020	0.020	0.130	100%			0.15	0.15		
	interior surface							0.17	0.17		
		0.230	**0.230**					**4.69**	**4.69**	**0.21**	**0.21**
window				**glass/LE/gas**							
	pane	0.004	0.004	low emissivity							
	gas	0.016	0.016	Argon							
	pane	0.004	0.004	clear							
		0.024	**0.024**							**1.5**	**1.20**
	frame			wood						**1.70**	**1.70**

References

Duffie, J. A. and Beckman, W. A. (1991) *Solar Engineering of Thermal Processes*, John Wiley and Sons, New York

GEMIS (2004) *Gemis: Global Emission Model for Integrated Systems*, Öko-Institut, Darmstadt, Germany

TRNSYS (2005) *A Transient System Simulation Program*, Solar Energy Laboratory, University of Wisconsin, Madison, WI

Weiss, W. (ed) (2003) *Solar Heating Systems for Houses: A Design Handbook for Solar Combi-systems*, James and James Ltd, London

8.6 Apartment building in the Cold Climate Conservation Strategy

Johan Smeds

Table 8.6.1 *Targets for apartment building in the Cold Climate Conservation Strategy*

	Targets
Space heating	15 kWh/m²a
Non-renewable primary energy: (space heating + water heating + electricity for mechanical systems)	60 kWh/m²a

This section presents two solutions for the apartment building in the cold climate. As a reference for the cold climate, the city of Stockholm is considered. The solutions are based on energy conservation, minimizing the heat losses of the building. A balanced mechanical ventilation system with heat recovery is used to reduce the ventilation losses.

8.6.1 Solution 1a: Conservation with electric resistance heating and solar DHW

Building envelope

In this solution the envelope and ventilation system were designed so that the peak space heating load is limited to approximately 10 W/m². This allows one central heating element for each apartment and heat distribution by the supply air, which is more economical than individual room heaters. This solution requires a well-insulated and air-tight building envelope.

The apartment building has reinforced concrete structure and wooden frame façades with mineral wool. The windows have a frame ratio of 30 per cent, triple glazing, one low-e coating and they are filled with krypton. The U-values are shown in Table 8.6.2. Construction data are shown in Table 8.6.6.

Table 8.6.2 *The building components*

Building component	U-value (W/m²K)
Walls	0.13
Roof	0.09
Floor (excluding ground)	0.12
Windows (frame + glass)	0.92
Window frame	1.20
Window glass	0.80
Whole building envelope	0.21

Mechanical systems

A balanced mechanical ventilation system with heat recovery and a bypass for summer ventilation is used. The heat exchanger has an efficiency of 80 per cent. The supply air is used for the distribution of heat. Domestic hot water is supplied by a solar domestic hot water system in combination with an electric furnace. The solar system consists of 60 m^2 collectors, a storage tank of 6 m^2 and an electric furnace of 8 kW.

Energy performance

Space heating demand and peak load: simulation results from DEROB-LTH show that the space heating demand is 10,300 kWh/a (6.5 kWh/m^2a). The monthly space heating demand of solution 1a in comparison to the reference case (building code 2001) is shown in Figure 8.6.1.

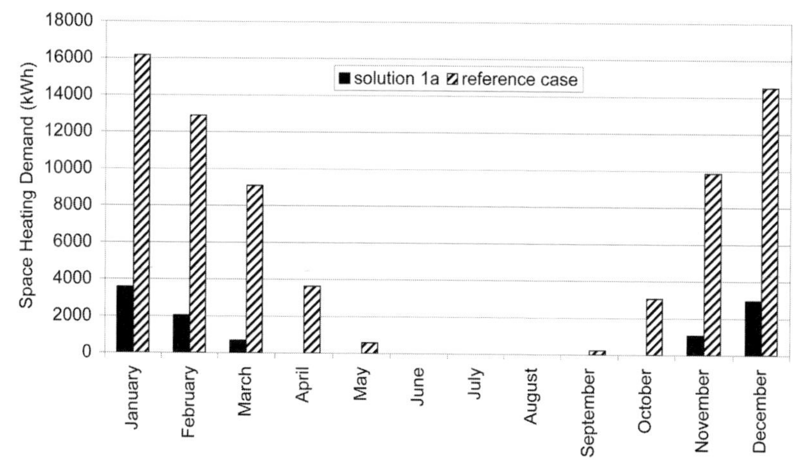

Source: Johan Smeds

Figure 8.6.1 *Space heating demand of solution 1a*

Hourly loads of the heating system are calculated from the DEROB-LTH simulation results without direct solar radiation in order to simulate a totally shaded building. The annual peak load is 16 kW or 10 W/m^2 and occurs in January.

General assumptions for these simulations were:

- heating set point: 20°C;
- maximum room temperatures: 23°C during winter, 26°C during summer (assumes use of shading devices and window ventilation);
- ventilation rate: 0.45 ach;
- infiltration rate: 0.05 ach; and
- heat recovery: 80 per cent efficiency.

Domestic hot water demand: the net heat demand for DHW is approximately 37,800 kWh/a (23.6 kWh/m^2a). Thus, the DHW demand is much larger than the space heating demand.

The system losses consist mainly of losses from the hot water storage tank, but also from pipe losses in the distribution system. The tank and circulation losses are 1.5 and 4.4 kWh/m^2a, respectively. The circulation losses are taken into account as internal gains in the simulations with DEROB-LTH. The tank losses and losses from the collector loop, on the other hand, cannot be used as internal gains since the tank and boiler are placed in the basement outside the thermal envelope of the living area. The conversion losses of the electric boiler are set to zero.

The choices of system solutions are based on simulations of the space heating demand for the building performed with the programme DEROB-LTH. Simulations of the active solar energy systems are performed with Polysun 3.3. A modified version of Polysun, the Polysun-Larsen version, was used in order to enable the use of the data files of space heating demand from heat load simulations performed in DEROB-LTH.

Household electricity: the use of household electricity for each apartment (two adults and one child) is 2190 kWh or 21.9 kWh/m^2. The primary energy target does not include household electricity since this factor can vary considerably depending on the occupants' behaviour.

Non-renewable primary energy demand and CO_2 emissions: the total energy use for DHW and space heating, system losses and mechanical systems is 63,200 kWh/a. After taking the solar contribution for DHW into account, the delivered energy is 34,200 kWh/a, which is provided by electricity. This results in a total use of non-renewable primary energy of 80,300 kWh/a (55 kWh/m^{2}a) and CO_2 equivalent emissions of 10 kg/m^{2}a (see Table 8.6.3).

Table 8.6.3 *Total energy demand, non-renewable primary energy demand and CO_2 equivalent emissions for the apartment building with electric resistance space heating and solar DHW system with electrical backup*

Net Energy (kWh/m²a)		Total Energy Use (kWh/m²a)				Delivered energy (kWh/m²a)		Non renewable primary energy		CO_2 equivalent emissions	
		Energy use		Energy source				factor (-)	(kWh/m²a)	factor (kg/kWh)	(kg/m²a)
Mechanical systems	5.0	Mechanical systems	5.0	Electricity	5.0	Electricity	5.0	2.35	11.8	0.43	2.2
Space heating	6.5	Space heating	6.5	Electricity	6.5	Electricity	18.3	2.35	43.0	0.43	7.9
DHW	23.6	DHW	23.6	Electricity	11.8						
		Tank and circulation losses	5.5	Solar	17.3						
		Conversion losses	0.0								
Total	35.1		40.6		40.6		23.3		54.8		10.0

8.6.2 Solution 1b: Energy conservation with district heating

Building envelope

Using a district heating system with hot water radiant heating allows a less insulated building envelope than solution 1a with supply air heating.

The apartment building has reinforced concrete structure and wooden frame façades with mineral wool. The windows have a frame ratio of 30 per cent, triple glazing, one low-e coating and are filled with krypton. U-values are shown in Table 8.6.4. Construction data are shown in Table 8.6.7.

Table 8.6.4 *U-values of the building components*

Building component	U-value (W/m²K)
Walls	0.24
Roof	0.25
Floor (excluding ground)	0.30
Windows (frame + glass)	0.92
Window frame	1.20
Window glass	0.80
Whole building envelope	0.34

Mechanical systems

A balanced mechanical ventilation system with heat recovery and a bypass for summer ventilation is used. The heat exchanger has an efficiency of 80 per cent. District heating is distributed by a hot water radiant heating system. A wastewater counter-flow pipe heat exchanger for preheating of water to the space heating system can be installed; but this is not included in any of the calculations for energy performance shown below.

Energy performance

Space heating demand: simulation results from DEROB-LTH show that the space heating demand of the building is 21,500 kWh/a (13.4 kWh/m^{2}a). The monthly space heating demand of the building in comparison to the reference case is shown in Figure 8.6.2. The heating season for the proposed solution 1b extends from November to March.

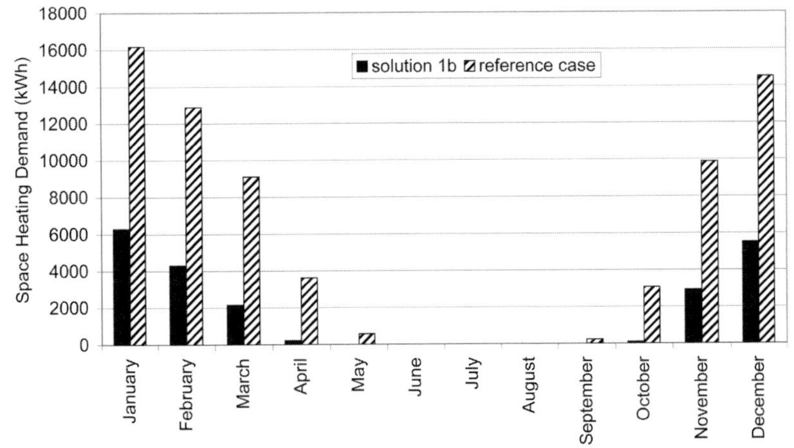

Source: Johan Smeds

Figure 8.6.2 *Space heating demand*

Hourly loads of the heating system are calculated with results from DEROB-LTH simulations without direct solar radiation in order to simulate a totally shaded building. The annual peak load is 21.2 kW or 13 W/m^2 and occurs in January.

Domestic hot water demand: the net domestic hot water demand for the 48 inhabitants is 37,800 kWh/a (23.6 kWh/m^{2}a).

System losses: the system losses consist mainly of circulation losses in the distribution system. The circulation losses are taken into account as internal gains in the simulations with DEROB-LTH. The conversion losses of the heat exchanger connected to the district heating system are set to zero.

Household electricity: the household electricity use for each apartment is 2190 kWh (21.9 kWh/m^2) for two adults and one child. The primary energy target does not include household electricity.

Non-renewable primary energy demand and CO_2 emissions: the total energy use for DHW and space heating, system losses and mechanical systems is 74,240 kWh/a. The use of district heating results in a non-renewable primary energy demand of 69,760 kWh/a (44 kWh/m^{2}a) and CO_2 equivalent emissions of 12 kg/m^{2}a in one year (see Table 8.6.5).

Table 8.6.5 *Total energy demand, non-renewable primary energy demand and CO_2 equivalent emissions for the apartment building with district heating*

Net Energy (kWh/m²a)		Total Energy Use (kWh/m²a)				Delivered energy (kWh/m²a)		Non renewable primary energy		CO_2 equivalent emissions	
		Energy use		Energy source				factor (-)	(kWh/m²a)	factor (kg/kWh)	(kg/m²a)
Mechanical systems	5.0	Mechanical systems	5.0	Electricity	5.0	Electricity	5.0	2.35	11.8	0.43	2.2
Space heating	13.4	Space heating	13.4	District heating	41.4	District heating	41.4	0.77	31.9	0.24	10.0
DHW	23.6	DHW	23.6								
		Circulation losses	4.4								
		Conversion losses	0.0								
Total	42.0		46.4		46.4		46.4		43.6		12.1

8.6.3 Design advice

Both solutions here fulfil the primary energy target. When comparing the construction of the building envelope to current building standards, it is clear that today's apartment buildings could live up

Table 8.6.6 *Solution 1a: Conservation with electric resistance heating and solar DHW – building envelope construction*

	Material	Thickness	Conductivity	Per cent	Studs	Studs	Resistance without Rsi, Rse	Resistance with Rsi, Rse	U-value
		m	λ (W/mK)	%	λ (W/mK)	%	(m²K/W)	(m²K/W)	(W/m²K)
wall	exterior surface							0.04	
	wooden panel	0.045		100%					
	air gap	0.025		100%					
	mineral wool hd	0.050	0.030	100%			1.67		
	mineral wool	0.250	0.036	85%	0.14	15%	4.84		
	plastic foil			100%					
	mineral wool	0.050	0.036	85%	0.14	15%	0.97		
	plaster board	0.013	0.220	100%			0.06		
	interior surface							0.13	
		0.433					**7.54**	**7.71**	**0.130**
roof	exterior surface							0.04	
	roof felt	0.003		100%					
	mineral wool hd	0.050	0.030	100%			1.67		
	mineral wool	0.350	0.036	100%			9.72		
	plastic foil			100%					
	concrete	0.15	1.700	100%			0.09		
	interior surface							0.10	
		0.553					**11.48**	**11.62**	**0.086**
floor	exterior surface								
	mineral wool	0.300	0.036	100%			8.33		
	concrete	0.150	1.700	100%			0.09		
	interior surface							0.17	
		0.450					**8.42**	**8.59**	**0.116**
window				emissivity					
	pane	0.004	low emissivity	5%	reversed				
	gas	0.012	krypton						
	pane	0.004	clear	83.70%					
	gas	0.012	krypton						
	pane	0.004	clear	83.70%					
		0.036							**0.800**
	frame	0.093	wood		0.14		0.66	0.83	**1.20**

to the targets set by IEA Task 28 if they were built very air tight and if they used efficient ventilation heat exchangers.

If the building is situated in an urban area where district heating systems use large quantities of renewable fuels (as, for example, in Scandinavia, where 80 per cent of the fuel is renewable), solution 1b is preferable. In rural areas where district heating is not available, solution 1a, with a solar domestic hot water system and electrical auxiliary heating, is a good alternative, especially since the solar system can cover the energy demand in summer. Of course, the production method of electricity is an important issue that significantly affects the resulting primary energy demand of a building using a high share of electricity – for example, as in solution 1a. An average value for the primary energy factor for electricity based on 17 European countries is used in all calculations. An environmentally friendly choice of electricity production would have a great impact on the primary energy use.

Table 8.6.7 *Solution 1b: Energy conservation with district heating – building envelope construction*

	Material	Thickness	Conductivity	Per cent	Studs	Studs	Resistance without Rsi, Rse	Resistance with Rsi, Rse	U-value
		m	λ (W/mK)	%	λ (W/mK)	%	(m^2K/W)	(m^2K/W)	(W/m^2K)
wall	exterior surface							0.04	
	wooden panel	0.045		100%					
	air gap	0.025		100%					
	mineral wool hd	0.030	0.030	100%			1.00		
	mineral wool	0.100	0.036	85%	0.14	15%	1.94		
	plastic foil			100%					
	mineral wool	0.050	0.036	85%	0.14	15%	0.97		
	plaster board	0.013	0.220	100%			0.06		
	interior surface							0.13	
		0.263					**3.97**	**4.14**	**0.242**
roof	exterior surface							0.04	
	roof felt	0.003		100%					
	mineral wool hd	0.030	0.030	100%			1.00		
	mineral wool	0.100	0.036	100%			2.78		
	plastic foil			100%					
	concrete	0.15	1.700	100%			0.09		
	interior surface							0.10	
		0.283					**3.87**	**4.01**	**0.250**
floor	exterior surface								
	mineral wool	0.110	0.036	100%			3.06		
	concrete	0.150	1.700	100%			0.09		
	interior surface							0.17	
		0.260					**3.14**	**3.31**	**0.302**
window				emissivity					
	pane	0.004	low emissivity	5%	reversed				
	gas	0.012	krypton						
	pane	0.004	clear	83.70%					
	gas	0.012	krypton						
	pane	0.004	clear	83.70%					
		0.036							**0.800**
	frame	0.093	wood		0.14		0.66	0.83	**1.20**

Table 8.6.8 *Design parameters of the solar DHW system in solution 1a*

Parameter	Value
Electrical heater efficiency	8 kW
The pipes of the solar circuit	
Pipe material	Copper
ø inside	16 mm
ø outside	18 mm
Length indoors	24 m
Length outdoors	6 m
Thermal conductivity, λ, of pipe	0.040 W/mK
Thermal insulation thickness	25 mm
Collector circuit	
Pump power	210 W
Collector circuit flow	525 l/h
Specific throughput	7 l/h,m^2
Heat transfer medium	Water (50%), glycol (50%)
Power output to heat transfer medium	60%
Heat transfer rate k*A for heat exchanger of stratifying device	5000 W/K
Tank	
Volume	6 m^3
Height	4 m
Temperature in the tank room	15°C

References

Kvist, H. (2005) *DEROB-LTH for MS Windows, User Manual Version 1.0–20050813*, Energy and Building Design, Lund University, Lund, Sweden, www.derob.se

Meteotest (2004) *Meteonorm 5.0 – Global Meteorological Database for Solar Energy and Applied Meteorology*, Bern, Switzerland, www.meteotest.ch

Polysun 3.3 (2002) *Polysun 3.3: Thermal Solar System Design, User's Manual*, SPF, Institut für Solartechnik, Rapperswil, Switzerland, www.solarenergy.ch

8.7 Apartment building in the Cold Climate Renewable Energy Strategy

Helena Gajbert and Johan Smeds

Table 8.7.1 *Targets for apartment building in the Cold Climate Renewable Energy Strategy*

	Targets
Space heating	20 kWh/m^2a
Non-renewable primary energy: (space heating + water heating + electricity for mechanical systems)	60 kWh/m^2a

This section presents a solution for the apartment building in the cold climate. As a reference for the cold climate, the city of Stockholm is considered. A balanced mechanical ventilation system with heat recovery is used to reduce the ventilation losses. The solution is based on renewable energy supply.

8.7.1 Solution 2: Renewable energy with solar combi-system and biomass boiler

Building envelope

The apartment building has reinforced concrete structure and wooden frame façades with mineral wool. The U-values are shown in Table 8.7.2. The double-glazed windows have a frame ratio of 30 per cent, one low-e coating and are filled with air. Construction data are found in Table 8.7.7.

Table 8.7.2 *U-values of the building components*

Building component	U-Value (W/m^2K)
Walls	0.27
Roof	0.29
Floor (excluding ground)	0.30
Windows (frame + glass)	1.34
Window frame	1.20
Window glass	1.40
Whole building envelope	0.41

Mechanical systems

Ventilation: a balanced mechanical ventilation system with heat recovery and a bypass for summer ventilation is installed. The heat exchanger has an efficiency of 80 per cent. The ventilation rate is 0.045 ach and the infiltration rate is 0.05.

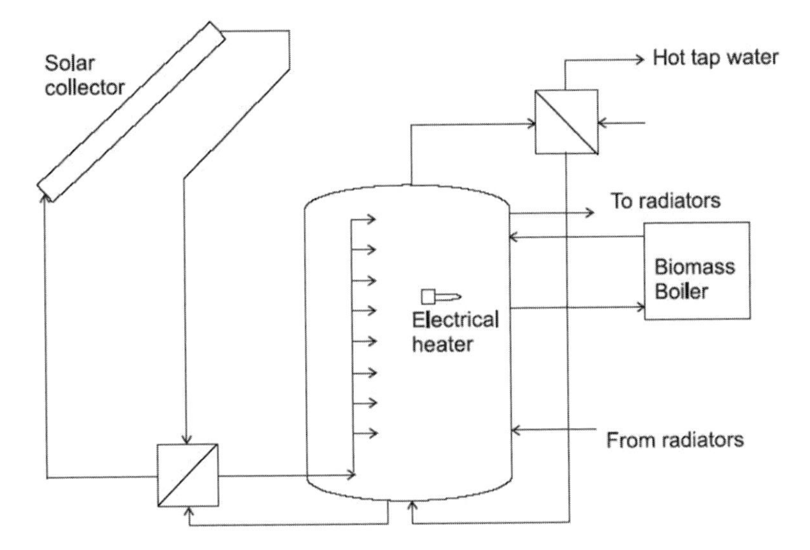

Source: Helena Gajbert and Johan Smeds

Figure 8.7.1 *The design of the suggested solar combi-system with a pellet boiler and an electrical heater as auxiliary heat sources and two external heat exchangers, one for DHW and one for the solar circuit; the latter is attached to a stratifying device in the tank*

Space heating and DHW: a solar combi-system is chosen as the energy supply system for the building – that is, a system where the space heating system and the domestic hot water (DHW) system are combined, here by a joint storage tank, to which both solar collectors and auxiliary energy sources are connected.

The solar thermal heating system consists of 50 m^2 collectors placed on the south-facing roof, tilted 40°. The storage tank is 4 m^2 and there is a 35 kW pellet boiler, as well as an electrical heater of 5 kW, connected to it for increased operational flexibility. The heat from the collector circuit is delivered to the tank via an external heat exchanger coupled to a stratifying device, which improves the important thermal stratification in the tank. Another external heat exchanger is used for heating the DHW, taking water from the top of the tank and returning it at the bottom. In the collector circuit, the liquid heat transfer medium is a mixture of 50 per cent water and 50 per cent glycol. Water from the tank is also used for the radiant hot water space heating system in the building. Other important design parameters of this system can be found in Table 8.7.8. An illustration of the system is shown in Figure 8.7.1.

The electrical heater and the biomass boiler largely working alternately, heating the upper part of the tank when necessary. When only very little auxiliary heat is required during short time intervals, as is often the case in the summer, the biomass boiler operation mode with frequent starts and stops causes unnecessary energy losses. It is therefore suitable to shut off the furnace during summer, when solar energy can cover most of the energy demand. The electrical heater can then be a good complement during cold and cloudy days. The electrical heater can also be used in tandem with the boiler if needed in extremely cold winter days.

An alternative system design for increased efficiency could also include a possibility of delivering solar energy directly to a low temperature radiator heating system when there is prevailing heating demand. Since the collector temperature then can be lowered, heat losses of the system are reduced and the efficiency is increased.

The building is designed according to simulations with DEROB-LTH (Kvist, 2005). Simulations of the active solar energy systems are performed with a modified version of Polysun 3.3 (Polysun, 2002): the Polysun-Larsen version, which uses the data files of space heating demand from the DEROB-LTH simulations. Stockholm climate data from Meteonorm (Meteotest, 2004) are used in both simulation programs. The dimensioning of the solar thermal heating system was performed primarily with a concern for low auxiliary energy use in summer, avoidance of stagnation and overheating, and economical aspects. For general assumptions and parameters used in the DEROB-LTH and Polysun simulations, see Table 8.7.7 and 8.7.8.

Energy performance

Space heating demand: the space heating demand of the building is 30,400 kWh/a (19 kWh/m^{2}a) according to simulation results from DEROB-LTH. The assumed heating set point was 20°C. Maximum room temperature set points were 23°C during winter and 26°C during summer (this assumes the use of shading devices and window ventilation).

The monthly space heating demand for this high-performance case in comparison to a reference building (current building codes year 2001) are shown in Figure 8.7.2. Hourly peak loads of the heating system are calculated with results from DEROB-LTH simulations without direct solar radiation in order to simulate a shaded building. The annual peak load occurs in January and is 26.8 kW (17 W/m^2).

Domestic hot water demand: the net energy demand for domestic hot water is assumed to be 37,800 kWh/a (23.6 kWh/m^{2}a). The DHW demand is thus larger than the space heating demand.

System losses: the system losses consist mainly of losses from the hot water storage tank, but also from circulation losses in the distribution system for DHW. The tank and circulation losses are 1.2 and 4.4 kWh/m^{2}a, respectively. The circulation losses are taken into account as internal gains in the building heat load simulations. The tank losses and losses from the collector circuit can, however, not be used as internal gains since the tank and boiler are placed in the basement outside the thermal envelope of the building. The efficiency of the biomass boiler is set to 85 per cent, resulting in conversion losses of 5.5 kWh/m^{2}a.

Household electricity: the household electricity use is 2190 kWh/a (21.9 kWh/m^{2}a) for two adults and one child in each apartment. The primary energy target does not include household electricity since this factor can vary considerably depending on the occupants' behaviour.

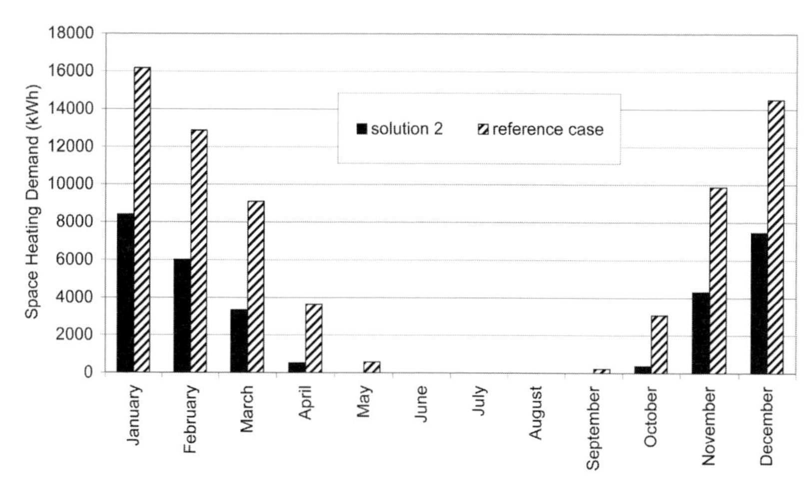

Source: Helena Gajbert and Johan Smeds

Figure 8.7.2 *Monthly values of the space heating demand during one year; the annual total space heating demand is 30,400 and 70,000 kWh/a, respectively, for the high-performance building and the reference building*

Total energy use, non-renewable primary energy demand and CO_2 emissions: as shown in Table 8.7.3, the net energy required for DHW, space heating and mechanical systems is 47.6 kWh/m^{2}a (76,200 kWh/a). Adding tank, boiler and system losses gives a total energy use of 58.7 kWh/m^{2}a (93900 kWh/a).The total auxiliary energy demand (electricity and pellets), including the conversion losses in the pellets boiler, is 36.4 kWh/m^{2}a (58300 kWh/a), where 36.2 kWh/m^{2}a (57900 kWh/a) are provided by biomass in the pellet boiler and approximately 0.3 kWh/m^{2}a (400 kWh/a) by the electrical heater.

The active solar gains are 17.3 kWh/m^{2}a (27600 kWh/a) and the solar fraction of the heating system is 35 per cent since the simulation result of the same building but without solar collector would require 77,400 kWh/a (described in section 8.7.2). The electricity use for mechanical systems is 5.0 kWh/m^{2}a (8000 kWh/a).

Due to the solar contribution, the delivered energy is reduced to 41 kWh/m^{2}a (66,200 kWh/a). The total use of non-renewable primary energy is approximately 17 kWh/m^{2}a (27,900 kWh/a) and the CO_2 equivalent emissions are 3.8 kg/m^{2}a (6100 kg/a). Compared to the reference building, the non-renewable energy use and CO_2 equivalent emissions are considerably lower (see Table 8.7.4 and Figure 8.7.3).

Table 8.7.3 *Total energy demand, non-renewable primary energy demand and CO_2 equivalent emissions for the apartment building*

Net Energy (kWh/m²a)		Total Energy Use (kWh/m²a)				Delivered energy (kWh/m²a)		Non renewable primary energy		CO_2 equivalent emissions	
		Energy use		Energy source				factor (-)	(kWh/m²a)	factor (kg/kWh)	(kg/m²a)
Mechanical systems	5.0	Mechanical systems	5.0	Electricity	5.0	Electricity	5.0	2.35	11.8	0.43	2.2
Space heating	19.0	Space heating	19.0	Electricity	0.3	Electricity	0.3	2.35	0.6	0.43	0.1
				Solar	17.3						
DHW	23.6	DHW	23.6	Bio pellets	36.2	Bio pellets	36.2	0.14	5.1	0.04	1.6
		Tank and circulation losses	5.6								
		Boiler conversion losses	5.5								
Total	47.6		58.7		58.7		41.4		17.4		3.8

Table 8.7.4 *A comparison between the high-performance house and the reference house regarding energy use and CO_2 equivalent emissions*

Building	Net energy (kWh/m²a)	Total energy use (kWh/m²a)	Delivered energy (kWh/m²a)	CO_2 equivalent emissions (kWh/m²a)
High performance building	47.6	58.7	41.4	3.8
Reference building	72.4	81.2	81.2	20.5

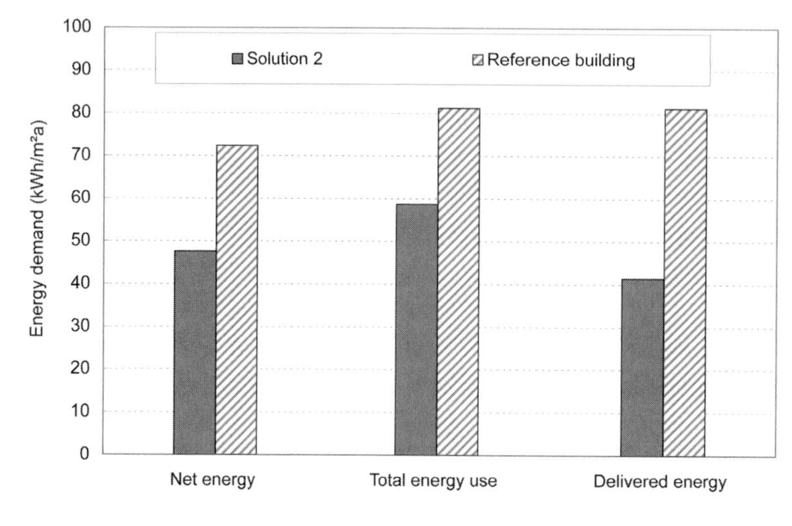

Source: Helena Gajbert and Johan Smeds

Figure 8.7.3 *An overview of the net energy, the total energy use and the delivered energy for the high-performance building and the reference building*

8.7.2 Sensitivity analysis

Since the performance of the solar energy system is strongly affected by the system design parameters (for example, collector area, tank volume, slope and direction of the collectors and collector type of the solar heating system), some of the most important results from the Polysun simulations shown during the designing of the system are presented here. If nothing else is mentioned, the parameters of the simulated systems are the same as for the suggested system. The 'auxiliary energy', which is shown in many of the figures, is the heat output from the auxiliary energy system, thus including heat for DHW heating, space heating and also heat losses. It does not include the conversion losses from the heat production (i.e. combustion and heat exchanger losses in a wood pellet furnace).

Collector area

The collector area and tank volume should preferably be considered both individually and as a unit. The collector area is, however, the more important feature to optimize in order to minimize the auxiliary energy use. The collector area was varied in the simulation while keeping a constant storage tank volume. The aim was to find a collector size that is large enough to almost cover the summer heating demand (i.e. the DHW demand), while still maintaining a relatively low risk of stagnation in the collectors.

The solar fraction (SF) is calculated according to Equation 8.1. The parameters Aux and Aux_0 represent the auxiliary energy required with and without the solar system, respectively. In Polysun, the value of Aux_0 was obtained by setting the collector area to zero, which resulted in an annual auxiliary energy demand of 77,400 kWh per year and 11,600 kWh during the summer months (June, July and August):

$$SF = 1 - \frac{Aux}{Aux_0} \qquad [8.1]$$

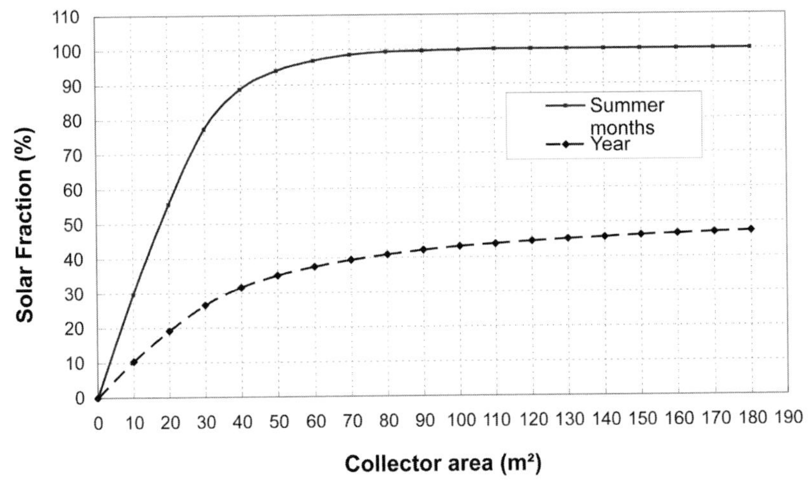

Source: Helena Gajbert and Johan Smeds

Figure 8.7.4 *The solar fraction of the system for the whole year and for the summer months*

In Figure 8.7.4 it is shown how the calculated solar fractions for the whole year and also for the summer months, June, July and August, increase with increasing collector area.

In order to lower the risk of stagnation and overheating, 50 m^2 seems a suitable collector size. The pellet boiler can still be turned off during summer, although a small amount of electricity may be needed during a few colder days. The solar fraction during the three summer months will then be 95 per cent. If a solar fraction of 100 per cent in the summer is to be obtained, the risk of overheating and stagnation would be large. Even if good control strategies can mitigate these problems, it is still desirable to avoid the problem completely. Should a heated swimming pool for summer use be included in the heating system, the problems of overheating and stagnation could be reduced as this increases the heat load during summer. Figure 8.7.5 shows monthly values of the auxiliary energy demand during summer for different system dimensions.

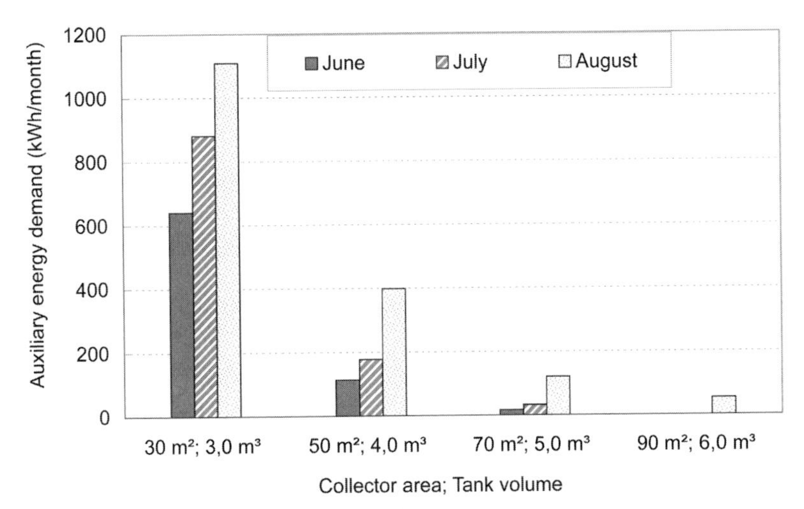

Source: Helena Gajbert and Johan Smeds

Figure 8.7.5 *Monthly values of auxiliary energy demand for solar systems of different dimensions*

In Figure 8.7.6, the energy saved due to the installation of the collectors is shown for the margin collector area (i.e. the extra savings made with an increase in collector area by 10 m^2). A rough estimation is that the collector area preferably should be between 40 m^2 and 60 m^2 to provide a reasonable amount of energy per margin collector area, thus giving sufficient energy yield in relation to the total investment costs.

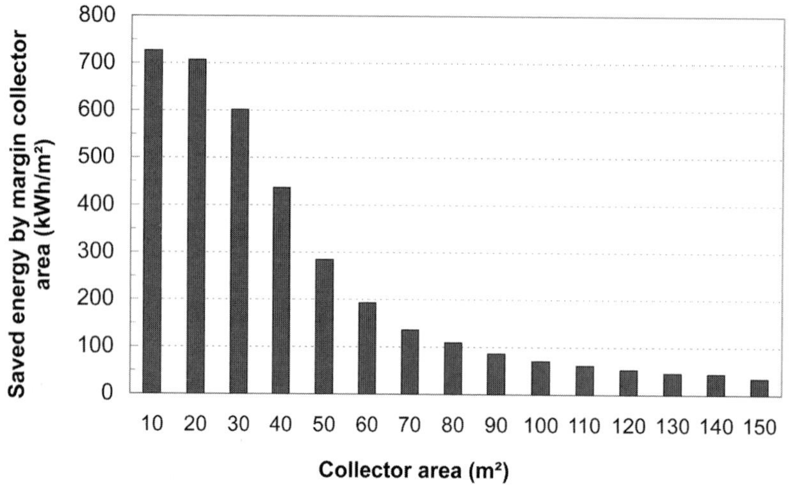

Source: Helena Gajbert and Johan Smeds

Figure 8.7.6 *The energy savings per margin collector area (i.e. how much additional energy is saved if 10 m^2 is added to the collector area, read from left to right)*

The resulting non-renewable primary energy demand and CO_2 equivalent emissions when solar combi-systems of different dimensions are used can be seen in Figure 8.7.7. The tank volumes were optimized for each case.

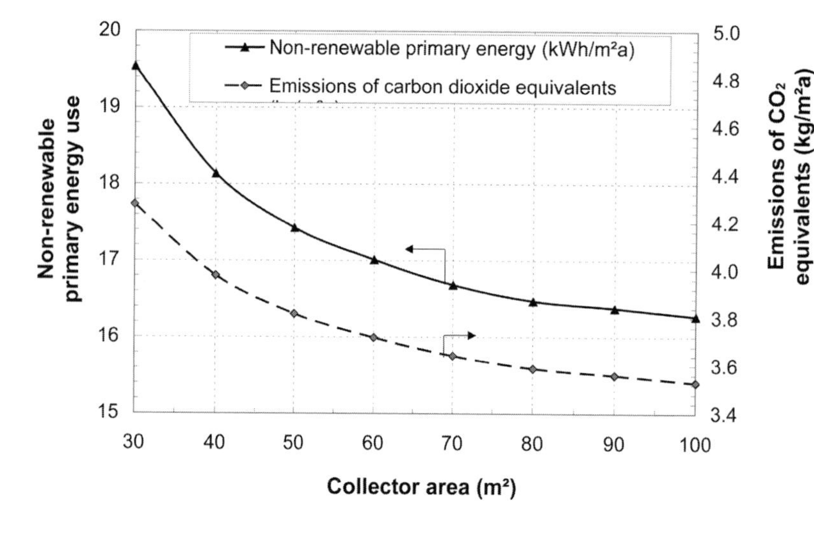

Source: Helena Gajbert and Johan Smeds

Figure 8.7.7 *The resulting non-renewable primary energy demand and CO_2 equivalent emissions for different collector areas*

If the solar system had been installed in the reference building (described in section 8.1) instead, the auxiliary energy demand would be much higher, as can be seen in Figure 8.7.8, where the auxiliary energy demand of the two buildings is shown for different solar system dimensions. Again, the tank volumes were optimized for each case.

In Figure 8.7.9 the overview of the energy demands of the high-performance building and for the reference building are shown, as well as the solar gains of collectors of 30, 50 and 70 m^2. Observe that the heat for DHW, circulation losses and tank losses is valid for a 50 m^2 collector area. For a larger collector, the tank losses will be slightly higher.

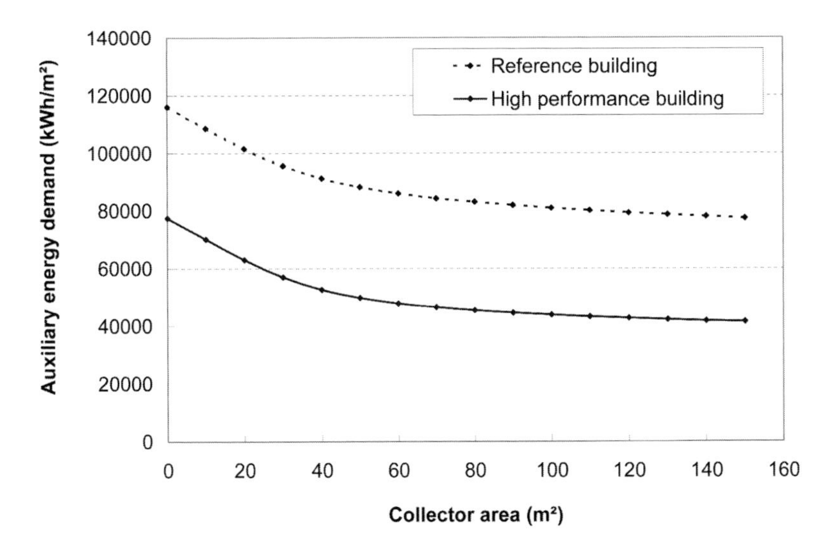

Source: Helena Gajbert and Johan Smeds

Figure 8.7.8 *Annual auxiliary energy demand per living area for different system dimensions based on Polysun simulations*

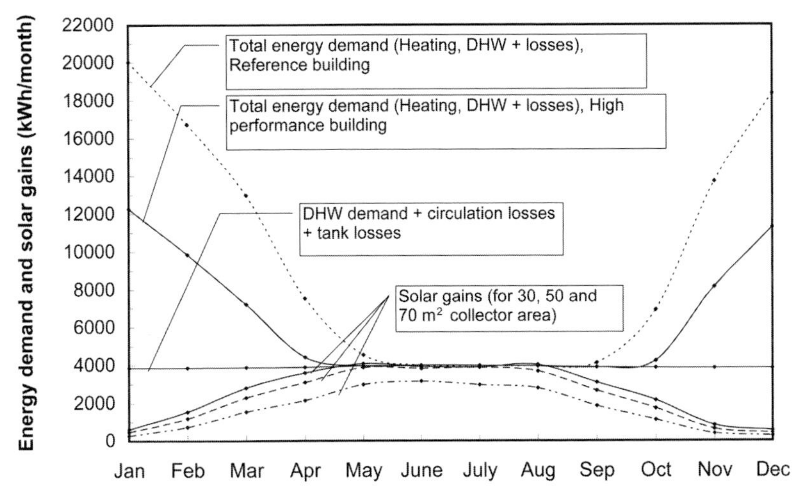

Source: Helena Gajbert and Johan Smeds

Figure 8.7.9 *The energy demands and the solar gains of the high-performance building and the reference building; the solar gains are shown for different collector areas*

Storage tank volume

For economical reasons it is important to consider the storage tank volume. A larger tank is much more expensive, although the impact on the auxiliary energy use is rather small, as shown in Figure 8.7.10. The tank could preferably be designed based on the daily energy demand due to the limited storage time capacity. The tank should not be too small due to the risk of overheating and because of lowered storage capacity. On the other hand, an oversized tank might require higher auxiliary energy demand in order to heat the tank and also because of the higher heat losses. A high insulation level is therefore important and it becomes more important the larger the tank is, as shown in Figure 8.7.10.

Simulation results for collector areas between 50 m^2 and 100 m^2, with varying tank volumes, are shown in Figure 8.7.11. The chosen system is marked with a circle. These results also show that the collector area has a much larger influence on the auxiliary energy use than the tank volume. Although the tank size has a relatively low influence on the auxiliary energy use, it is important not to choose too small a tank since there will be a risk of overheating and as it might be difficult to provide the required heat for DHW and space heating. For the larger collectors, some energy can be saved by choosing a larger tank. The suggested system with 50 m^2 collector area and a storage volume of 4 m^3 would neither imply overheating nor excessive tank losses, according to the simulation results.

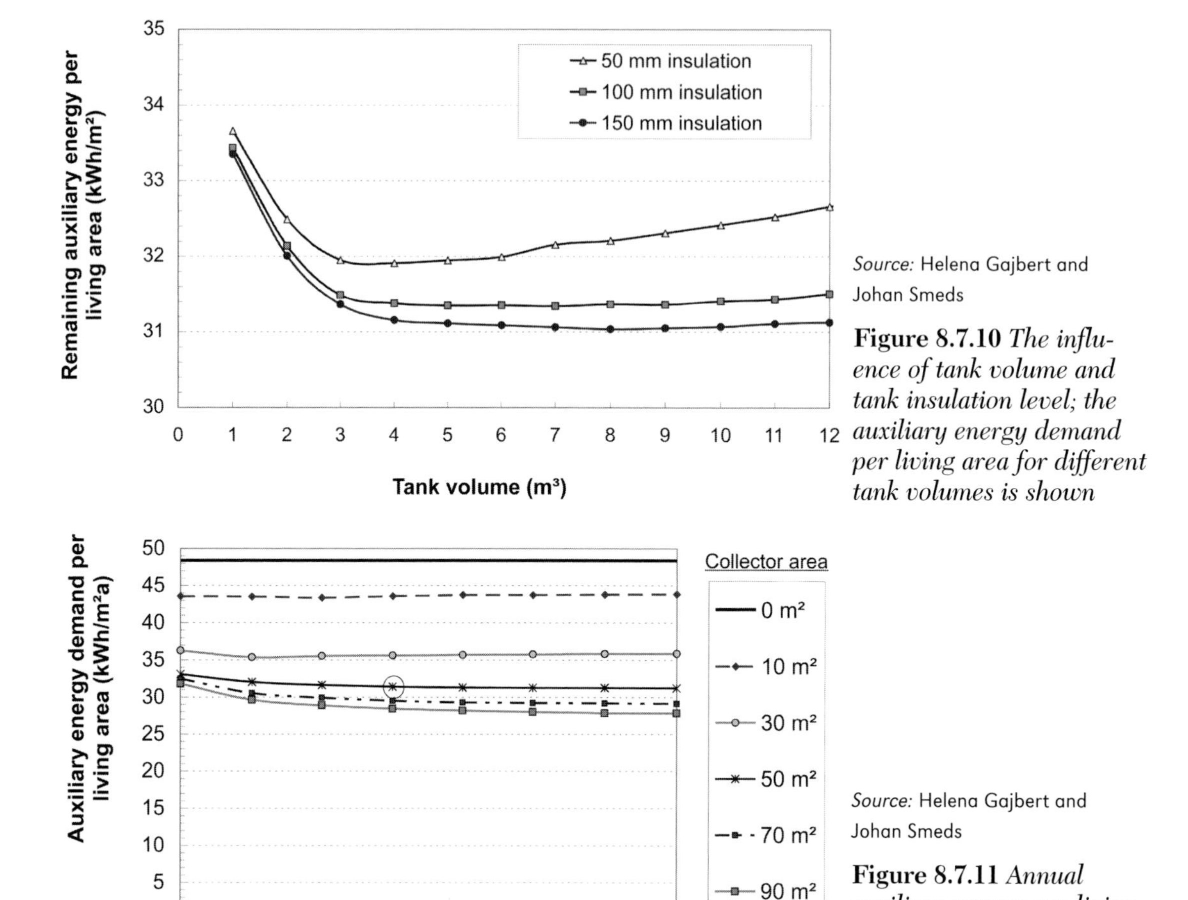

Source: Helena Gajbert and Johan Smeds

Figure 8.7.10 *The influence of tank volume and tank insulation level; the auxiliary energy demand per living area for different tank volumes is shown*

Source: Helena Gajbert and Johan Smeds

Figure 8.7.11 *Annual auxiliary energy per living area for different system dimensions based on Polysun simulations*

Collector tilt

The tilt of the collector determines how large the collector should be in order to cover the summer demand. In order to maintain a solar fraction of 95 per cent during the summer months, the required collector size increases as the collector tilt is increased or decreased from 30° to 40°, which is the tilt where the smallest collector area (50 m^2) is needed (see Figure 8.7.12).

When using highly tilted or vertical collectors, the solar gains will be better balanced to the seasonal heating demand since more of the winter space heating demand will be covered. However, this requires a larger collector area in order to cover the summer demand. For a very large collector with a high tilt angle, energy can be saved in total over the year, thus implying a higher annual solar fraction. A good, yet very expensive, solution would therefore be to use a vertical wall-mounted collector of very large size. The benefits of this are naturally much higher for combi-systems than for DHW systems. There are generally no benefits in using a horizontal collector or a collector with a low tilt.

Looking further at two mounting options for good building integration to either mount the collectors on the 40° tilted roof or vertically on the wall, tilted 90°, it can be seen in Figure 8.7.13 how the auxiliary energy demand decreases with increasing system size. The 40° tilt implies a lower demand of auxiliary energy on an annual basis and the difference is more significant in smaller systems.

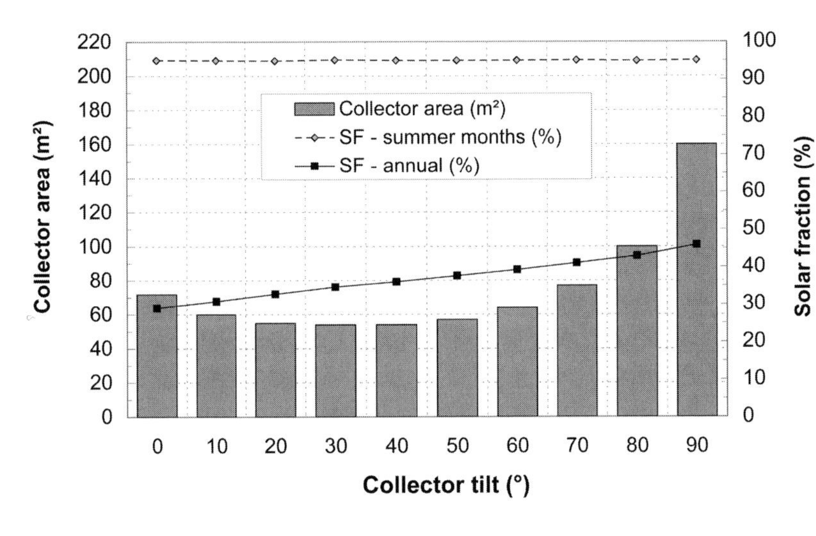

Source: Helena Gajbert and Johan Smeds

Figure 8.7.12 *The collector area required for differently tilted collectors in order to obtain a solar fraction of 95% during summer (this is thus an appropriate area that does not imply boiling in the tank); the solar fraction for the whole year and the constant solar fraction in summer are also shown*

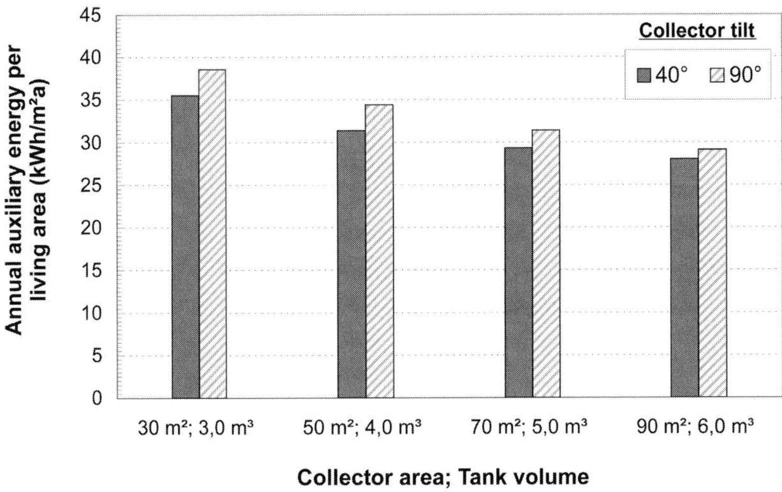

Source: Helena Gajbert and Johan Smeds

Figure 8.7.13 *Results from simulations of systems with differently tilted collectors: 40° for the roof-mounted and 90° for the wall-mounted collectors*

Furthermore, the risk of snow cover should be considered if a very low tilt angle is used. The risk of snow cover is even larger for evacuated tube collectors.

Azimuth angle

If the deviation from south is varied (i.e. the azimuth angle), the auxiliary energy demand and solar fraction changes according to Figure 8.7.14.

Collector performance

The choice of collector is very important for the system performance. Simulations have been performed with collectors of different types, from older collector types without a selective surface to the most advanced evacuated tube collectors available, with characteristics and simulation results listed in Table 8.7.5. The results are also shown in Figure 8.7.15. In the suggested system, an advanced flat-plate collector with a selective surface, anti-reflection treated glass and highly efficient thermal insulation, number 3 in Table 8.7.5, is used. When compared to an older collector type without selective surface, the chosen collector increases the solar fraction from 21 per cent to 30 per cent. If an evacuated tube collector is used, the solar fraction increases even more. For this solar

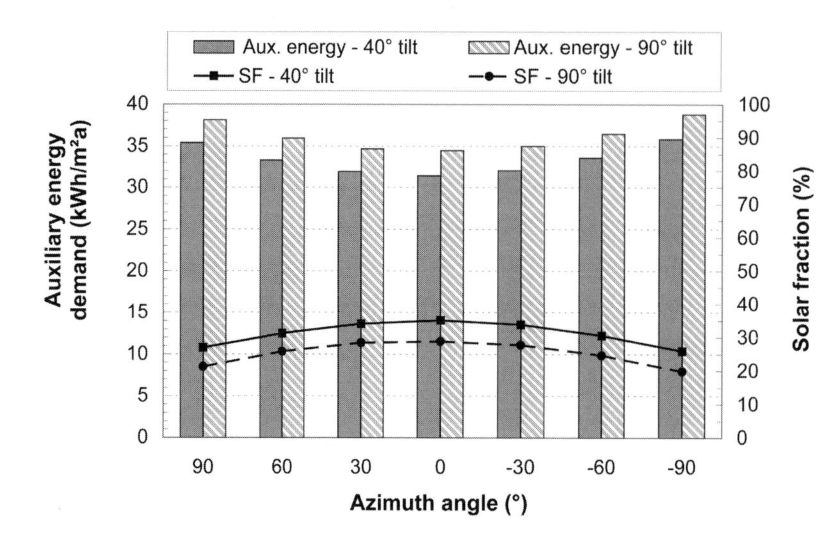

Source: Helena Gajbert and Johan Smeds

Figure 8.7.14 *The auxiliary energy demand and solar fractions of systems with different azimuth angles*

Table 8.7.5 *Collector parameters and corresponding solar fraction and auxiliary energy*

Collector type	?0 (-)	c_1 (W/m²K)	c_2 (W/m² K²)	KCH1 (-)	KCH2 (-)	Specific heat capacity (kJ/m²K)	Solar fraction (%)	Remaining auxiliary energy (kWh/a)	Auxiliary energy (kWh/m² a)
1. Advanced evacuated tube collector	0.88	1.41	0.013	0.92	1.15	7.84	40,5	46053	28.8
2. Evacuated tube collector	0.77	1.85	0.004	0.9	1.00	5.71	36.7	49004	30.6
3. Advanced flat plate collector	0.85	3.7	0.007	0.91	0.91	6.32	35.1	50232	31.4
4. Flat plate collector	0.8	3.5	0.015	0.9	0.9	6.32	33.1	51767	32.4
5. Flat plate collector - no selective surface	0.75	6,00	0.03	0.9	0.9	7,00	26.4	56976	35.6

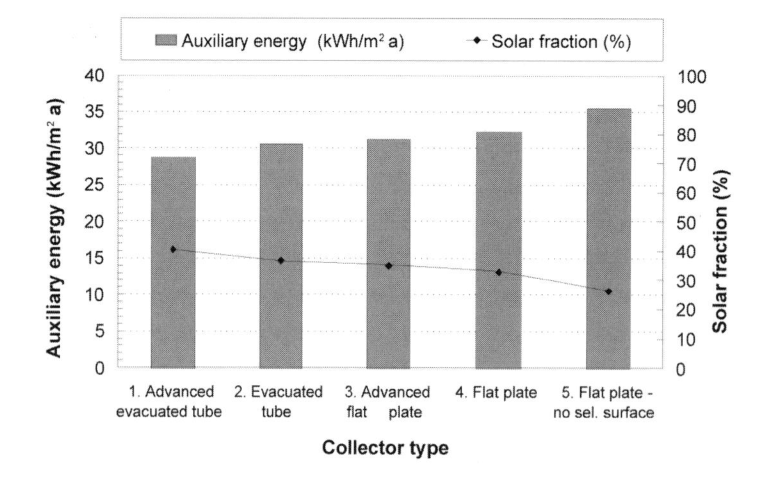

Source: Helena Gajbert and Johan Smeds

Figure 8.7.15 *Results from simulations of systems with different collector types*

heating system an evacuated tube collector of 35 m² would be sufficient. Evacuated tube collectors are very efficient, but are also often very expensive.

Concentrating collectors

By using concentrating reflectors to enhance the irradiance on the absorber, the output from the collectors can be increased. If the geometry of the collectors excludes irradiation from higher solar angles, the collector could improve the seasonal balance by increasing the winter performance and

Table 8.7.6 *The auxiliary energy demand for an evacuated tube collector with and without reflectors*

	70°, 42 m^2		90°, 56 m^2	
	Without reflector	With reflector	Without reflector	With reflector
Total annual auxiliary energy (kWh/a)	47,538	42,297	45,643	40,570
Solar fraction (%)	38.6	45.4	41.0	47.6

suppressing the summer performance. It is interesting to apply this strategy to an evacuated tube collector, which has the potential to deliver energy during lower irradiance levels and which also, currently, is relatively expensive. An estimation of the effect that this would imply was investigated for an evacuated tube collector (with collector characteristics according to Table 8.7.5, collector type 1). Polysun simulations were performed assuming a double collector area from October to March and with the collector tilted 70° and 90°. The area was chosen so as to imply a solar fraction of 95 per cent during summer. The results, shown in Table 8.7.6, indicate that the reflector implies energy savings of 11 per cent if it is tilted either 70° or 90°.

Design of the collector loop heat exchanger and flow rate

The design of the storage tank and its connected components is also important for the system efficiency since different types of heat exchangers work optimally at different flow rates. The flow rate in the collector loop should not be too low so as not to cause laminar flow, air traps and/or increased heat losses. Therefore, if the heat from the collectors is transferred to the tank through an internal heat exchanger, a high flow rate implies lower auxiliary energy demand, as shown in Figure 8.7.16.

The efficiency of the solar heating system is improved if an external heat exchanger is used instead and even better if a stratifying device is employed. In a smaller tank, the importance of a stratifying device is more significant. In these cases, a low flow rate is better since the fluid inserted into the tank otherwise can cause mixing of the stratified water. However, the flow rate should not be too low. The optimum can be seen in Figure 8.7.16. Lower flow rates also contribute to a lower electricity use for the pump and the dimensions of the pipes are reduced and less expensive.

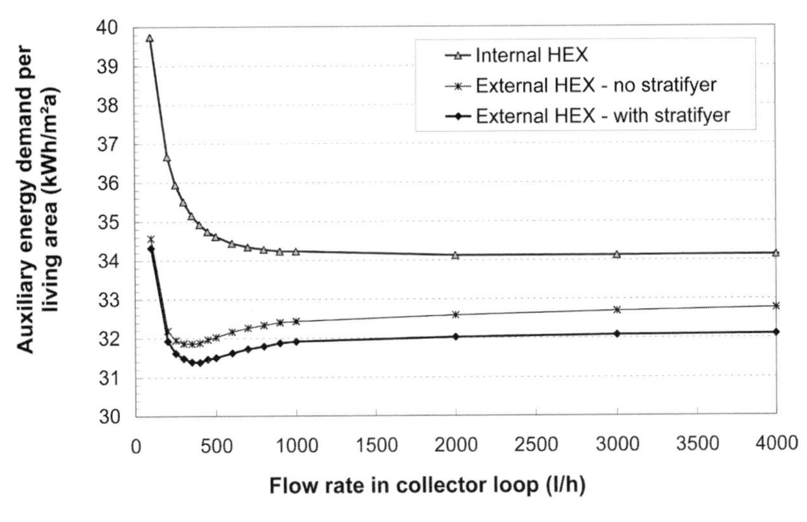

Source: Helena Gajbert and Johan Smeds

Figure 8.7.16 *The auxiliary energy demand, with varied flow rate and type of heat exchanger in the solar circuit*

Non-renewable energy use for solar DHW systems and combi-systems

If a solar DHW system would be used instead of the combi-system, the target of non-renewable primary energy would not be reached if combined with electricity for space heating. If combined with district heating for space heating, the solar collectors would have to be at least 40 m^2 in order to reach the target of 60 kWh/m^2a (living area), as shown in Figure 8.7.17. By looking at Figure 8.7.18, where the emissions of CO_2 equivalents are shown, the most environmentally friendly design also appears to be with a solar combi-system with a pellet boiler and an electrical heater.

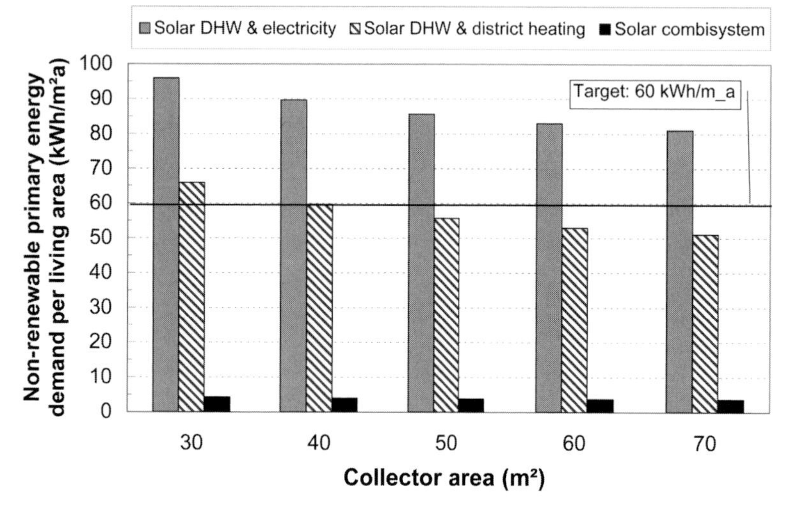

Source: Helena Gajbert and Johan Smeds

Figure 8.7.17 *The use of non-renewable primary energy for three different energy system designs: the combi-system in question, with a pellet boiler and an electrical heater; a solar DHW system combined with district heating; and a solar DHW system combined with electrical heating*

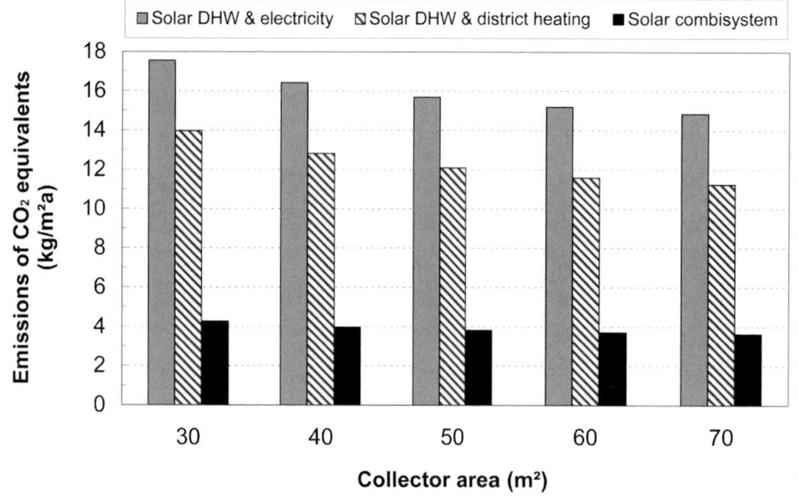

Source: Helena Gajbert and Johan Smeds

Figure 8.7.18 *The emissions of CO_2 equivalents for three different energy system designs: the combi-system in question, with a pellet boiler and an electrical heater; a solar DHW system combined with district heating; and a solar DHW system combined with electrical heating*

8.7.3 Design advice

The solution presented in this section fulfils the primary energy target. When comparing the construction of the building envelope to current building standards, it is clear that today's apartment buildings could live up to the targets set if they were built very air tight and with efficient ventilation heat exchangers.

In cold climates where the heating demand is high, the solar energy from a solar combi-system can provide a significant part of the total energy demand of a high-performance house. If looking specifically at the space heating demand, solar energy can also contribute, to some extent, to the space heating supply even for a well-insulated building. However, it may be easier to justify a solar DHW system economically if the energy demand is low enough to warrant a space heating system based only, for example, on heat recovery and electrical resistance heating of the incoming air.

If a solar combi-system is used, the combined storage tank can give beneficial effects due to the higher energy withdrawal from the tank, which can lower the temperatures in the collector circuit and thus increase the collector efficiency somewhat during the heating season. By combining the tank for space heating and hot water and thus improving the system flexibility, the heat losses might be reduced in comparison to using two separate tanks for DHW and space heating.

The collector solar system dimensions should preferably be designed so as to cover most of the energy demand in summer, both for solar combi-systems and solar DHW systems. The highest solar fraction on an annual basis is achieved with a collector tilt of approximately 30° to 50°. If a larger collector is affordable, it is beneficial to design a combi-system, over-dimension the collector area and to place it vertically on the wall since this increases the solar gains during the heating season and reduces the risk of stagnation in summer.

For the technical facility management of the building, a solar combi-system does, of course, mean some additional costs in comparison to the district heating system; but since the biomass boiler can be shut down during summer, service costs should be at a reasonable level.

The solution presented here implies very low use of fossil fuels and low emissions of CO_2 equivalents to the atmosphere. Therefore, it is a very good solution for a building in rural areas where district heating is not available, especially as the solar system can cover the energy demand in summer and the furnace can be switched off.

District heating systems use a large share of renewable fuels in Scandinavia; approximately 80 per cent of the fuel is renewable. If the building is situated in an urban area with such a district heating system, the district heating is to be preferred to the use of biomass boilers as biomass boilers give rise to some emissions from the furnaces, something that may be forbidden in certain urban areas. Today, the purification of the exhaust gases is very good and modern furnaces are often

Table 8.7.7 *General assumptions for simulations of the building in DEROB-LTH – construction according to solution 2, space heating target 20 kWh/m²a*

	Material	Thickness	Conductivity	Per cent	Studs	Studs	Resistance without Rsi, Rse	Resistance with Rsi, Rse	U-value
		(m)	λ (W/mK)	(%)	λ (W/mK)	(%)	(m²K/W)	(m²K/W)	(W/m²K)
wall	exterior surface							0.04	
	wooden panel	0.045		100%					
	air gap	0.025		100%					
	mineral wool hd	0.030	0.030	100%			1.00		
	mineral wool	0.100	0.036	85%	0.14	15%	1.94		
	plastic foil			100%					
	mineral wool	0.030	0.036	85%	0.14	15%	0.58		
	plaster board	0.013	0.220	100%			0.06		
	interior surface							0.13	
		0.243					**3.58**	**3.75**	**0.267**
roof	exterior surface							0.04	
	roof felt	0.003		100%					
	mineral wool hd	0.015	0.030	100%			0.50		
	mineral wool	0.100	0.036	100%			2.78		
	plastic foil			100%					
	concrete	0.15	1.700	100%			0.09		
	interior surface							0.10	
		0.268					**3.37**	**3.51**	**0.285**
floor	exterior surface								
	mineral wool	0.110	0.036	100%			3.06		
	concrete	0.150	1.700	100%			0.09		
	interior surface							0.17	
		0.260					**3.14**	**3.31**	**0.302**
window				emissivity					
	pane	0.004	clear	83.70%					
	gas	0.015	air						
	pane	0.004	low emissivity	4%					
		0.023							**1.400**
	frame	0.093	wood		0.14		0.66	0.83	**1.20**

designed to use optimal oxygen amounts in the combustion process. If a biomass boiler is to be used, it should be an 'environmentally approved' boiler and it should be coupled with a thermal heat storage tank in order to achieve an effective and controlled combustion process and to minimize the emissions of volatile organic compounds (VOCs) and particles (Johansson et al, 2003).

Table 8.7.8 *Design parameters of the solar combi-system used in the Polysun simulations*

Parameter	Value
Pellet boiler	
Pellet boiler peak power	35 kW
Electrical heater peak power	5 kW
The pipes of the solar circuit	
Pipe material	Copper
Ø inside	16 mm
Ø outside	18 mm
Length indoors	24 m
Length outdoors	6 m
Thermal conductivity, λ, of pipe	0.040 W/mK
Thermal insulation thickness	25 mm
Collector circuit	
Pump power	210 W
Collector circuit flow	525 l/h
Specific throughput	7 l/h,m^2
Heat transfer medium	Water (50%), glycol (50%)
Power output to heat transfer medium	60%
Heat exchanger in solar circuit	
Heat transfer rate k*A for heat exchanger	5000 W/K
Tank	
Volume	4 m^3
Height	4 m
Temperature in the tank room	15°C
Thermal insulation of the tank	150 mm

References

Johansson, L., Gustafsson, L., Thulin, C. and Cooper, D. (2003) *Emissions from Domestic Bio-fuel Combustion – Calculations of Quantities Emitted*, SP Report 2003:08, SP Swedish National Testing and Research Institute, Borås, Sweden

Kvist, H. (2005) *DEROB-LTH for MS Windows, User Manual Version 1.0–20050813*, Energy and Building Design, Lund University, Lund, Sweden, www.derob.se

Meteotest (2004) *Meteonorm 5.0 – Global Meteorological Database for Solar Energy and Applied Meteorology*, Bern, Switzerland, www.meteotest.ch

Polysun (2002) *Polysun 3.3: Thermal Solar System Design, User's Manual*, SPF, Institut für Solartechnik, Rapperswil, Switzerland, www.solarenergy.ch

Swedish Environmental Protection Agency (2005) www.naturvardsverket.se

Swedish National Testing and Research Institute (2005) www.sp.se

Weiss, W. (ed) (2003) *Solar Heating Systems for Houses: A Design Handbook for Solar Combisystems*, James and James Ltd, London

8.8 Apartment buildings in cold climates: Sunspaces

Martin Reichenbach

8.8.1 Introduction

This section examines what can be expected from glazed balconies as part of apartment buildings with a very low energy demand in a northern climate. The motivation in including glazing in a balcony is to gain additional living space tempered from the outdoor climate by both solar gains and heat losses from the apartment, with no increase in heating costs. This is a purely passive concept.

An alternative concept is to actively extract sun-tempered air from the sunspace as a pre-heated fresh air source for the apartment. In this concept, the sunspace will be colder than a pure buffer space, a major disadvantage for its use as a sometimes habitable space.

Adding a sunspace to an apartment is more difficult than is the case with row houses or detached houses. The window area facing directly outside (and not facing the sunspace) will be reduced, which limits the daylight and the possibility of using window ventilation. For this reason, the sunspace framing should be as thin as possible to minimize light loss. Considering the high insulation standard of the studied apartment building, a rather high standard of glazing and very air-tight construction was assumed.

The first part of this study examines the effects which different sunspace geometries in combination with varying material parameters have on the heating demand of the apartment. This geometry study is based on the simplified calculations commonly used in the early design phase. In a second step, comfort in the sunspace is studied in detail using rigorous modelling to assess the effects of different ventilation and shading strategies.

8.8.2 Performance of different sunspace types

Each apartment and each of the respective sunspaces was analysed as an own thermal zone in a dynamic simulation model. Details on the assumptions made are given in section 8.8.5. Apartment positions were chosen to cover the full range of heating demand resulting from different amounts of exterior exposure (see Figure 8.8.1). Unit A is of special interest since one would expect extreme results due to its top corner position.

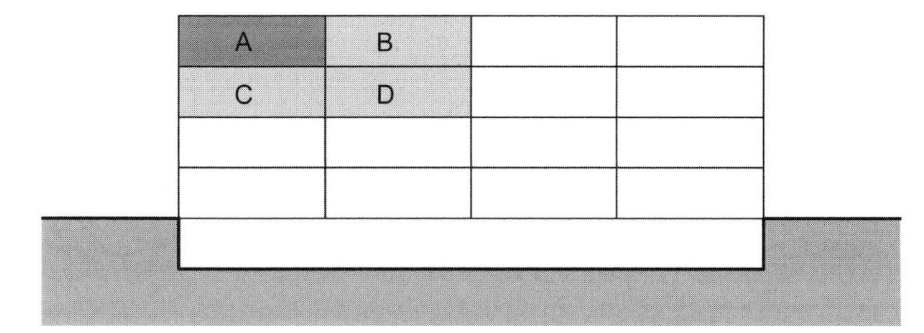

Source: Martin Reichenbach, Reinertsen Engineering AS, N 0216 Oslo, Norway

Figure 8.8.1 *Apartment units selected for simulation in the study*

Figure 8.8.2 shows the sunspace geometries that were studied. Types 1 and 2 occur in front of a flush façade; types 3, 3A and 4 are partly or completely inset into the building.

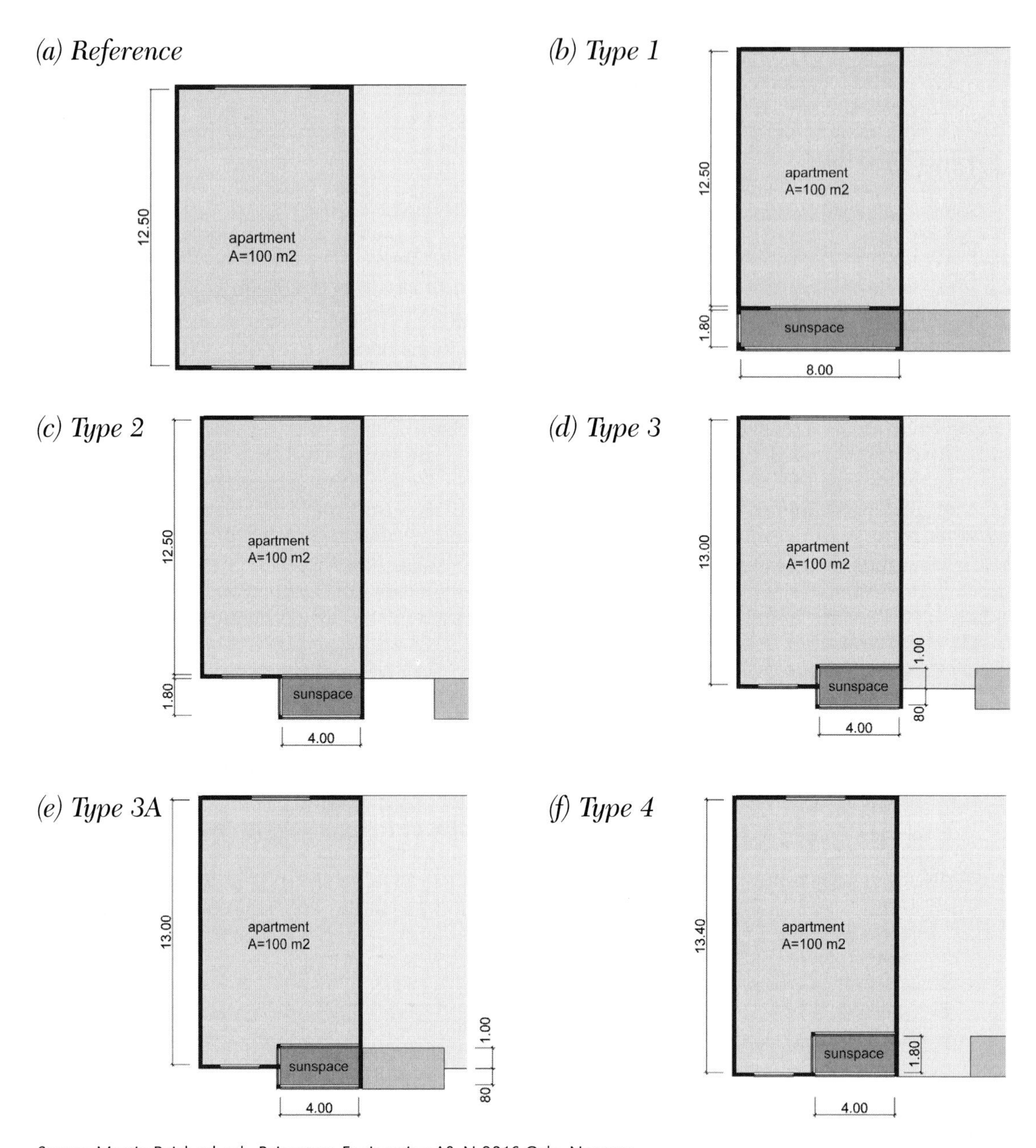

Source: Martin Reichenbach, Reinertsen Engineering AS, N 0216 Oslo, Norway

Figure 8.8.2 *Sunspace types: (a) reference: no sunspace (A/V = 0.285); (b) sunspace covering the entire southern façade: depth = 1.80 m (A/V = 0.265); (c) sunspace covering 50% of the southern façade: depth = 1.80 m (A/V = 0.282); (d) sunspace covering 50% of the southern façade and partly withdrawn into the apartment volume: depth = 0.80 m out/1.00 m in (A/V = 0.285); (e) sunspace geometry and orientation as type 3, turned 'back-to-back' with the neighbouring unit (A/V = 0.278); (f) sunspace covering 50% of the southern façade and completely withdrawn into the apartment volume: depth = 1.80 m (A/V = 0.274)*

Cross-comparison of simulation results

Simulations showed the range in heating demands according to the apartment's position, when the average of all units is 15 kWh/m²a. The space heating demand of a top corner unit is nearly triple the demand of a middle apartment. This pattern remained the same for all types of sunspaces. This cross-comparison concentrates on unit A, the worst load case.

				Q_h [kWh/m²a]			
glazed area, south facade [%] *	U glazing, south [W/m²K]	g glazing , south	Total glazed area (S+N) to floor area [%]	**A** top floor, corner unit	**B** top floor, mid unit	**C** 3rd floor, side unit	**D** 3rd floor, mid unit
29.2	0.84	0.52	10.0	**22.4**	**16.9**	**13.7**	**9.0**

Table 8.8.1 *Space heating demand: Reference case without sunspace*

Note: * Gives ratio glazed area (without window frames) to façade area.

In Table 8.8.2 as well as in Figure 8.8.3, the results are given for all cases simulated with the common wall glazing with a U=1.2 W/m²K. None of the sunspace cases strongly affected the apartment's space heating demand. Only case 1 – the complete double façade solution – can be considered an energy saving design. It is the most compact geometry of the studied variations. The other cases are, however, beneficial as they add semi-conditioned living space at no added heating costs (assuming the occupants don't try to heat the sunspace by leaving the door open). Increased heat losses from the enlarged windows (compared to the reference case) are offset by the sunspaces buffer effect. It is evident that the more a sunspace is inset into the building, the higher its minimum temperature will be.

cases	space heating demand	worst case sunspace temperature
	[kWh/m2a]	[°C]
reference	**22.4**	
1	**21.2**	**0.4**
2	**22.9**	**-1.3**
3	**22.8**	**1.5**
3A	**22.7**	**1.7**
4	**22.6**	**3.7**

Table 8.8.2 *Simulation results for the studied sunspace types (unit A)*

Side-by-side sunspaces also have the higher minimum sunspace temperatures and somewhat reduce the apartment space heating demand (compare type 3 to type 3A). For all variations, a fully glazed common wall tends to outperform the common wall being glazed only above windowsill height.

The least space heating demand occurs when the sunspace outside glazing and common glazing have an identical and good U-value (0.84 W/m²K). However, reducing the common wall U-value to let the sunspace be indirectly heated by the apartment results in its minimum temperatures being

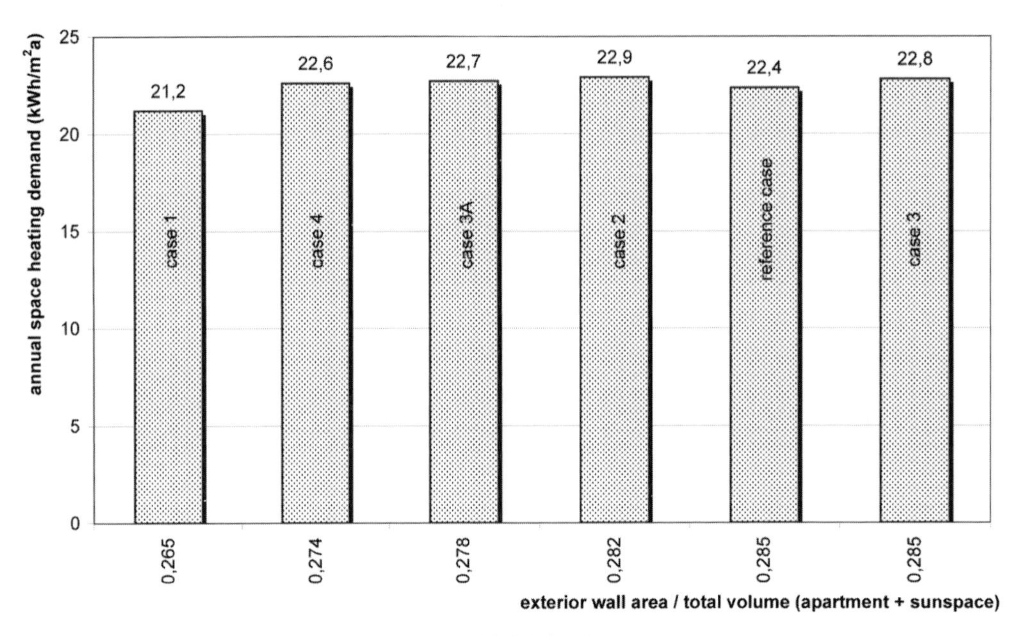

Source: Martin Reichenbach, Reinertsen Engineering AS, N 0216 Oslo, Norway

Figure 8.8.3 *Space heating demand in relation to area to volume (A/V) ratio*

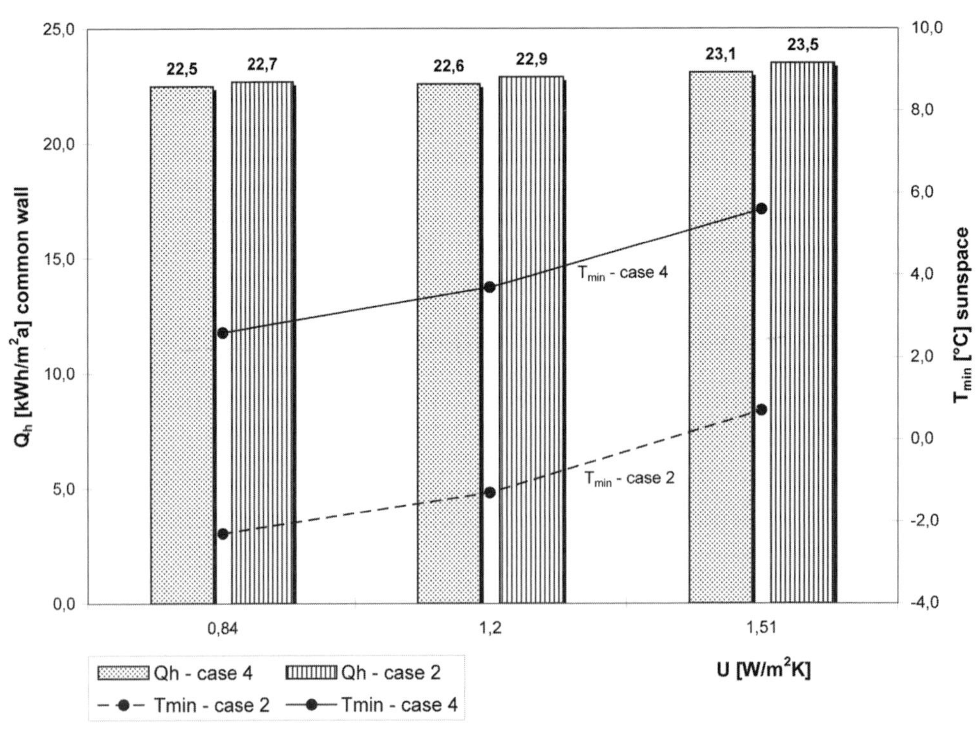

Source: Martin Reichenbach, Reinertsen Engineering AS, N 0216 Oslo, Norway

Figure 8.8.4 *Space heating demand and sunspace minimum temperature in relation to U-value of the common wall glazing*

above 0°C. This leads to only a very moderate increase in space heating demand with a great benefit of frost protection for plants. The relationship between common wall and sunspace glazing quality, and sunspace minimum temperature and space heating demand, is shown in Figure 8.8.4.

A variation exploring a poor outside glazing (U = 1.51 W/m^{2}K) in combination with the standard, highly insulating, common wall still gives reasonable space heating demand, but sunspace minimum temperatures drop below 0°C. An advantage is a lower cost and easily moveable construction – for example, folding glass panels. Here the goal is simply to extend autumn and to experience spring earlier.

8.8.3 Detailed study: Comfort in the sunspace throughout the year

Shading and ventilation strategies

The second part of this study focuses on frost protection of the sunspace in winter and overheating protection in summer. The study is based on case 3, a partly withdrawn sunspace covering 50 per cent of the southern façade and slightly projecting out from the façade. The glazed area of the apartment's remaining south façade was increased from 29 per cent to 44 per cent after a simulation showing that this does not increase heating demand. Horizontal shading elements above all windows both in the sunspace and in the exposed façade were assumed – to some extent even in winter. This models the shading caused by windows being slightly inset in the façade.

Simulations of case 3 with and without western glazing show only very small differences. The conclusion is that there is no reason to refrain from glazing to the west in the specific case under study here.

The constant parameters in the simulation model – with the noticeable exception of ventilation and shading schedules of the sunspace – remained unchanged (see section 8.8.5).

To reasonably model the temperature conditions in the sunspace, shading and ventilation schedules for the sunspace were defined both on a daily and on an annual basis. In a trial- and-error process, the parameters to ensure acceptable temperature conditions in the sunspace throughout the year were found.

Three ventilation strategies for the sunspace were studied:

1. From November to February, the sunspace is ventilated as little as possible (i.e. by infiltration only). During this period, reducing heat losses by using the sunspaces as a buffer is the goal. The sunspace can only occasionally be occupied. The sunspace need not be ventilated or shaded most of the time since it is not occupied.
2. From March to May and mid September to October, ventilation is needed to prevent the sunspace from overheating. Increased mid-day ventilation together with moderate sun shading achieved acceptable conditions. During nighttime the ventilation is kept minimal in order to preserve the sunspace temperature. During this period the sunspace serves as an effective extension to the living space of the apartment.
3. From June to mid September, the aim is to avoid overheating of the glazed space and the apartment. Increased night ventilation and solar shading proved effective. The air change rates assumed here can be achieved by natural ventilation. Daytime ventilation is reduced to avoid the intake of hot outside air off the façade.

Predicted space heating demand and sunspace temperatures

These sunspace variations result in an increase in the apartment heating load compared to the reference case by as much as 10 per cent. This indicates that the estimated performance by the early design tool approach may be too optimistic. Comfort performance is impressive, however. The sunspace maintains a mean temperature approximately 10K to 15K higher than the ambient through the winter (see Figure 8.8.5). In summer, the sunspace temperatures more closely approximate the ambient temperatures given that the sunspace is effectively ventilated and sun shaded. Indeed,

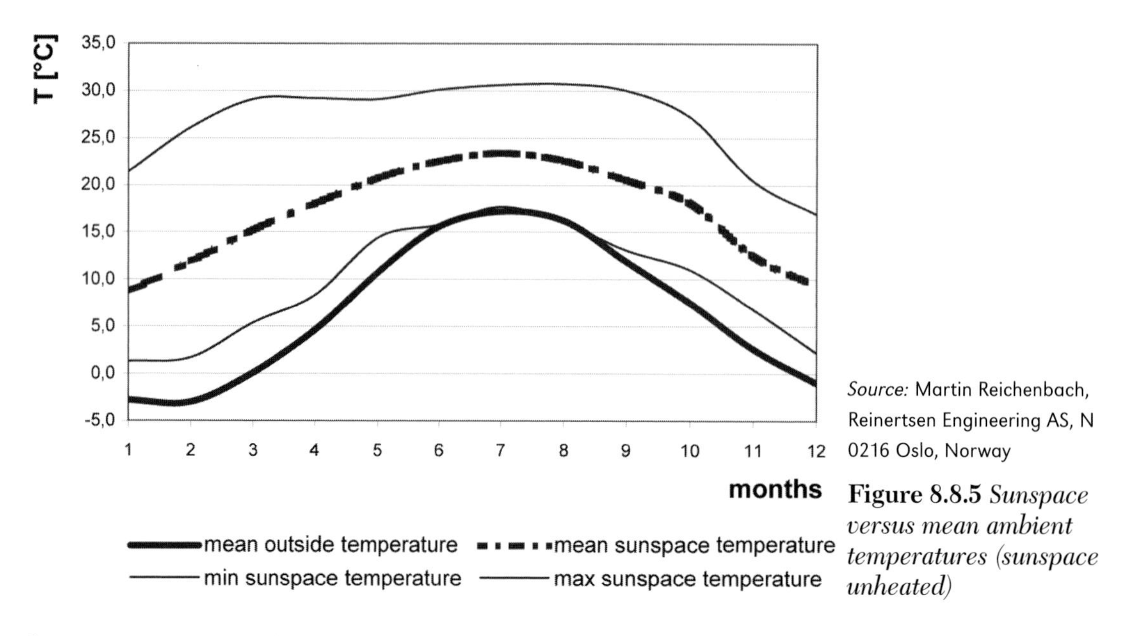

Source: Martin Reichenbach, Reinertsen Engineering AS, N 0216 Oslo, Norway

Figure 8.8.5 *Sunspace versus mean ambient temperatures (sunspace unheated)*

during 62 per cent of the year, the sunspace temperature is within the temperature range from 15°C to 28°C.

Heating the sunspace or not?

Two variations were modelled: the temperature in the sunspace free floating, or heated to a minimum of +5°C. Simulations predict that the sunspace will not cool down below zero, even without heating (T_{min} = 1.3°C in January). The heating load to safely ensure a minimum temperature of +5°C in winter is very small – in fact, as little as 0.1 kWh/m^2a (of m^2 apartment area). This justifies heating the sunspace at such a set point.

Influence of additional mass

Adding exposed brick to the 80 cm high windows sill height did not significantly affect space heating demand relative to a fully glazed common wall between the sunspace and apartment.

Results do show slightly higher minimal temperatures and minimally reduced peak loads due to

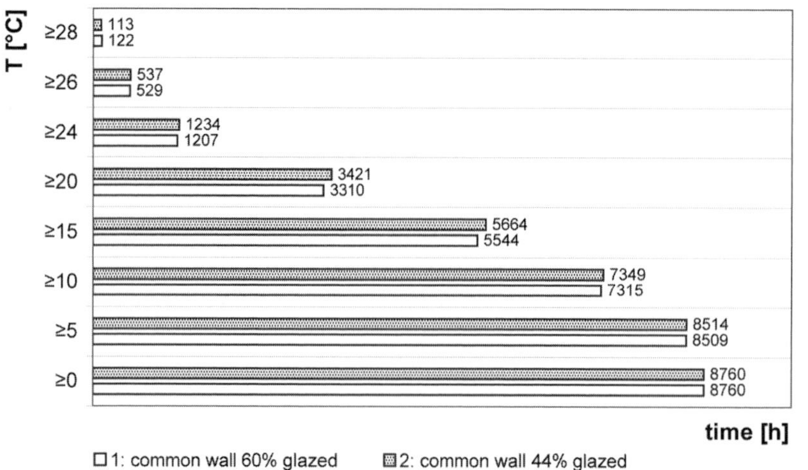

Source: Martin Reichenbach, Reinertsen Engineering AS, N 0216 Oslo, Norway

Figure 8.8.6 *Temperature frequency in the sunspace*

the increased heat storage. Figure 8.8.6 compares the temperature frequencies of the fully glazed common wall (series 1) versus the exposed brick to window sill height (series 2). There are approximately 120 more hours between 15°C and 28°C due to the thermal storage effect. In all simulations the sunspace floor is modelled as exposed concrete. Therefore, apparently there is little benefit from adding more thermal storage capacity.

Preheating ventilation air in the sunspace

In this last alternative the sunspace is used to preheat intake air for the mechanical ventilation system of the apartment. A constant air flow of approximately 11m3/h from the sunspace to the apartment from November to April was assumed. In October, approximately 22 m^3/h was assumed. This winter air change rate of around 0.5 ach for the sunspace supplies approximately 10 per cent of the hygienically necessary ventilation air in the apartment. The remaining supply air is assumed to be taken directly from the ambient via a heat exchanger with 80 per cent efficiency. In March and April, the sunspace is ventilated additionally to avoid too high temperature peaks, while in winter the air flow through the sunspace into the apartment is sufficient as sunspace ventilation. To check the comfort temperature of the ventilation, supply air from the sunspace was allowed to float above 20°C during the heating season. Temperatures occasionally exceed the maximum room temperature of 23°C in April and October, but there is no comfort problem.

This sunspace preheating of ventilation supply air does save energy without overheating the apartment or excessively cooling down the sunspace. The overall space heating demand and peak demand are reduced by this strategy. It is advisable to place the intake for the sunspace low in the exterior wall, while the extraction opening to the apartment should be high in the common wall. The ventilation air then crosses the sunspace diagonally with the maximum temperature rise. Trying to control the air exchange between sunspace and apartment by opening the door or windows in the common wall would certainly result in increased heat losses and condensation in the sunspace during the winter.

8.8.4 Lessons learned

The following lessons may be learned from these simulations:

- The construction of the sunspace should reflect its use: a comparatively poor outside glazing can be successful as long as the comfort limitations are respected. A high-performance sunspace glazing maximizes its use in winter also and provides frost protection for plants, even in this northern climate.
- With high insulation sunspace glazing, a large window area between the sunspace and apartment is possible with daylighting benefits.
- The common wall should be less well insulated than the sunspace exterior walls if a frost-free sunspace is desired.
- Sunspaces should not be deep: a long and relatively narrow sunspace along the southern façade outperforms a short and deep sunspace, and it offers better daylighting of the apartment.
- Solutions with the sunspace fully or partly inset into the building outperform sunspaces fully projecting out from the façade.
- A sunspace in this climate is not a heating source; rather, it should be viewed as adding living quality at no or little additional cost for space heating. A sunspace makes it possible to open the south façade of the apartment without significant increase in demand for heating energy or peak capacity. This, in itself, is a significant accomplishment.

8.8.5 Assumptions for the dynamic simulations

The following parameters were assumed as being constant and remained unchanged throughout all simulations with DEROB-LTH:

- reference case: apartment house, TSS 1, cold climate;
- strict orientation to south, no shading from other buildings or vegetation;
- internal loads according to TSS1, high-performance case;
- U-value, all opaque outside walls: 0.21 W/m^{2}K;
- U-value, roof: 0.24 W/m^{2}K;
- U-value apartment glazing: 0.84 W/m2K (triple glazing, one low-e coating, krypton fill);
- g-value apartment glazing: 0.52;
- frame ratio: 30 per cent;
- U-value window frames: 1.2 W/m^{2}K;
- glazed area, northern façade: 16 per cent;
- glazed area, southern façade apartment: 29 per cent;
- glazing ratio sunspace outside walls (south and west): 60 per cent;
- ventilation rate apartment: 0.45 ach;
- infiltration rate apartment: 0.05 ach;
- heat exchanger with 80 per cent efficiency;
- infiltration rate, sunspace: 0.5 ach (constant), no mechanical ventilation;
- no shading devices simulated;
- heating set point apartment: 20°C;
- no heating in sunspaces;
- cooling set point in apartment at 23°C/ 26°C in summer (to simulate the necessary additional ventilation and/or shading) and
- cooling set point in sunspace at 28°C (to avoid an overestimation of the sunspace's buffer effect in the simulation).

All sunspace types were studied through variations of the remaining relevant parameters (i.e. glazing quality and glazing ratio). Three different glass qualities were applied in the common wall between sunspace and apartment (U = 0.84/1.2/1.51) Each of these variations was studied both with an almost completely glazed wall between sunspace and apartment and with the same wall only glazed above an 80 cm high window sill. In all of these variations the sunspaces outside glazing is at least as well insulating as the glazing between sunspace and apartment (U = 0.84). In addition, one final variation studies the reversed situation with a relatively light glazing to the outside (double glazing, one low-e coating, air filled, U = 1.51) and the highly insulating glazing (U = 0.84) to the inside.

Next to a solution's space heating demand, the minimal sunspace temperature occurring in the course of one year was included in the set of simulation results in order to give a rough idea what thermal comfort in winter one could expect for the respective variation. The figure is not actually reliable in itself since the simulations in this first set are run with very simplified assumptions concerning ventilation and shading of the sunspace.

References

Kvist, H. (2005) *DEROB-LTH for MS Windows, User Manual Version 1.0–20050813*, Energy and Building Design, Lund University, Lund, Sweden, www.derob.se

9

Temperate Climates

9.1 Temperate climate design

Maria Wall and Johan Smeds

The example solutions for conservation and renewables in this chapter are compared to reference buildings that fulfil average building codes from the year 2001 in Austria, Belgium, Germany, The Netherlands, Switzerland, England and Scotland. The example solutions have been designed to fulfil the energy targets of this book while achieving superior comfort.

9.1.1 Temperate climate characteristics

The climate of Zurich was used to present the temperate climate region. Zurich, located at 47.3° N latitude, has an average yearly temperature of 9.1°C (Meteotest, 2004). A comparison with the other two climate regions studied in this book is shown in Figure 9.1.1. The countries which were used as a base for the temperate region have heating degree days ranging from 3086 Kelvin days (England/Nottingham) to 3639 Kelvin days (Glasgow/Scotland).

The temperate climate is characterized by a heating season with a moderate number of degree days, low direct solar gains in winter and a summer season with no extreme temperatures. Figure 9.1.2 shows the monthly average outdoor temperature and global solar radiation for Zurich. The monthly global solar radiation on a horizontal plane varies between 22 kWh/m^2 (December) and 173 kWh/m^2 (July).

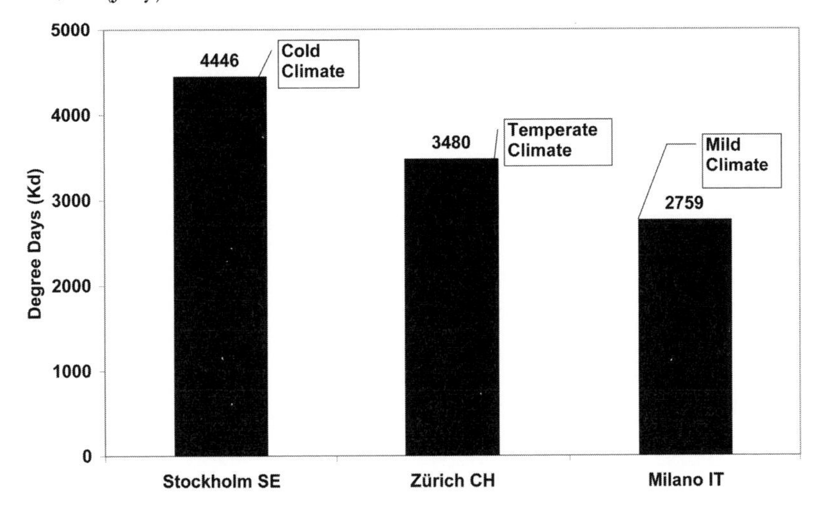

Source: Maria Wall and Johan Smeds

Figure 9.1.1 *Degree days (20/12) in cold, temperate and mild climate cities*

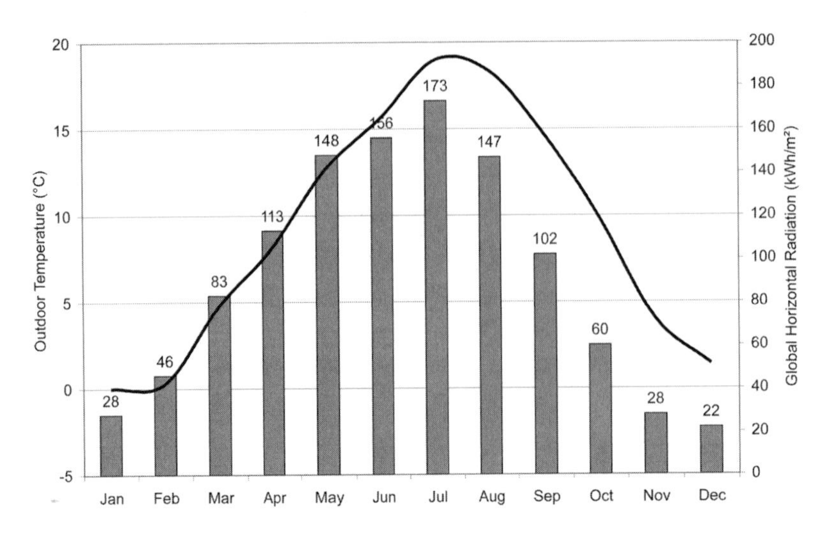

Source: Maria Wall and Johan Smeds

Figure 9.1.2 *Monthly average outdoor temperature and solar radiation (global horizontal) for Zurich*

Zurich is typical in degree days for the temperate region; but the solar gains are below the mean value for this region. The given strategies and example solutions will therefore perform even better in more sunny locations.

Due to the temperate climate, high-performance houses have been established first in this region. The reasons are that heat losses can be relatively easily and drastically reduced by means of extending the insulation thickness and including mechanical ventilation with heat recovery. On the energy supply side, there is good solar availability to cover a substantial part of the annual domestic hot water heating demand. During the heating season, passive solar gains can frequently eliminate the momentary heating demand. Extreme measures are not necessary in this temperate climate. Many paths are available to achieve high performance, each with benefits offsetting disadvantages in different ways. This allows cost-effective solutions adapted to the location and gives some freedom in architectural design.

The main design goals, a low space heating demand and primary energy demand, can be easily reached in buildings with a low envelope area to volume ratio (A/V) (i.e. row houses and apartment buildings). For detached single family houses, a greater effort is required to achieve such performance; but it is also possible. Critical, however, is supplying the needed greater heating capacity. This is mainly a problem when the heat is supplied by a small heat pump connected to the exhaust air, and the heat is delivered by the ventilation air.

The challenge is therefore not to achieve the energy goals, but rather to keep the required heating capacity low and to do so economically. The choice of components should be done with respect to lifetime and simple operation. Environmental aspects of materials and production should be taken into account.

9.1.2 Single family house

Single family reference design

The reference single family house with 150 m^2 floor area has an average U-value of the building envelope of 0.47 W/m^{2}K. This reflects the average insulation standard according to the building code 2001 for the above mentioned countries. A condensing gas boiler for DHW and space heating is assumed. The combined ventilation and infiltration rate is 0.6 ach and no heat recovery is assumed. This results in a space heating demand of 70.4 kWh/m^2 a according to simulations with SCIAQ Pro (ProgramByggerne, 2004). The DHW heating is 21 kWh/m^{2}a. Finally, household electricity is assumed to be 29 kWh/m^{2}a.

The total energy use for DHW and space heating, mechanical systems and system losses is approximately 104 kWh/m^{2}a. Using electricity and gas as the energy sources results in a non-renewable primary energy demand of 124 kWh/m^{2}a and CO_2 equivalent emissions of 26.5 kg/m^{2}a.

Single family house example solutions

The space heating targets of 20 kWh/m^{2}a for strategy 1 (energy conservation) and 25 kWh/m^{2}a for strategy 2 (renewable energy) mean that the energy demand for the single family house must be very small. Different solutions are, of course, possible depending on available energy supply systems and the local energy costs. All solutions have mechanical ventilation with 80 per cent heat recovery.

The examples are as follows.

CONSERVATION: SOLUTIONS 1A AND 1B

U-value of the whole building:	0.21 W/m^{2}K
Space heating demand:	19.8 kWh/m^{2}a
Heating distribution:	hot water radiant heating
Heating system:	condensing gas boiler (solution 1a) or wood pellet stove (solution 1b)
DHW heating system:	solar collectors and condensing gas boiler (solution 1a), or solar collectors and wood pellet stove (80%) and electricity (20%) (solution 1b)

RENEWABLE ENERGY: SOLUTION 2

U-value of the whole building:	0.25 W/m^{2}K
Space heating demand:	25.0 kWh/m^{2}a
Heating distribution:	hot water radiant heating
DHW and space heating system:	solar combi-system and biomass boiler

In Figure 9.1.3 the total energy demand, delivered energy and non-renewable primary energy demand are shown for the different solutions compared to the reference single family house. The different solutions reduce total energy demand to 50 per cent or less than that of the reference house. The demand for non-renewable primary energy is reduced to a mere 15 per cent to 24 per cent of the reference house. Solution 2 achieves a very low primary energy demand due to the high supply of renewable energy from solar and biomass. The CO_2 equivalent emissions are shown in Figure 9.1.4.

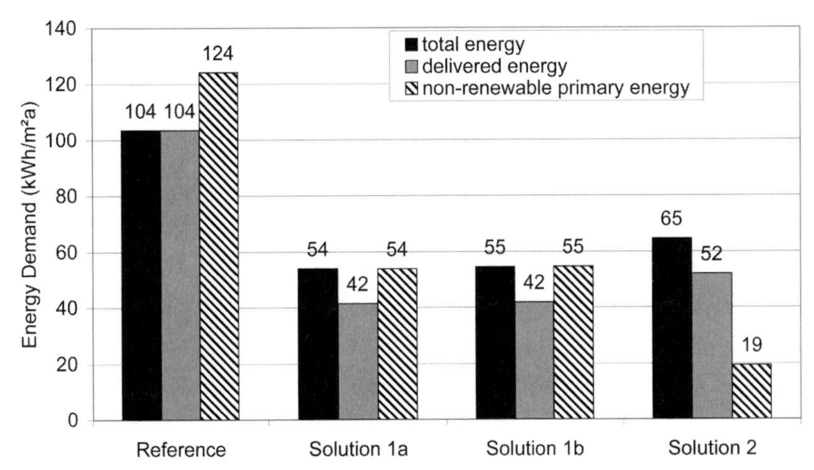

Source: Maria Wall and Johan Smeds

Figure 9.1.3 *Overview of the total energy use, the delivered energy and the non-renewable primary energy demand for the single family houses; the reference building has a condensing gas boiler for heating*

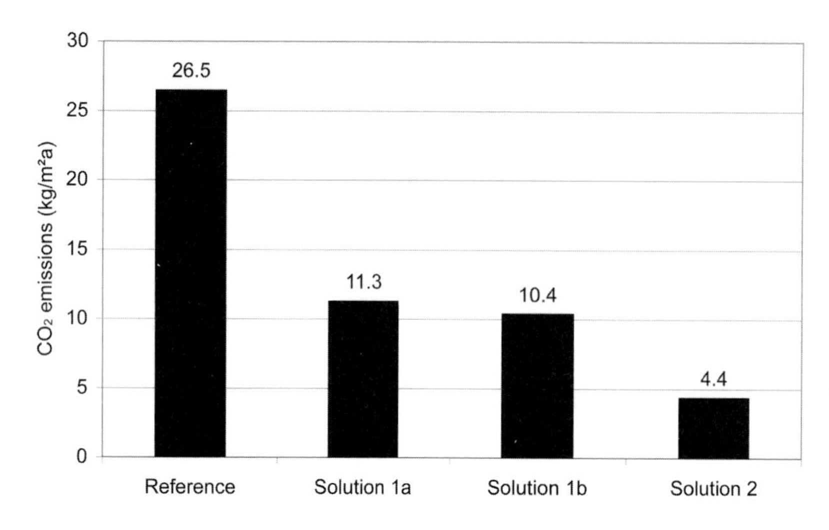

Source: Maria Wall and Johan Smeds

Figure 9.1.4 *Overview of the CO_2 equivalent emissions for the single family houses; the reference building has a condensing gas boiler for heating*

The CO_2 emissions are only 4.4 kg/m²a for solution 2, which is only 17 per cent of the emissions from the reference house (26.5 kg/m²a).

The results show that the choice of heating system drastically affects the primary energy demand and CO_2 equivalent emissions. Large reductions in energy use and emissions are possible with conservation measures.

9.1.3 Row houses

Row house reference design

A row of six houses is assumed, each 120 m². The reference row house has an average U-value of the building envelope of 0.55 W/m²K. A condensing gas boiler for DHW and space heating is assumed. The combined ventilation and infiltration rate is 0.6 ach and no heat recovery is assumed. The space heating demand for this reference house is 60.8 kWh/m² according to TRNSYS simulations. The DHW heating is 25.6 kWh/m²a. Household electricity is assumed to be 36 kWh/m²a for the reference house.

The total energy use for DHW and space heating, mechanical systems and system losses is approximately 100 kWh/m²a. Energy supply with electricity and gas results in a non-renewable primary energy demand of 120 kWh/m²a and CO_2 equivalent emissions of 26 kg/m²a.

Row house example solutions

The space heating targets of 15 kWh/m²a for strategy 1 (energy conservation) and 20 kWh/m²a for strategy 2 (renewable energy) require that the energy demand be kept very small. An economical and efficient option for a row of houses is therefore to have a common heating system. This solution is acceptable in some regions. All solutions shown have mechanical ventilation with 80 per cent heat recovery. The main examples are as follows.

Figure 9.1.5 shows the total energy use, delivered energy and the use of non-renewable primary energy for the reference row house and the example solutions. Solution 1b with an outdoor air heat pump has the lowest delivered energy demand. However, the primary energy demand will be slightly higher than for solutions 1a and 2, which profit from using solar energy. The energy demand is overall half of what is needed for the reference row house. The CO_2 equivalent emissions are approx 40% of the reference house (see Figure 9.1.6).

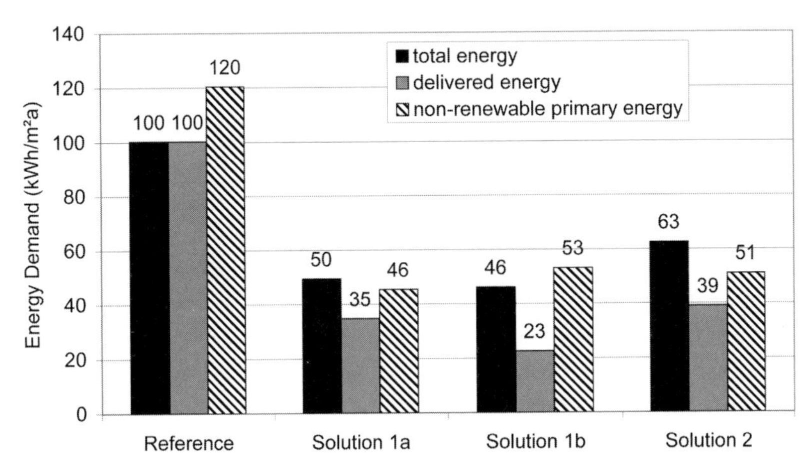

Source: Maria Wall and Johan Smeds

Figure 9.1.5 *Overview of the total energy use, the delivered energy and the use of non-renewable primary energy for the row houses; the reference house has a condensing gas boiler*

CONSERVATION: SOLUTION 1A

U-value of the whole building:	0.32 W/m^2K
Space heating demand:	15 kWh/m^2a
Heating distribution:	hot water radiant heating
Heating system:	oil or gas boiler
DHW system:	solar collectors and oil or gas boiler

CONSERVATION: SOLUTION 1B

U-value of the whole building:	0.32 W/m^2K
Space heating demand:	15 kWh/m^2a
Heating distribution:	hot water radiant heating
DHW and space heating system:	ambient air heat pump

RENEWABLE ENERGY: SOLUTION 2

U-value of the whole building:	0.38 W/m^2K
Space heating demand:	19.4 kWh/m^2a
Heating distribution:	hot water radiant heating, fresh air heating
DHW and space heating system:	solar combi-system and gas boiler

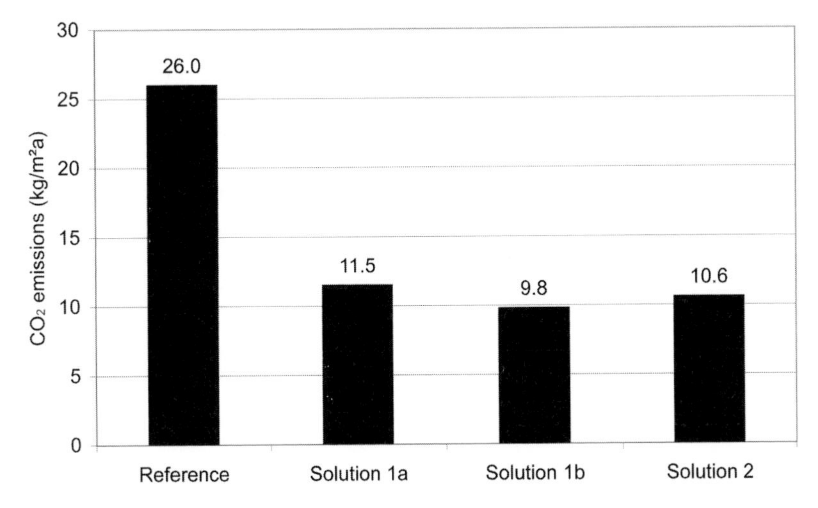

Source: Maria Wall and Johan Smeds

Figure 9.1.6 *Overview of the CO_2 equivalent emissions for the row houses; the reference house has a condensing gas boiler*

9.1.4 Apartment building

Apartment building reference design

The apartment building has a floor area of 1600 m^2. The reference apartment building has an average U-value of the building envelope of 0.60 W/m^2K. A condensing gas boiler for DHW and space heating is assumed. The combined ventilation and infiltration rate is 0.6 ach and no heat recovery is assumed. The space heating demand for this reference house is 56 kWh/m^2a according to calculations based on EN832 (Heidt, 1999). The DHW heating is 23.1 kWh/m^2a. Household electricity is assumed to be 38 kWh/m^2a for the reference apartment with two adults and one child.

The total energy use for DHW and space heating, mechanical systems and system losses are approximately 98 kWh/m^2a for the reference apartment building. Delivered energy is the same since no renewable energy sources are used. Using electricity and gas as the energy sources results in a non-renewable primary energy demand of 118 kWh/m^2a and CO_2 equivalent emissions of 25 kg/m^2a.

Apartment building example solutions

The space heating targets of 15 kWh/m^2a for strategy 1 (energy conservation) and 20 kWh/m^2a for strategy 2 (renewable energy) are not difficult to achieve for an apartment building due to the favourable relation between envelope area and building volume. The solutions have mechanical ventilation with 80 per cent heat recovery. The examples are as follows.

CONSERVATION: SOLUTION 1	
U-value of the whole building:	0.40 W/m^2K
Space heating demand:	10 kWh/m^2a
Heating distribution:	hot water radiant heating
Heating system:	condensing gas boiler
DHW heating:	solar collectors and condensing gas boiler
RENEWABLE ENERGY: SOLUTION 2	
U-value of the whole building:	0.40 W/m^2K
Space heating demand:	10 kWh/m^2a
Heating distribution:	hot water radiant heating
Heating system:	biomass boiler
DHW heating:	solar collectors and biomass boiler

Even without an extremely insulated building envelope, the space heating demand will be low for this type of building. For apartment buildings, it is easy to reduce the energy demand only with modest measures. In Figure 9.1.7, the total energy use, delivered energy and the use of non-renewable primary energy are shown for the reference apartment building and the example solutions. The solutions presented only need approximately 20 per cent of the delivered energy compared to the reference building. The non-renewable primary energy demand can be reduced even further. The CO_2 equivalent emissions are approximately 12 per cent to 25 per cent of the emissions from the reference building (see Figure 9.1.8).

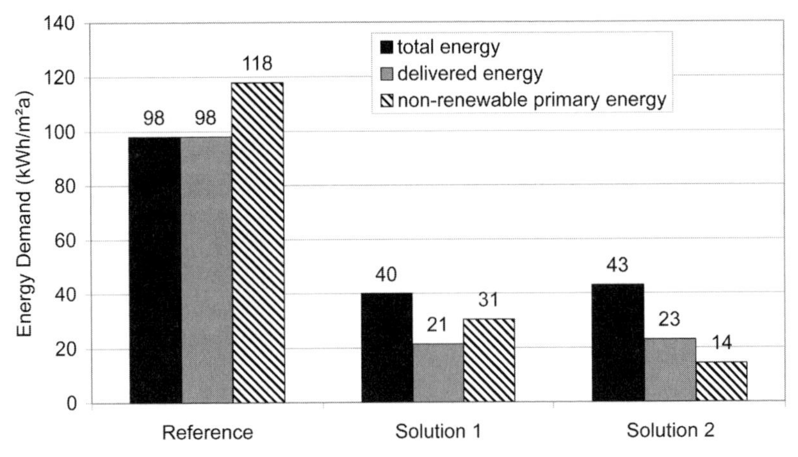

Source: Maria Wall and Johan Smeds

Figure 9.1.7 *Overview of the total energy demand, the delivered energy and the non-renewable primary energy demand for the apartment buildings; the reference building uses a condensing gas boiler*

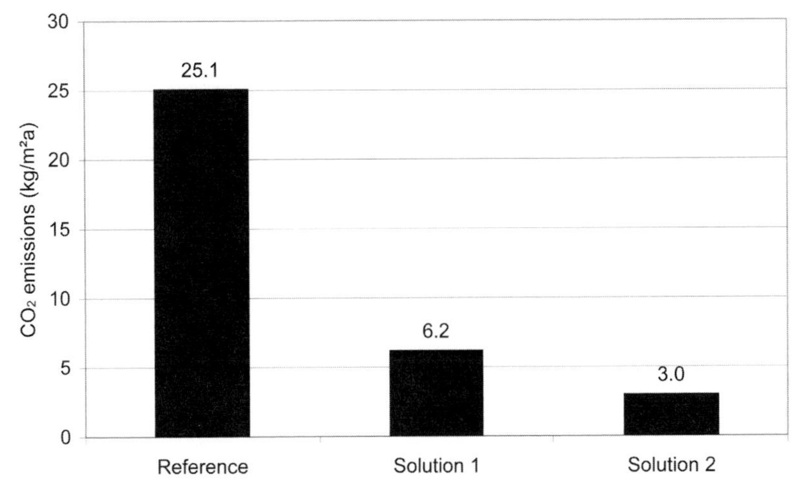

Source: Maria Wall and Johan Smeds

Figure 9.1.8 *Overview of the CO_2 emissions for the apartment buildings; the reference building uses a condensing gas boiler*

References

Heidt, F. D. (1999) *Bilanz Berechnungswerkzeug, NESA-Datenbank*, Fachgebiet Bauphysik und Solarenergie, Universität-GH Siegen, Siegen, Germany

Kvist, H. (2005) *DEROB-LTH for MS Windows, User Manual Version 1.0–20050813*, Energy and Building Design, Lund University, Lund, Sweden, www.derob.se

Meteotest (2004) *Meteonorm 5.0 – Global Meteorological Database for Solar Energy and Applied Meteorology*, Bern, Switzerland, www.meteotest.ch

ProgramByggerne (2004) *ProgramByggerne ANS, SCIAQ Pro 2.0 – Simulation of Climate and Indoor Air Quality: A Multizone Dynamic Building Simulation Program*, www.programbyggerne.no

TRNSYS (2005) *A Transient System Simulation Program*, Solar Energy Laboratory, University of Wisconsin, Madison, WI

9.2 Single family house in the Temperate Climate Conservation Strategy

Tor Helge Dokka

Table 9.2.1 *Targets for single family house in the Temperate Climate Conservation Strategy*

	Targets
Space heating	20 kWh/m²a
Non-renewable primary energy:	
(space heating + water heating + electricity for mechanical systems)	60 kWh/m²a

This section presents a solution for the single family house in the temperate climate. As a reference for the temperate climate, the city of Zurich is considered. Climatic data are taken from the meteorological software database Meteonorm (Meteotest, 2004). The solution is based on energy conservation, minimizing the heat losses of the building. A balanced mechanical ventilation system with heat recovery is used to reduce the ventilation losses.

9.2.1 Solution 1a: Conservation with a condensing gas boiler and solar DHW and Solution 1b: Conservation with a wood pellet stove and solar DHW

Building envelope

The opaque part of the building envelope is a wooden lightweight frame with mineral wool. U-values and areas are shown in Table 9.2.2. The floor construction is a slab on the ground with expanded polystyrene insulation (EPS). The windows have a frame area ratio of 30 per cent, triple glazing, one low-e coating and are filled with krypton gas.

Table 9.2.2 *Building envelope components*

Component	U-Value (W/m²K)	Area (m²)
Walls	0.20	113.6
Roof	0.13	129.7
Floor (excluding ground)	0.17	96.4
Windows (frame + glass)	0.92	22.0
Window frame	1.20	–
Window glass	0.80	–
Entrance door	1.00	2.0
Whole building envelope	0.21	–

Mechanical systems

A balanced mechanical ventilation system with 80 per cent heat recovery and a bypass for summer ventilation is used. A condensing gas boiler with a water-based radiator system is employed for space heating.

Approximately 50 per cent of the DHW demand is supplied by a flat-plate solar collector and the rest is covered with the condensing gas boiler. The south-facing solar collectors are mounted at an angle of 40° and a storage tank of 0.4 m^3 is assumed. A circulation pump of 40 W operates the solar collector loop for about 2000 hours during a year. The corresponding annual electricity needed for the pump per living area is therefore approximately 0.5 kWh/m^2a.

An alternative energy system can be a local wood pellet stove covering 80 per cent of the space heating demand. The rest is then covered with electricity (electric radiators). Fifty per cent of the DHW demand is covered by the same solar collector system described above, and the rest is covered by electricity. Both of these energy systems meet the primary energy target of 60 kWh/m^2a.

Energy performance

Space heating demand: simulation results from the Norwegian energy simulation software SCIAQ Pro (ProgramByggerne, 2004) give the monthly space heating demand of the building as shown in Figure 9.2.1. The heating season extends approximately from November to March:

- space heating demand: 2960 kWh/a (19.8 kWh/m^{2}a);
- heating set point: 20°C;
- ventilation rate: 0.45 ach;
- infiltration rate: 0.05 ach; and
- heat recovery: 80 per cent efficiency.

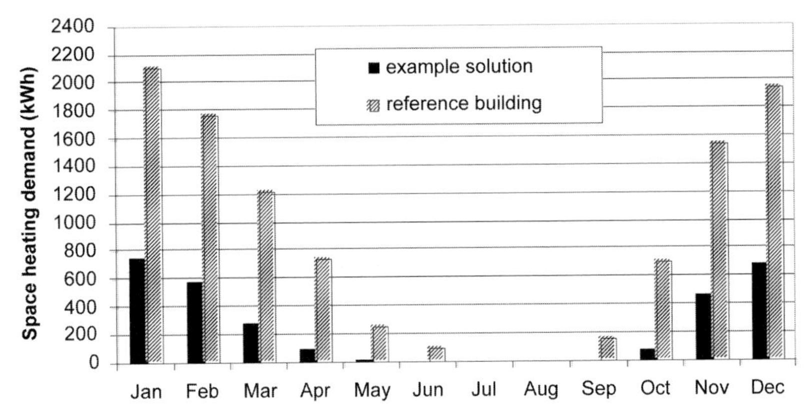

Source: Tor Helge Dokka

Figure 9.2.1 *Monthly space heating demand for the proposed solution (19.8 kWh/m^{2}a) and the reference building (70.4 kWh/m^{2}a)*

Hourly loads of the heating system are calculated with results from SCIAQ Pro simulations without direct solar radiation. The annual peak load is approximately 2200 W or 14.5 W/m^2 and occurs in January. The annual peak load for the reference building is approximately 5300 W or 35 W/m^2 and occurs in January.

Domestic hot water demand: the net DHW heat demand is approximately 3150 kWh/a (21 kWh/m^{2}a). Two adults and two children live in a typical single family detached house. The average DHW consumption per person is 40 litres per person per day of 55°C water. Consequently, the single family house consumes 160 litres of DHW per day. The average temperature of the cold water over the year is set to 8.5°C. The temperature of the hot water was set to 50°C at the faucet. The on/off temperature set points for the thermostat in the tank were set to 55°C/57°C.

System losses: the system losses consist mainly of losses from the hot water storage tank, but also from pipe losses in the distribution system and conversion losses in the boiler. The losses provided by the solar collector simulation program are the tank losses, which include the total heat losses through the tank wall, base and cover, and the connection losses. The tank losses become larger with an increase in tank size and/or an increase in solar collector area. The tank losses for a solar collector system with 4 m^2 of collectors and a 400 litre tank are about 630 kWh per year or 4.2 kWh/m^{2}a (living area). The annual coefficient of performance (COP) for the wood pellet stove is 80 per cent, resulting in conversion losses of 5.0 kWh/m^{2}a. The annual COP of the condensing gas boiler including distribution losses in the water-based radiant heating system is set to 90 per cent.

Household electricity: household electricity is set to 2500 kWh or 16.6 kWh/m^2 (two adults and two children). The primary energy target does not include household electricity since this factor is very much dependent on the occupants' behaviour.

Total energy use: the calculated total energy use for DHW, space heating, system losses, conversion losses and electricity for fans and pumps is approximately 8300 kWh/a. Including the assumed household electricity, the total energy use is 10,800 kWh/a.

Table 9.2.3 *Total energy use*

Total energy use	kWh/m²a	kWh/a
Space heating	19.8	2964
DHW heating	21.0	3150
System losses	4.2	630
Conversion losses	5.0	743
Electricity, ventilation fans	5.0	750
Electricity, circulation pump	0.5	75
Total	55.4	8312

Non-renewable primary energy demand and CO_2 emissions: the computer program Polysun has been used to calculate the remaining amount of energy use for DHW and space heating, taking solar gains into account, and to calculate tank and pipe losses. The solar DHW system consists of 4.0 m² flat-plate collectors tilted 40° and a 400 litre storage tank. The results for the condensing gas solution for total energy use, delivered energy, non-renewable primary energy and CO_2 emissions for this system are given in Table 9.2.3.

Results for the alternative solution with the wood pellet stove is given in Table 9.2.4.

Table 9.2.4 *Total energy use, non-renewable primary energy demand and CO_2 emissions for the solar domestic hot water (DHW) system with condensing gas boiler*

Net Energy (kWh/m²a)		Total Energy Use (kWh/m²a)				Delivered energy (kWh/m²a)		Non renewable primary energy factor (-)	Non renewable primary energy (kWh/m²a)	CO_2 factor (kg/kWh)	CO_2 equivalent emissions (kg/m²a)
		Energy use		Energy source							
Mechanical systems	5.5	Mechanical systems	5.5	Electricity	5.5	Electricity	5.5	2.35	12.9	0.43	2.4
Space heating	19.8	Space heating	19.8	Gas	36.0	Gas	36.0	1.14	41.0	0.25	8.9
DHW	21.0	DHW	21.0								
		Tank and circulation	4.2								
		Conversion losses	3.6	Solar	12.6						
Total	46.3		54.1		54.1		41.5		54.0		11.3

Source: Tor Helge Dokka

Table 9.2.5 *Total energy use, non-renewable primary energy demand and CO_2 emissions for the solar DHW system and wood pellet stove*

Net Energy (kWh/m²a)		Total Energy Use (kWh/m²a)				Delivered energy (kWh/m²a)		Non renewable primary energy factor (-)	Non renewable primary energy (kWh/m²a)	CO_2 factor (kg/kWh)	CO_2 equivalent emissions (kg/m²a)
		Energy use		Energy source							
Mechanical systems	5.5	Mechanical systems	5.5	Electricity	22.1	Electricity	22.1	2.35	51.9	0.43	9.5
Space heating	19.8	Space heating	19.8	Wood pellets	19.8	Wood pellets	19.8	0.14	2.8	0.04	0.9
DHW	21.0	DHW	21.0								
		Tank and circulation	4.2								
		Conversion losses	4.0	Solar	12.6						
Total	46.3		54.5		54.5		41.9		54.7		10.4

9.2.2 Summer comfort

Two strategies are tested by simulations in SCIAQ Pro to achieve acceptable thermal comfort in summer. The first strategy uses a bypass of the heat recovery with enforced ventilation at 1.5 ach between 7 pm and 6 am from June to August (ventilation strategy). External fixed shading for windows facing west, south and east is used, with 50 per cent transmittance and 10 per cent absorption. The second strategy uses the same ventilation schedule, but south-facing windows have automatically controlled external Venetian blinds with only 15% transmittance and 5% absorption (solar shading and ventilation strategy). The automatic Venetian blinds are activated when the external solar irradiation exceeds 150 W/m^2.

As shown in Figures 9.2.2 and 9.2.3, with only the ventilation strategy, the temperature exceeds 26°C during many hours of the year, with a maximum temperature of 33.5°C in July. One way of characterizing the summer comfort is by the sum of degree hours for temperatures exceeding 26°C. The amount of overheating in this case is 1989 Kh/a.

Results from simulation of the combined solar shading and ventilation strategy are shown in Figures 9.2.4 and 9.2.5. The maximum temperature, occurring in July, is reduced to 32.3°C. The degree hours above 26°C is reduced to 996 Kh/a, but is still high.

To achieve even better thermal comfort in summer, other passive cooling techniques such as effective cross-ventilation in the night (night ventilation) and adding some thermal mass (concrete or masonry construction) have to be considered. These measures are not studied in this chapter.

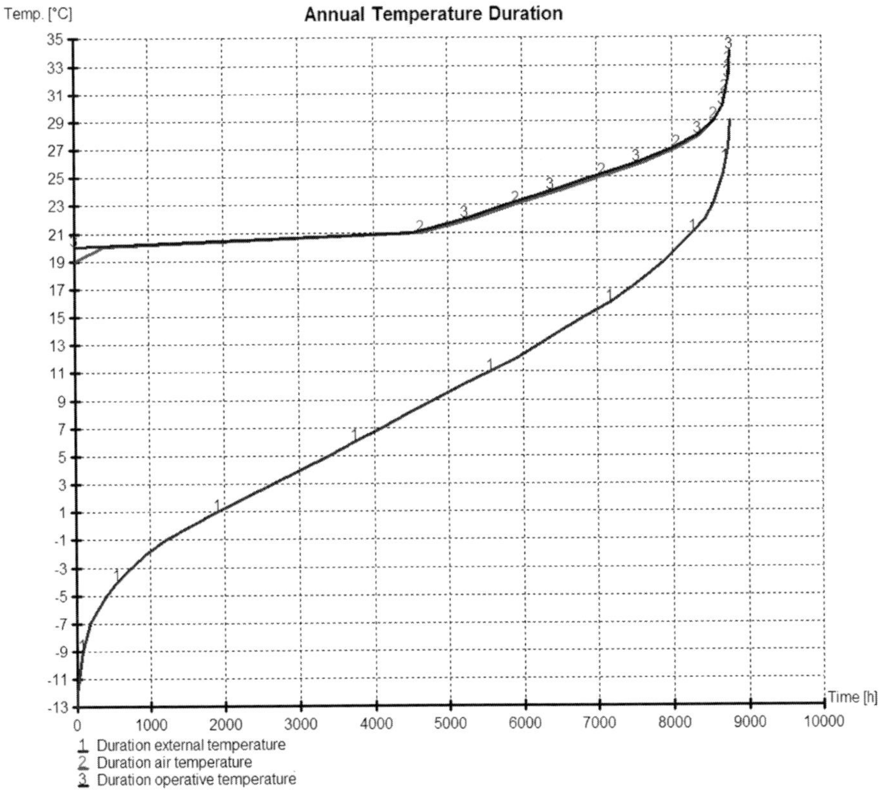

Source: Tor Helge Dokka

Figure 9.2.2 *The annual temperature duration with only the ventilation strategy*

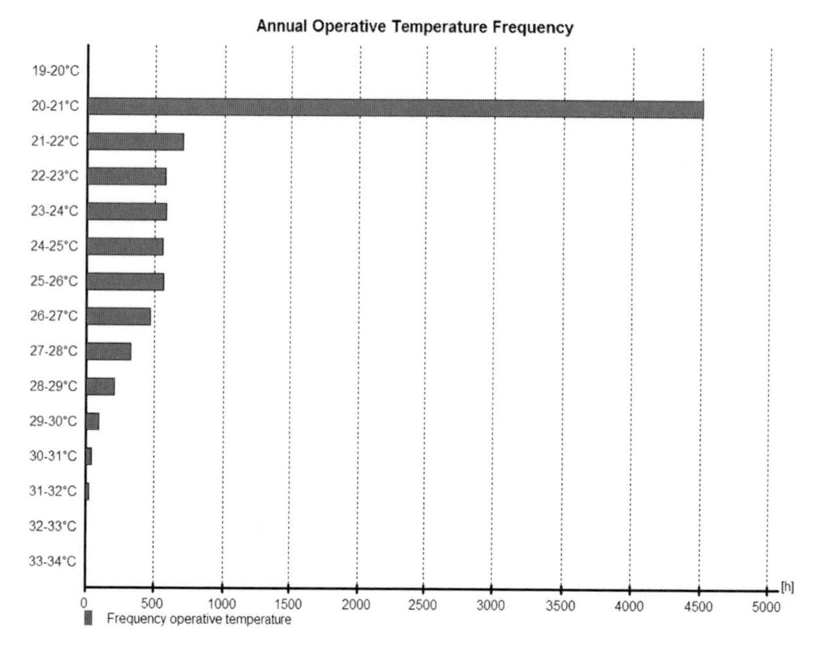

Source: Tor Helge Dokka

Figure 9.2.3 *The annual temperature frequency with only the ventilation strategy*

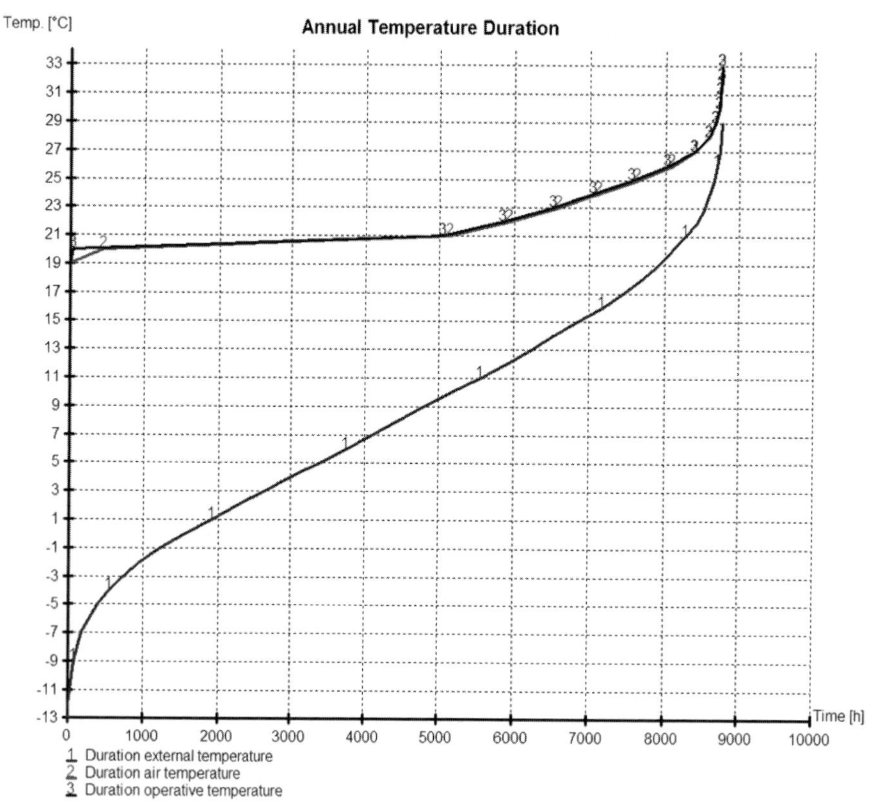

Source: Tor Helge Dokka

Figure 9.2.4 *The annual temperature duration with both the solar shading and the ventilation strategy*

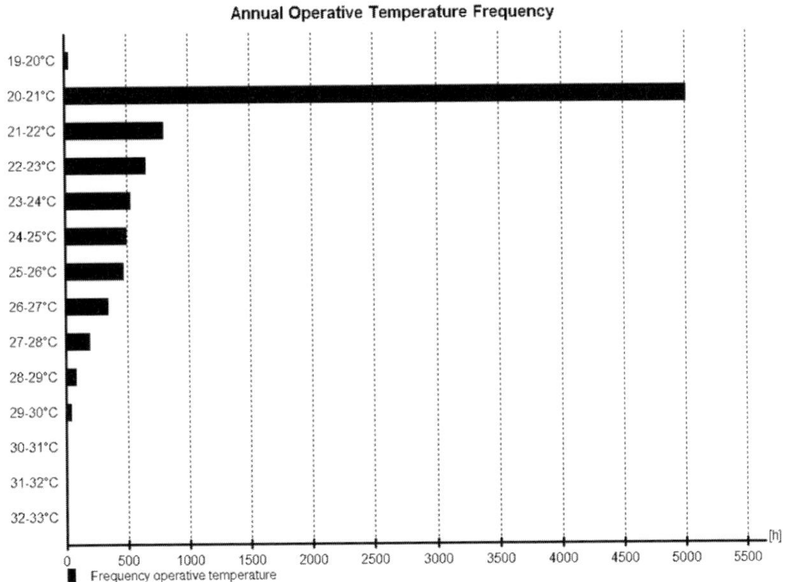

Source: Tor Helge Dokka

Figure 9.2.5 *The annual temperature frequency with both the solar shading and the ventilation strategy*

9.2.3 Sensitivity analysis

In the conservation strategy, the most important measure is to reduce the heat losses of the building and, secondly, to increase and utilize passive solar gains. A sensitivity study with respect to the most important parameters that affect the heating demand is presented in the following sections.

Opaque building envelope

The space heating demand is calculated for five different standards of the opaque building envelope: poor, fair, good, very good and excellent. The U-values for the different building elements and the corresponding mean U-value of the building envelope are given in Table 9.2.6. The space heating demand for the different building envelope standards is given in Figure 9.2.6. All other inputs influencing the heating demand, as described earlier in this chapter, are kept constant.

As expected, the mean U-value of the building envelope influences the space heating demand significantly. In fact, down to approximately a mean U-value of 0.15 W/m^{2}K, a reduction in the mean U-value of 0.1 W/m^{2}K gives a reduction in space heating demand of about 15 kWh/m^{2}a. A further reduction in U-value from 0.15 W/m^{2}K gives less reduction in space heating demand.

Usually, the extra cost for reducing the U-value is lowest for the roof and floor construction, and highest for the external walls. But this may vary significantly for different construction and building types, and has to be optimized for each project in order to achieve a cost-effective low energy building.

Table 9.2.6 *Mean U-value for different standards of the building shell*

Building shell standard	U-value wall (W/m^{2}K)	U-value floor (W/m^{2}K)	U-value roof (W/m^{2}K)	Mean U-value (W/m^{2}K)
Poor	0.40	0.30	0.30	0.333
Fair	0.30	0.20	0.20	0.233
Good	0.20	0.17	0.13	0.165
Very good	0.15	0.13	0.11	0.129
Excellent	0.10	0.10	0.08	0.092

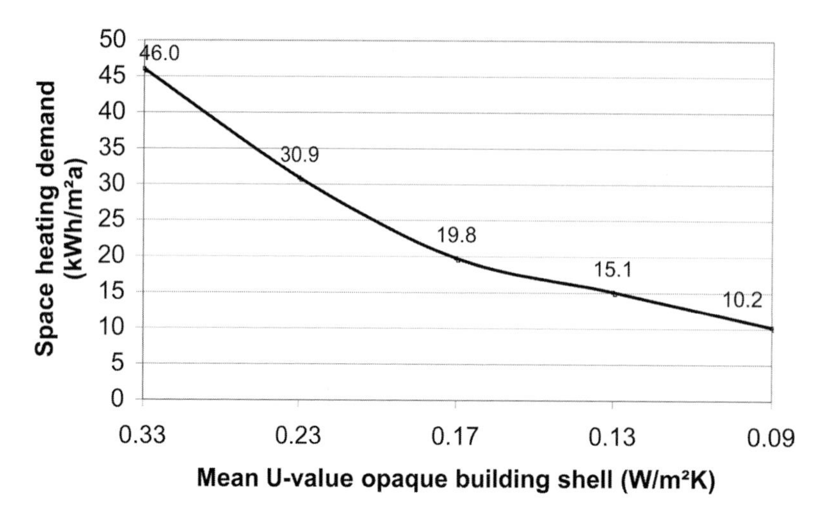

Source: Tor Helge Dokka

Figure 9.2.6 *The influence of the mean U-value of the opaque building envelope on the space heating demand*

Window type

To investigate the effect of different window constructions on the space heating demand, four different window types are considered, as shown in Table 9.2.7. Highly insulating glazing reduces the U-value significantly, but also reduces the passive solar gain due to the lower g-value.

The space heating demand is reduced considerably by going from the double glazing to the triple glazing with U-value 0.92 W/m^2K. A further reduction in U-value to 0.75 W/m^2K gives an insignificant reduction in space heating demand in this case. It has to be mentioned that this house has a large share of the windows facing south, making the higher g-value of the conservation strategy window (U = 0.92) more significant than in other cases with more uniform window distribution. This result is also climate dependent and cannot directly be extended to other climate conditions.

Table 9.2.7 *Calculated space heating demand for different window constructions; the triple glazing with one low-e coating and krypton is used in the solution*

Window construction (kWh/m²a)	U-value (W/m²K)	g-value (–)	Space heating demand
Double glazing	2.00	0.73	29.9
Double glazing, one low-e, argon	1.40	0.64	24.0
Triple glazing, one low-e, krypton	0.92	0.55	19.8
Triple glazing, two low-e, krypton, insulated frame	0.75	0.47	19.0

Window orientation

A passive solar gain strategy is to have a large portion of the windows facing in a southerly direction. Five different window distributions, ranging from extreme southerly to extreme northerly orientation, have been simulated. Results are shown in Table 9.2.8. The influence on the space heating demand is limited; the difference between the two extreme distributions of windows is below 4 kWh/m^2a. This small influence on the heating demand is mostly due to the fact that the low heat losses using the conservation strategy make the heating season short and therefore limit the usefulness of the solar heat gain in autumn and spring. A building with standard thermal insulation will have a significant heating demand in spring and autumn, and can therefore benefit more from a passive solar strategy.

When considering a passive solar strategy, the risk of excessive temperatures has to be studied and necessary design measures have to be taken. Effective external solar shading is often necessary, together with effective airing.

Table 9.2.8 *Calculated space heating demand for different window distributions*

Window distribution	Window area (m^2) East	South	West	North	Space heating demand (kWh/m^2a)
Extreme southerly distribution	3	15	3	1	18.8
Southerly distribution	3	9	9	1	19.8
Normal distribution	3	8	3	8	20.6
Northerly distribution	3	1	9	9	21.9
Extreme northerly distribution	3	1	3	15	22.6

Air tightness

Air tightness is normally measured by a pressurization test, with a pressure difference of 50 Pa over the building fabric. This measured value is denoted N_{50} and may, in practice, vary from 20 ach (very leaky building) to 0.2 ach (extremely air-tight building). The Norwegian building code requires that the air tightness for a single family house should be below 4.0 ach at 50 Pa, and the Passivhaus standard (the Passive House Institute) requires a value below 0.6 ach. A low N_{50} value gives low air infiltration, and a large value gives high air infiltration and high infiltration losses. In a high-performance building the necessary air flow rate is provided by the ventilation system, and a high air infiltration only leads to unnecessary heat losses.

Five different levels of air tightness, with corresponding infiltration rate calculated according to the EN 832 standard, have been simulated. The results from the simulations, shown in Table 9.2.9, demonstrate that good air tightness is crucial for a high-performance building. However, the difference in space heating demand between a good air tightness level (1.0 ach) and the Passivhaus standard (0.6 ach) is rather small (1.7 kWh/m^2a).

Table 9.2.9 *Calculated space heating demand for different air tightness standards*

Air tightness standard	N_{50} (ach)	Infiltration (ach)	Space heating demand (kWh/m^2a)
Leaky	6.0	0.42	43.3
Normal air tightness	4.0	0.28	34.0
Good air tightness	2.0	0.14	25.2
Very good air tightness	1.0	0.07	20.9
Extremely air tight (Passivhaus standard)	0.6	0.04	19.2

Ventilation heat recovery

The efficiency of a ventilation heat recovery unit is normally measured by the temperature efficiency on the exhaust air side. The efficiency should be given as an annual mean value, taking into consideration energy use for defrosting the exchanger. Typical efficiencies of ventilation heat exchangers range from 50 per cent to nearly 90 per cent. Simulated space heating demand for different heat exchangers with corresponding typical efficiencies are shown in Table 9.2.10.

The space heating demand is approximately reduced by 2.5 kWh/m^2 for each 10 per cent increase in heat exchanger efficiency. Energy efficient heat exchangers are often a profitable measure since the cost of high efficiency heat exchangers are normally just slightly higher than less effective exchangers.

Table 9.2.10 *Calculated space heating demand for different heat exchangers*

Heat exchangers	Temperature efficiency (%)	Space heating demand (kWh/m^2a)
Cross-flow heat exchanger	55	26.2
Counter-flow heat exchanger	70	22.3
Rotary wheel exchanger or optimized counter-flow exchanger	80	19.8
Optimized rotary wheel exchanger	88	17.8

Super conservation level

To see how low a space heating demand it is possible to achieve, a 'super' conservation level is simulated based on the best available solution from the above sensitivity analyses. The space heating demand, with the solutions given in Table 9.2.11, will then be extremely low: 965 kWh/a (6.4 kWh/m^{2}a) (see Figure 9.2.7). In fact, this very low heating demand can be covered by electricity and still meet the primary energy target of 60 kWh/m^{2}a (see Table 9.2.12).

Table 9.2.11 *Description of the super conservation level*

Optimized building envelope and ventilation	U-Value (W/m^{2}K)	g-value (-)	Infiltration rate (ach)	Heat recovery efficiency (%)
Walls	0.10			
Floor	0.10			
Roof	0.08			
Windows: triple glazing, two low-e coatings, krypton, insulated frame *Distribution:* east 3 m^2, south 15 m^2, west 3 m^2, north 1 m^2	0.75	0.47		
Air tightness: N50 = 0.6 ach			0.04	
Heat exchanger: optimized rotary wheel exchanger				88

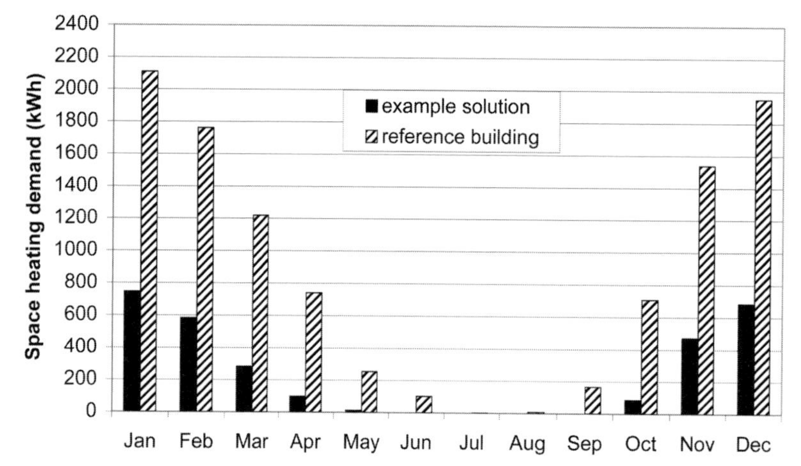

Source: Tor Helge Dokka

Figure 9.2.7 *Monthly space heating demand (annual total 965 kWh/a) for the super conservation solution*

Table 9.2.12 *Primary energy demand and CO_2 emissions for the super conservation solution combined with a DHW solar system*

Net Energy (kWh/m^{2}a)		Total Energy Use (kWh/m^{2}a): Energy use		Total Energy Use (kWh/m^{2}a): Energy source		Delivered energy (kWh/m^{2}a)		Non renewable primary energy factor (-)	Non renewable primary energy (kWh/m^{2}a)	CO_2 factor (kg/kWh)	CO_2 equivalent emissions (kg/m^{2}a)
Mechanical systems	5.5	Mechanical systems	5.5	Electricity	24.5	Electricity	24.5	2.35	57.6	0.43	10.5
Space heating	6.4	Space heating	6.4								
DHW	21.0	DHW	21.0								
		Tank and circulation	4.2	Solar	12.6						
		Conversion losses	0.0								
Total	32.9		37.1		37.1		24.5		57.6		10.5

9.2.4 Design advice

It is quite easy to meet the requirement set for space heating (20 kWh/m^{2}a) in the temperate climate. Important design parameters to achieve this requirement are as follows:

- A well-insulated building envelope: a mean U-value for the opaque building envelope should be in the range of 0.15–0.20 W/m^{2}K.
- Well-insulated windows, with a U-value in the range of 0.90–1.10 W/m^{2}K and a g-value higher than 0.5. A further reduction of the U-value down to the Passivhaus standard (0.80 W/m^{2}K) or below is not necessary to meet the space heating demand of 20 kWh/m^{2}a. The distribution and orientation of windows is not of great importance for a well-insulated house like this one.
- A high level of air tightness is a prerequisite for a high-performance building and should be in the range 0.5 to 1.0 ach at 50 Pa pressure difference.
- A ventilation system with a high efficiency heat exchanger is also crucial. The efficiency should be at least 75 per cent, preferably higher.

Table 9.2.13 *Constructions according to the space heating target of 20 kWh/m^{2}a*

	Material	Thickness (m)	Conductivity (W/mK)	Per cent (%)	Conductivity Studs (W/mK)	Percentage of studs	Resistance (m^{2}K/W)	U-value (W/m^{2}K)
Wall	Exterior surface	0.108					0.04	
	Brick	0.025	0.75	100			0.14	
	Air gap	0.009					0.14	
	Gypsum	0.225	0.23	100			0.04	
	Insulation/studs		0.037	85	0.13	15	4.42	
	Plastic foil						0.05	
	Internal siding	0.013	0.12	100			0.11	
	Interior surface						0.13	
	Total	**0.38**					**4.90**	**0.20**
Roof	Exterior surface							
	Roof tiles	0.108	0.75	100			0.14	
	Roofing paper						0.05	
	Wooden panel	0.015	0.12	100			0.13	
	Air gap	0.05					0.14	
	Mineral wool/rafter	0.35	0.037	90	0.13	0.15	6.63	
	Plastic foil						0.05	
	Internal siding	0.013	0.12	100			0.11	
	Interior surface							
	Total	**0.536**					**7.25**	**0.13**
Floor	Exterior surface							
	EPS insulation	0.2	0.036	100			5.56	
	Plastic foil						0.05	
	Concrete	0.07	1.5	100			0.05	
	Parquet	0.014	0.12	100			0.12	
	Interior surface							
	Total	**0.284**					**5.77**	**0.17**

To meet the primary energy requirement of 60 kWh/m^{2}a, several solutions can be used:

- A combination of a condensing gas boiler and a DHW flat-plate solar collector covering a minimum of 50 per cent of the DHW demand. A water-based radiator system for space heating is used for this solution.
- A combination of a wood pellet stove covering 80 per cent of the space heating demand and a DHW flat-plate solar collector covering a minimum of 50 per cent of the DHW demand.

- A 'super' conservation solution with an extremely low space heating demand in combination with a DHW flat-plate solar collector covering a minimum of 50 per cent of the DHW demand. A cost-effective direct electric space heating system is used for this solution.

Other solutions, such as heat pump systems and/or district heating, if available, are also possible in order to meet the primary energy target of 60 kWh/m^{2}a.

References

Meteotest (2004) *Meteonorm 5.0 – Global Meteorological Database for Solar Energy and Applied Meteorology*, Bern, Switzerland, www.meteotest.ch

ProgramByggerne (2004) *ProgramByggerne ANS, SCIAQ Pro 2.0 – Simulation of Climate and Indoor Air Quality: A Multizone Dynamic Building Simulation Program*, www.programbyggerne.no

9.3 Single family house in the Temperate Climate Renewable Energy Strategy

Tor Helge Dokka

Table 9.3.1 *Targets for single family house in the Temperate Climate Renewable Energy Strategy*

	Targets
Space heating	25 kWh/m^{2}a
Non-renewable primary energy: (space heating + water heating + electricity for mechanical systems)	60 kWh/m^{2}a

This section presents a solution for the single family house in the temperate climate. As a reference for the temperate climate, the city of Zurich is considered. Climatic data are taken from the meteorological software database Meteonorm (Meteotest, 2004). The solution is mainly based on renewable energy. A balanced mechanical ventilation system with heat recovery is used to reduce the ventilation losses. The DHW and space heating system is based on a combi-system with solar collectors and a bio-pellet burner.

9.3.1 Renewable energy with solar combi-system and biomass boiler

Building envelope

The opaque part of the building envelope is a wooden lightweight frame with mineral wool. The floor construction is a slab on the ground with expanded polystyrene insulation (EPS). U-values and areas are shown in Table 9.3.2.

Table 9.3.2 *Building envelope components*

Component	U-Value (W/m^{2}K)	Area (m^2)
Walls	0.22	113.6
Roof	0.18	129.7
Floor (excluding ground)	0.22	96.4
Windows (frame + glass)	0.92	22.0
Window frame	1.20	-
Window glass	0.80	-
Entrance door	1.00	2.0
Whole building envelope	0.25	

Mechanical systems

A balanced mechanical ventilation system with 80 per cent heat recovery and a bypass for summer ventilation is used. A system with a combined solar collector and biomass boiler is used for the water-based space heating system and for DHW heating. The space heating system has water-based radiators. The solar collector system supplies approximately 50 per cent of the DHW load. The remaining energy demand is covered with the biomass boiler. The flat-plate solar collector to the south has an area of 4 m^2 and is mounted at an angle of 40° directly south. A storage tank of 0.4 m^3 is assumed. A circulation pump of 40 W operates the solar collector loop for about 2000 hours during a year. The corresponding annual electricity needed for the pump per living area is therefore 0.5 kWh/m^2a.

Energy performance

Space heating demand: simulation results from the Norwegian energy simulation software SCIAQ Pro (ProgramByggerne, 2004) give the monthly space heating demand of the building as shown in Figure 9.3.1. The heating season extends approximately from November to March:

- space heating demand: 3750 kWh/a (25 kWh/m^2a);
- heating set point: 20°C;
- ventilation rate: 0.45 ach;
- infiltration rate: 0.05 ach; and
- heat recovery: 80 per cent efficiency.

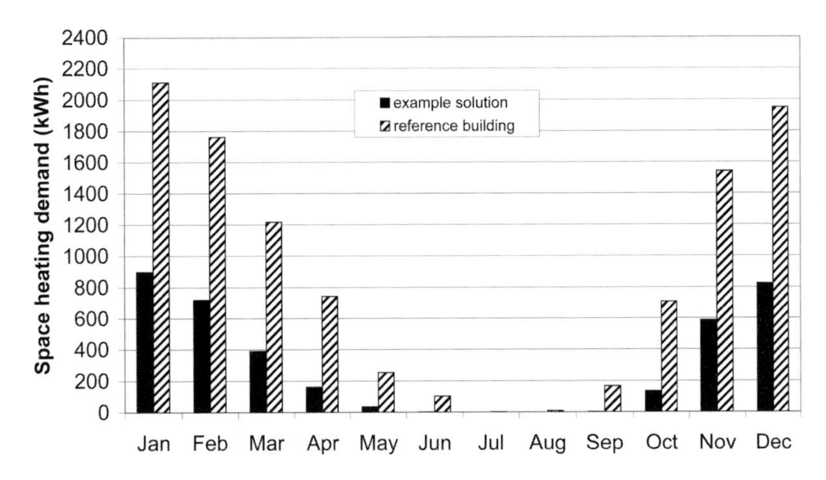

Source: Tor Helge Dokka

Figure 9.3.1 *Monthly space heating demand for the proposed solution (25 kWh/m^2a) and the reference building (70.4 kWh/m^2a)*

Hourly loads of the heating system are calculated based on results from SCIAQ Pro simulations without direct solar radiation. The annual peak load is approximately 2500 W or 16.6 W/m^2 and occurs in January. The annual peak load for the reference building is approximately 5300 W or 35 W/m^2 and occurs in January.

Domestic hot water demand: the net DHW heat demand is approximately 3150 kWh/a or 21 kWh/m^2a. Two adults and two children occupy a typical single family detached house. The average DHW consumption is 40 litres per person per day. Consequently, the single family house consumes 160 litres of domestic hot water per day. The average temperature over the year of the cold water is set to 8.5°C. The temperature of the hot water was set to 50°C at the faucet. The on/off temperature set points for the thermostat in the tank were set to 55/57°C.

System losses: the system losses consist mainly of losses from the hot water storage tank, but also from pipe losses in the distribution system and conversion losses in the boiler. The losses provided by the solar collector simulation program are the tank losses, which include the total heat losses through the tank wall, base and cover, and the connection losses. The tank losses become larger with increasing tank size and/or increasing solar collector area. The tank losses for a solar collector system with 4 m^2 of collectors and a 400 litre tank are about 630 kWh per year or 4.2 kWh/m^2a (living area). The annual COP for the biomass boiler is estimated to be 80 per cent, resulting in conversion losses of 8.9 kWh/m^2a.

Household electricity: the assumed household electricity is 2500 kWh or 16.6 kWh/m^2 (two adults and two children). The primary energy target does not include household electricity since this factor is very much dependent on the occupants' behaviour.

Non-renewable primary energy demand and CO_2 emissions: the computer program Polysun (Polysun, 2004) has been used to calculate the contribution from the solar collector system. The results with respect to total energy use, delivered energy, non-renewable primary energy and CO_2 emissions for this system are given in Table 9.3.3. The primary energy use is very low, below one third of the target (60 kWh/m^2a), and the CO_2 emissions are also very modest.

Table 9.3.3 *Energy use, non-renewable primary energy demand and CO_2 emissions for the DHW solar system and biomass boiler*

Net Energy (kWh/m^2a)		Total Energy Use (kWh/m^2a)				Delivered energy (kWh/m^2a)		Non renewable primary energy factor (-)	Non renewable primary energy (kWh/m^2a)	CO_2 factor (kg/kWh)	CO_2 equivalent emissions (kg/m^2a)
		Energy use		Energy source							
Mechanical systems	5.5	Mechanical systems	5.5	Electricity	5.5	Electricity	5.5	2.35	12.9	0.43	2.4
Space heating	25.0	Space heating	25.0	Biomass	46.5	Biomass	46.5	0.14	6.5	0.04	2.0
DHW	21.0	DHW	21.0								
		Tank and circulation	4.2								
		Conversion losses	8.9	Solar	12.6						
Total	51.5		64.6		64.6		52.0		19.4		4.4

9.3.2 Sensitivity analysis of alternative renewable energy systems

Several alternatives to the combined solar collector and biomass boiler system are possible. Three alternative systems have been explored here:

1 an optimized solar collector system combined with a condensing gas boiler;
2 an optimized solar collector system combined with district heating; and
3 an integrated heat pump, ventilation and DHW system, coupled with an earth tube system for the preheating of fresh air in the winter.

Optimized solar combi-system with condensing gas boiler

This system consists of an optimized solar collector system that is designed to cover the total DHW demand in the summer months of June to August. A solar system with 7.5 m^2 collector area and a 600 litre tank will meet this design requirement. The collector is directed due south and is tilted 40°. This solar system has an annual solar fraction of 70 per cent (for example, covering 70 per cent of the DHW demand).

The peak load for DHW and space heating is covered by a condensing gas boiler when the solar system is insufficient (during winter months). The annual COP of the condensing gas boiler, including distribution losses in the water-based radiant heating system, is set to 90 per cent.

The results with respect to total energy use, delivered energy, non-renewable primary energy and CO_2 emissions for this system are given in Table 9.3.4. The calculated primary energy demand is well below the target of 60 kWh/m²a, but the CO_2 emissions are significantly higher than the combined solar and biomass solution.

Table 9.3.4 *Energy use, non-renewable primary energy demand and CO_2 emissions for a solar combi-system with a condensing gas boiler*

Net Energy (kWh/m²a)		Total Energy Use (kWh/m²a)				Delivered energy (kWh/m²a)		Non renewable primary energy factor (-)	Non renewable primary energy (kWh/m²a)	CO_2 factor (kg/kWh)	CO_2 equivalent emissions (kg/m²a)
		Energy use		Energy source							
Mechanical systems	5.5	Mechanical systems	5.5	Electricity	5.5	Electricity	5.5	2.35	12.9	0.43	2.4
Space heating	25.0	Space heating	25.0	Gas	36.7	Gas	36.7	1.14	41.8	0.25	9.1
DHW	21.0	DHW	21.0								
		Tank and circulation	6.3								
		Conversion losses	3.5	Solar	19.1						
Total	51.5		61.3		61.3		42.2		54.8		11.4

Optimized solar combi-system with district heating

This solution has the same solar collector as the previous systems, but the gas boiler is replaced with a district heating system (assumed available). Space heating is provided by a water-based radiator system. The annual system COP of the district heating installation and the distribution losses is assumed to be 95 per cent. The district heating is supplied from a combined heat and power (CHP) plant using coal (35 per cent) and oil (65 per cent).

The results with respect to total energy use, delivered energy, non-renewable primary energy and CO_2 emissions for this system are given in Table 9.3.5. The calculated primary energy use is lower than for the gas boiler solution and below the target of 60 kWh/m²a, but the CO_2 emissions are high, and even higher than for the condensing gas boiler solution.

Table 9.3.5 *Energy use, non-renewable primary energy demand and CO_2 emissions for a solar combi-system with district heating*

Net Energy (kWh/m²a)		Total Energy Use (kWh/m²a)				Delivered energy (kWh/m²a)		Non renewable primary energy factor (-)	Non renewable primary energy (kWh/m²a)	CO_2 factor (kg/kWh)	CO_2 equivalent emissions (kg/m²a)
		Energy use		Energy source							
Mechanical systems	5.5	Mechanical systems	5.5	Electricity	5.5	Electricity	5.5	2.35	12.9	0.43	2.4
Space heating	25.0	Space heating	25.0	District heating	34.8	District heating	34.8	1.12	39.0	0.32	11.3
DHW	21.0	DHW	21.0								
		Tank and circulation	6.3								
		Conversion losses	1.6	Solar	19.1						
Total	51.5		59.4		59.4		40.3		51.9		13.6

Combined earth tube system with an integrated heat pump system

This system is quite common in passive houses in Austria and Germany. The function of the earth tube system is primarily to preheat the fresh air in the winter before it enters the air handling unit, preventing frost in the heat exchanger. This is often a problem in counter-flow and cross-flow heat exchangers when the external temperature is below 0°C, combined with high moisture production in the house (for example, showering). If the heat pump extracts heat from the exhaust air after the heat exchanger, the earth tube system provides a somewhat higher exhaust air temperature than without an earth tube system. This results in a higher annual COP for the heat pump system. The earth tube system can also be used as a passive cooling system during summer, but the effect is modest due to the relatively low air flow rate of the mechanical ventilation system (0.45 ach).

The maximum power capacity of the heat pump system is 1500 W. The mean peak load for DHW, including tank losses, is 450 W, assuming a 250 litre storage tank to even out peaks in DHW demand. Remaining heating power capacity for space heating is 1050 W. The peak load for space heating is simulated to be approximately 2480 W and the auxiliary peak load is then 1430 W, which is covered by electric resistance heating.

An annual simulation shows that approximately 70 per cent of the space heating demand can be covered by the heat pump system, in addition to the DHW demand. The heat pump system has an annual COP of 2.7, including distribution losses in the water-based radiator system (space heating).

The results with respect to total energy use, delivered energy, non-renewable primary energy and CO_2 emissions for this system are given in Table 9.3.6. The calculated primary energy use is below the target of 60 kWh/m^{2}a and the CO_2 emissions are lower than for the gas boiler and district heating solutions.

Table 9.3.6 *Energy use, non-renewable primary energy demand and CO_2 emissions for a combined earth tube and heat pump system*

Net Energy (kWh/m²a)		Total Energy Use (kWh/m²a)				Delivered energy (kWh/m²a)		Non renewable primary energy factor (-)	Non renewable primary energy (kWh/m²a)	CO_2 factor (kg/kWh)	CO_2 equivalent emissions (kg/m²a)
		Energy use		Energy source							
Mechanical systems	5.0	Mechanical systems	5.0	Electricity	7.5	Electricity	7.5	2.35	17.6	0.43	3.2
Space heating	25.0	Space heating	25.0	Electricity Heat pump	16.2	Electricity	16.2	2.35	38.1	0.43	7.0
DHW	21.0	DHW	21.0								
		Tank and circulation	5.3	Exhaust air /ground	32.6						
		Conversion losses	0.0								
Total	51.0		56.3		56.3		23.7		55.7		10.2

9.3.3 Design advice

Based on environmental considerations, the system with the solar collector and biomass boiler is clearly the best. The primary energy use and CO_2 emissions are much lower than the alternative solutions (i.e. the CO_2 emissions are only one third of that of the district heating solution). The drawback with the solar collector and biomass solutions is that some running and maintenance costs have to be expected.

An attractive alternative is the earth tube and heat pump system, with a higher primary energy use and CO_2 emissions, but with a much lower delivered energy use (all electricity). The advantage with this system is that the system is delivered in a compact unit, which can be placed in a washroom or a store room. An optimized version of the earth tube and heat pump system could be a combined conservation and renewable strategy, reducing the space heating energy demand down to approximately 15 kWh/m^{2}a. The heat pump could then cover typically 90 per cent to 95 per cent of the annual space heating load in addition to the DHW load. This is, in fact, the strategy used in many passive houses.

Even if the condensing gas boiler solution and the district heating solution meets the primary energy targets, they are not recommended due to the high use of non-renewable energy and comparably high emissions of greenhouse gases. However, in some areas, district heating based on a higher fraction of renewable energy is available and could then be an option.

Table 9.3.7 *Construction according to the space heating target of 25 kWh/m^{2}a*

	Material	Thickness (m)	Conductivity (W/mK)	Per cent (%)	Conductivity Studs (W/mK)	Percentage of Studs	Resistance (m^{2}K/W)	U-value (W/m^{2}K)
Wall	Exterior surface						0.04	
	Brick	0.108	0.75	100			0.14	
	Air gap	0.025					0.14	
	Gypsum	0.009	0.23	100			0.04	
	Insulation/studs	0.2	0.037	85	0.13	15	3.93	
	Plastic foil						0.05	
	Internal siding	0.013	0.12	100			0.11	
	Interior surface						0.13	
	Total	**0.355**					**4.58**	**0.22**
Roof	Exterior surface						0.04	
	Roof tiles	0.108	0.75	100			0.14	
	Roofing paper						0.05	
	Wooden panel	0.015	0.12	100			0.13	
	Air gap	0.05					0.14	
	Mineral wool/rafter	0.25	0.037	90	0.13	0.15	4.73	
	Plastic foil						0.05	
	Internal siding	0.013	0.12	100			0.11	
	Interior surface						0.13	
	Total	**0.436**					**5.52**	**0.18**
Floor	Exterior surface						0.00	
	EPS insulation	0.15	0.036	100			4.17	
	Plastic foil						0.05	
	Concrete	0.07	1.5	100			0.05	
	Parquet	0.014	0.12	100			0.12	
	Interior surface						0.13	
	Total	**0.234**					**4.51**	**0.22**

References

Meteotest (2004) *Meteonorm 5.0 – Global Meteorological Database for Solar Energy and Applied Meteorology*, Bern, Switzerland, www.meteotest.ch

ProgramByggerne (2004) *ProgramByggerne ANS, SCIAQ Pro 2.0 – Simulation of Climate and Indoor Air Quality: A Multizone Dynamic Building Simulation Program*, www.programbyggerne.no

Polysun (2004) *Polysun Version 3.3*, Institut für Solartechnik, www.solarenergy.ch

9.4 Row house in the Temperate Climate Conservation Strategy

Udo Gieseler

Table 9.4.1 *Targets for row house in the Temperate Climate Conservation Strategy*

	Targets
Space heating	15 kWh/m^{2}a
Non-renewable primary energy: (space heating + water heating + electricity for mechanical systems)	60 kWh/m^{2}a

This section presents a solution for the row houses in the temperate climate. As a reference for the temperate climate, the city of Zurich is used. The solution is based on energy conservation minimizing the heat losses of the building. A balanced mechanical ventilation system with heat recovery is used to reduce the ventilation losses.

9.4.1 Solution 1a: Conservation with oil or gas burner and solar DHW and Solution 1b: Conservation with ambient air heat pump

Building envelope and space heating demand

In the temperate climate, the reference row house has a space heating demand of about 61 kWh/m^{2}a. To reach the space heating target of 15 kWh/m^{2}a, building losses must be significantly reduced. Performance indicators for the reference house as well as for the high-performance house (solution 1) are given in Table 9.4.2. The strategy to achieve the space heating target includes three types of energy saving measures:

1. *Insulation*: the walls and the roof are well insulated. The U-values are between U = 0.16 W/m^{2}K and U = 0.22 W/m^{2}K (see Table 9.4.2 and Table 9.4.7). The insulation between the ground floor and the ground is usually more expensive and, at the same time, the possible energy saving is smaller than for the walls (see Gieseler et al, 2004). Therefore, the thickness is only slightly increased.
2. *Windows*: for all façades, standard windows with U = 1.1 W/m^{2}K for the glazing are used. The south window size is slightly reduced. A side effect of this is the reduction of possible overheating problems in summer.
3. *Ventilation*: the improved air tightness of the high-performance house with $n_{50} \leq 1$ ach leads to a very small infiltration of 0.05 ach. The necessary fresh air is provided by a mechanical ventilation system with a heat recovery efficiency of 80 per cent. An efficiency of 80 per cent is necessary to reach the space heating target. Note that the energy demand for defrosting of the heat exchanger increases with increasing efficiency (see Gieseler et al, 2002).

These measures lead to a space heating demand of 12.8 kWh/m^{2}a for the mid row house unit, and 19.5 kWh/m^{2}a for the end unit. The average space heating demand for a row with four mid and two end units is 15.0 kWh/m^{2}a, which meets the target.

For a more detailed comparison of the reference and the high-performance case, the energy balance is shown in Table 9.4.3. Note that these simulation results were calculated for the heating period. The average values for the row with six units is also given and is shown in Figure 9.4.1 for the reference and the high-performance house.

The aim of the conservation strategy was to reduce the heat losses and the means of achieving this is primarily by reducing ventilation losses by 73 per cent. The transmission losses are reduced by 45 per cent. Altogether, the losses of the high-performance house (solution 1) are less than half of the reference building. These losses must be balanced by the gains. The available internal gains of solution 1 are somewhat smaller than for the reference building due to an assumed higher efficiency of the appliances. The solar gains are also reduced, mainly due to the lower g-value of the windows (see Table 9.4.2). However, these effects are small compared with the significant reduction in the losses resulting in a space heating demand of 15 kWh/m²a.

Table 9.4.2 *Comparison of key numbers for the construction and energy performance of the row house (areas are per unit)*

	Reference building	**Conservation strategy**
Walls:		
Area north and south (m²)	39.4	41.4
U-value north/south (W/m²K)	0.45	0.22
Area east or west (m²)	57.0	57.0
U-value east/west (W/m²K)	0.36	0.16
Roof (area: 60 m²)		
U-value (W/m²K)	0.28	0.18
Floor (area: 60 m²)		
U-value (W/m²K)	0.39	0.31
Windows		
South		
Area (m²)	14.00	12.00
U-value glazing (W/m²K), 70%	2.80	1.10
U-value frame (W/(m²K), 30%	1.80	1.80
g-value	0.76	0.59
North		
Area (m²)	3.00	3.00
U-value glazing (W/m²K), 60%	2.80	1.10
U-value frame (W/m²K), 40%	1.80	1.80
g-value	0.76	0.59
East/west		
Area (m²)	3.00	3.00
U-value glazing (W/m²K), 60%	2.80	1.10
U-value frame (W/m²K), 40%	1.80	1.80
g-value	0.76	0.59
Air change rate (air volume: 275 m³)		
Infiltration (ach)	0.60	0.05
Ventilation (ach)	0.00	0.45
Heat recovery (-)	0.00	0.80
Space heating demand		
(simulation for 1 January–31 December)		
Mid unit (kWh/m²a)	55.5	12.8
End unit (kWh/m²a)	71.3	19.5
Row of four mid and two end units (kWh/m²a)	60.8	15.0

Table 9.4.3 *Simulation results for the energy balance in the heating period (kWh/m²a)*

Energy balance Simulation period: 1 October–30 April	Space heating demand	Gains		Losses	
		Solar	Internal	Transmission	Ventilation
Reference					
Mid unit	55.5	20.5	22.5	60.0	38.5
End unit	71.3	23.2	22.5	78.8	38.2
Row house (four mid + two end units)	**60.8**	**21.4**	**22.5**	**66.3**	**38.4**
Conservation strategy					
Mid unit	12.8	12.1	19.2	33.3	10.8
End unit	19.5	13.9	19.2	42.8	9.9
Row house (four mid + two end units)	**15.0**	**12.7**	**19.2**	**36.5**	**10.5**

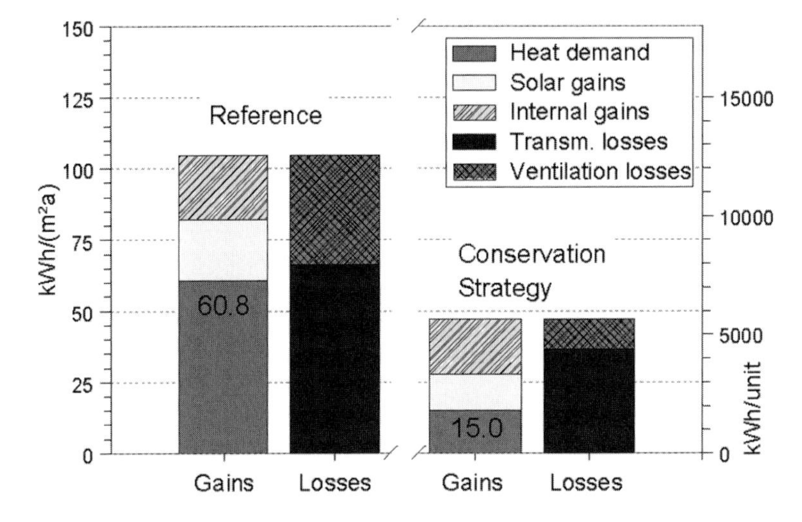

Source: Udo Gieseler

Figure 9.4.1 *Simulation results for the energy balance of the row house (six units) according to Table 9.4.3*

Mechanical systems

Space heating and DHW system: two alternatives are suggested for the row house. One possibility is a typical oil (or gas) burner, which is supplemented by a solar collector for DHW. Another possibility is a heat pump. Because the coefficient of performance (COP) of a heat pump is especially high in summer (high temperature heat source), this system is used without solar collector. Table 9.4.4 summarizes the two solutions.

For the actual sizing of the heating system, the maximum peak load is important. Figure 9.4.2 shows simulation results for the hourly heat load of an end unit. To model the worst case of weather conditions, only diffuse solar radiation is taken into account (overcast sky). The heat load is calculated for an occupied building (i.e. with internal gains). The resulting maximum heat load is 1900 W. This value has been used as the power limit of the heating system in the simulations. The monthly heat demand of an average row house unit and its distribution over the year is shown in Figure 9.4.3.

Domestic hot water system: a DHW demand of 160 litres per day at a temperature of 55°C for each row house unit is assumed. Each unit has its own 300 litre storage tank. The tank losses are calculated on the basis of a tank height of 1.6 m and thermal insulation of $U = 0.28$ W/m²K. Freshwater enters the system at a temperature of 9.7°C.

Electricity for mechanical systems: the electricity demand for fans, pumps and controls is estimated as 5 kWh/m²a. This value does not include the electricity demand for the heat pump.

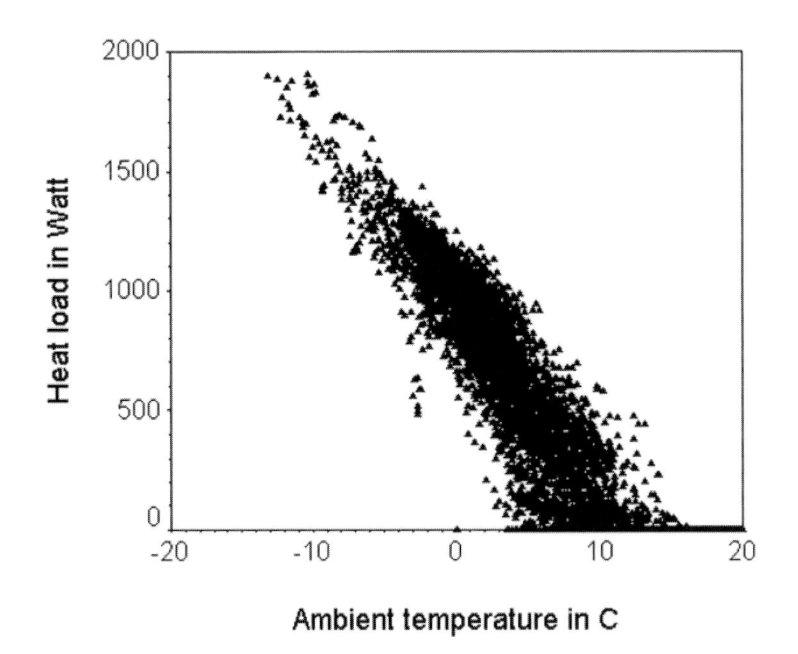

Source: Udo Gieseler

Figure 9.4.2 *Simulation results for the hourly heat load without direct solar radiation; the maximum heat load for an end unit is 1900 W*

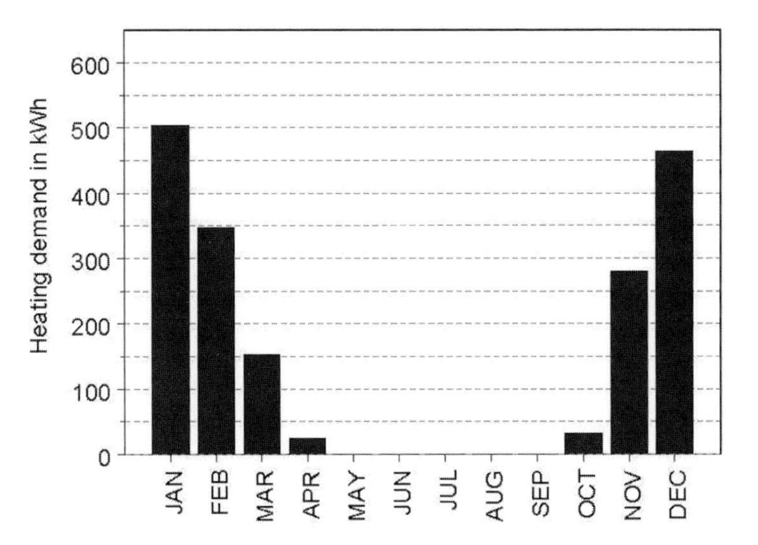

Source: Udo Gieseler

Figure 9.4.3 *Monthly space heating demand for a row house unit (average over four mid and two end units)*

Table 9.4.4 *Analysed example solutions*

Solution	Solution name	Energy supply for space heating	Energy supply for DHW
1a	Burner and solar	Oil (or gas or wood) burner	Oil (or gas or wood) burner 4 m^2 solar collector
1b	Heat pump	Electric heat pump (exhaust air/water)	Electric heat pump (exhaust air/water)

Solution 1a: Burner and solar: for the supply of DHW and space heating, a typical oil (or gas) burner is used. Since this system is located inside the heated volume, the efficiency is 100 per cent for space heating. For the supply of DHW, the efficiency is assumed as 85 per cent.

The energy supply for the DHW is supplemented by a 4 m^2 flat-plate solar collector for each row house unit. The collector is oriented towards south, with an inclination of 45°. The mass flow rate is

50 kg/h. The efficiency of a flat-plate collector depends on the mean fluid temperature T_i inside the collector, the ambient temperature T_a and on the total incident radiation I_T on the collector surface according to:

$$\eta = \eta_0 - a_1 \frac{T_i - T_a}{I_T} - \frac{(T_i - T_a)^2}{I_T} \qquad [9.1]$$

The coefficients of the collector efficiency, which were used here, are $\eta_0 = 0.8$, $a_1 = 3.5$ W/m²K, $a_2 = 0.015$ W/m²K². The efficiency of the heat exchanger to the DHW storage tank is 95 per cent.

Solution 1b: Heat pump: for the simulation of the heat pump, a representation of COP measurements from heat pumps is used. These measurements from more than 200 heat pumps are published in the *WPZ Bulletin* (Roth, 2000). The COP depends on type and temperatures of heat source and temperature of the supply side. For the solution presented here, an average air/water heat pump with electric power of 3 kW is used, with ambient air as the heat source. The supply temperature is 55°C for both DHW and space heating.

Energy performance

Non-renewable primary energy demand and CO_2 emissions: the total energy demand of the high-performance row house, as well as the corresponding primary energy demand and CO_2 emissions, are shown in Tables 9.4.5 and 9.4.6. The suggested high-performance house has a primary energy demand of 45 kWh/m²a for the oil burner (solution 1a) or 53 kWh/m²a for the heat pump (solution 1b). These values are significantly lower than the requested primary energy target.

Table 9.4.5 *Total energy demand, non-renewable primary energy demand and CO_2 emissions for the solution 1a with oil burner and solar DHW*

Net Energy (kWh/m²a)		Total Energy Use (kWh/m²a) Energy use		Energy source		Delivered energy (kWh/m²a)		Non renewable primary energy factor (-)	Non renewable primary energy (kWh/m²a)	CO_2 factor (kg/kWh)	CO_2 equivalent emissions (kg/m²a)
Mechanical systems	5.0	Mechanical systems	5.0	Electricity	5.0	Electricity	5.0	2.35	11.8	0.43	2.2
Space heating	15.0	Space heating	15.0								
DHW	25.6	DHW	25.6	Oil	29.8	Oil	29.8	1.13	33.7	0.31	9.3
		Tank losses	1.7								
		Conversion losses	2.2	Solar	14.7						
Total	45.6		49.5		49.5		34.8		45.5		11.5

Table 9.4.6 *Total energy demand, non-renewable primary energy demand and CO_2 emissions for the solution 1b with a heat pump*

Net Energy (kWh/m²a)		Total Energy Use (kWh/m²a): Energy use		Total Energy Use (kWh/m²a): Energy source		Delivered energy (kWh/m²a)		Non renewable primary energy factor (-)	Non renewable primary energy (kWh/m²a)	CO_2 factor (kg/kWh)	CO_2 equivalent emissions (kg/m²a)
Mechanical systems	5.0	Mechanical systems	5.0	Electricity	5.0	Electricity	5.0	2.35	11.8	0.43	2.2
Space heating	15.0	Space heating	15.0	Electricity Heat pump	17.6	Electricity	17.6	2.35	41.4	0.43	7.6
DHW	25.6	DHW	25.6								
		Tank losses	0.6	Ambient air	23.6						
		Conversion losses	0.0								
Total	45.6		46.2		46.2		22.6		53.2		9.8

9.4.2 Summer comfort

To evaluate summer comfort, separate simulations were performed. From 1 May to 30 September the ventilation is increased by 1 ach at night between 7 pm and 6 am. The heat recovery unit is not used. Additional shading reduces the direct and diffuse sunlight on windows by 50 per cent during the summer period. The indoor temperature is not allowed to fall below 20°C at any time in order not to overestimate the cooling effect. These assumptions represent a reasonably good passive cooling strategy.

Figures 9.4.4 and 9.4.5 show the number of hours with the average indoor temperature exceeding certain limits. The two shown cases in each diagram are the overheating hours with increased night ventilation only (dark columns) and the overheating hours with increased night ventilation and shading (light columns), as described above.

Figure 9.4.4 shows that the average indoor temperature is above 22°C during 2620 hours (71 per cent of the summer time) in the row house end unit if only increased night ventilation is used. With additional shading (of 50 per cent), the average indoor temperature is above 22°C during 1130 hours (31 per cent of the time). With this cooling strategy, the indoor temperature does not exceed 27°C. This would be quite acceptable in the temperate climate.

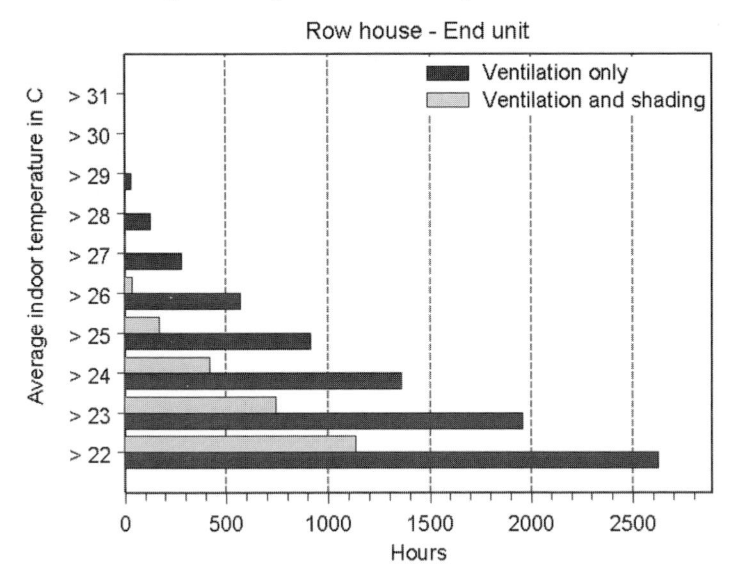

Source: Udo Gieseler

Figure 9.4.4 *Number of hours with average indoor temperature exceeding certain limits; the corresponding simulation period is 1 May to 30 September*

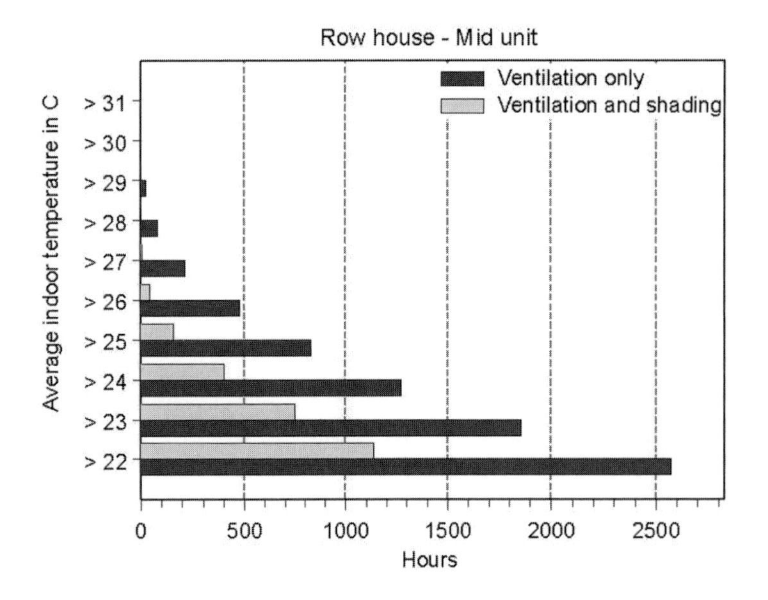

Source: Udo Gieseler

Figure 9.4.5 *Number of hours with average indoor temperature exceeding certain limits; the corresponding simulation period is 1 May to 30 September*

9.4.3 Sensitivity analysis

This section shows the importance of window type and window area to the space heating demand.

Figure 9.4.6 shows the space heating demand with varying south window area. The window area includes 30 per cent frame area. The U-value of the glazing for the two shown cases are Ug = 1.1 W/m^{2}K and U_g = 0.7 W/m^{2}K, with frames of U_F = 1.8 W/m^{2}K and U_F = 0.7 W/m^{2}K, respectively. The results show that for both window types the solar gains are roughly balanced by the transmission losses through the windows. Whereas the standard windows with Ug = 1.1 W/m^{2}K have a negative energy balance in a temperate climate, the high-performance windows with U_g = 0.7 W/m^{2}K lead to minimal energy savings with increased window size. For the 12 m^2 window area per unit of solution 1, the space heating demand can be reduced by 3.5 kWh/m^2 a down to 11.5 kWh/m^{2}a, if high-performance windows are used. The energy target could then be met with a lower insulation level for the walls. However, this is not cost efficient at the moment because of the higher costs for such windows.

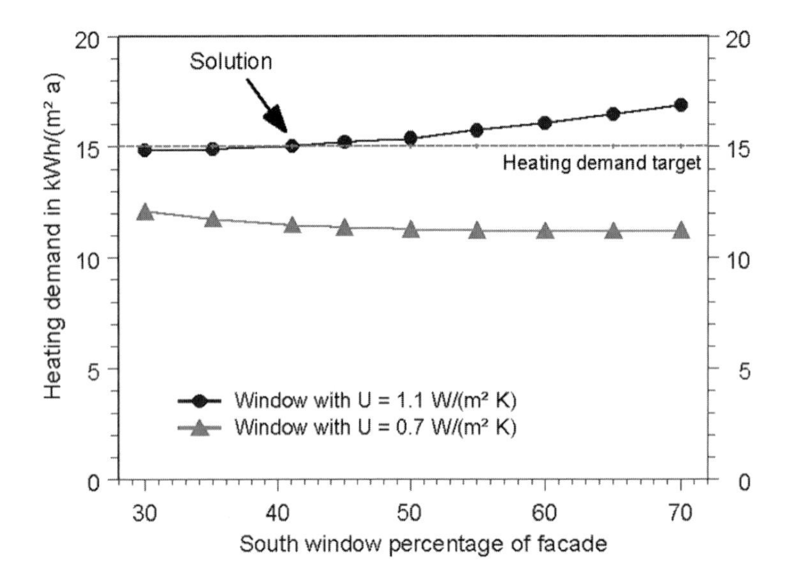

Source: Udo Gieseler

Figure 9.4.6 *Space heating demand for the south window variation in the row house; U-values are shown for the glazing*

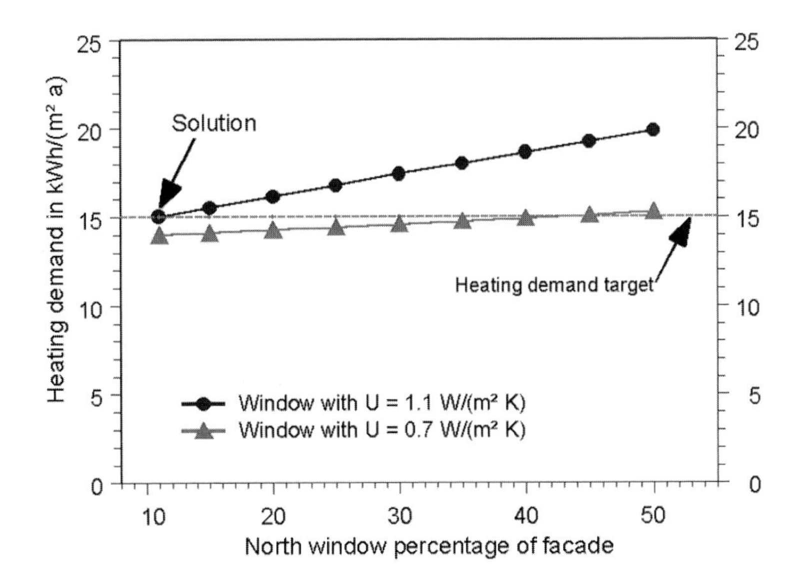

Source: Udo Gieseler

Figure 9.4.7 *Space heating demand for the north window variation in the row house; U-values are shown for the glazing*

In Figure 9.4.7, simulation results for the variation of the window on the north-facing façade are presented. Results for the same window types as in Figure 9.4.6 are shown. Here, both windows include 40 per cent frame area. Solution 1 uses 3 m^2 of standard type windows to the north for each unit (11 per cent window-to-wall area ratio). With high-performance windows (U_g = 0.7 W/m^2K), the space heating target can be met even with an extremely high window fraction.

The effect of shading is presented in Figure 9.4.8 for the south-facing façade. The space heating demand for the row house (all six units) is plotted for different shading coefficients. Shading is applied to direct sunlight on the south-oriented windows. Diffuse radiation is not reduced. This is a rough shading model for shading from buildings, trees or other objects. The heating demand can increase to more than 20 kWh/m^2a if no direct solar radiation reaches the south-facing façade. In such a case, the construction has to be improved to meet the space heating target.

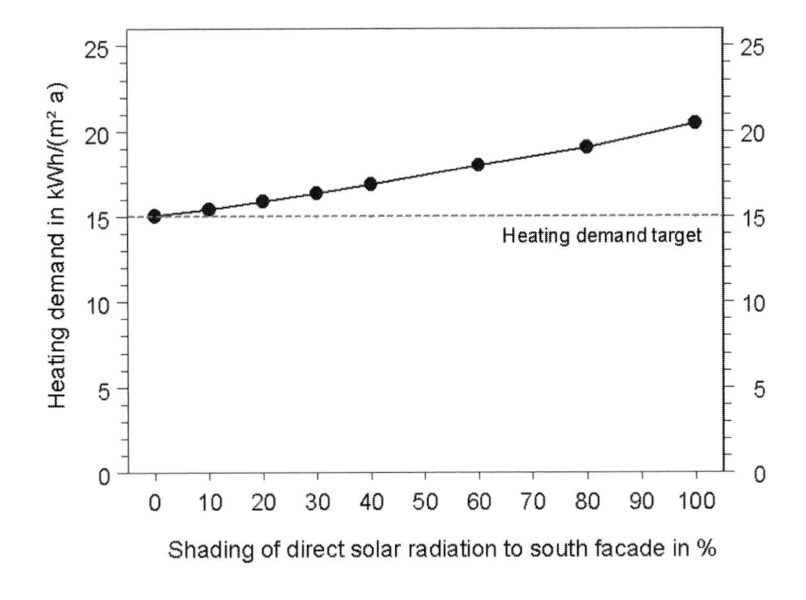

Source: Udo Gieseler

Figure 9.4.8 *Space heating demand for the row house for different shading coefficients*

9.4.4 Conclusions

In this section, the conservation strategy for a row house in a temperate climate has been presented. The conservation strategy is based on:

- ventilation system with heat recovery;
- high insulation level of the envelope; and
- reduced window area.

Both an example solution with a heat pump taking heat from the ambient air and solutions with oil or gas burner in combination with solar DHW heating are possible in order to fulfil the targets. The suggested conservation strategy fulfils the quite ambitious energy target of a space heating demand of 15 kWh/m^{2}a at reasonable construction costs. Both suggested solutions for the building services lead to non-renewable primary energy demands significantly below the target of 60 kWh/m^{2}a.

Table 9.4.7 *Details of the construction of the row house in the Temperate Climate Conservation Strategy (layers are listed from inside to outside)*

Element	Layer	Thickness (m)	Conductivity (W/mK)	Resistance (m^{2}K/W)	U-value (W/m^{2}K)
Wall south/north	Plaster	0.015	0.700	0.021	
	Limestone	0.175	0.561	0.312	
	Polystyrol	0.140	0.035	4.000	
	Plaster	0.020	0.869	0.023	
	Surface resistances	-	-	0.170	
	Total	**0.350**	-	**4.526**	**0.22**
Wall east/west	Plaster	0.015	0.700	0.021	
	Limestone	0.175	0.561	0.312	
	Polystyrol	0.200	0.035	5.714	
	Plaster	0.020	0.869	0.023	
	Surface resistances	-	-	0.170	
	Total	**0.410**	-	**6.240**	**0.16**
Roof 90%	Plaster board	0.013	0.211	0.062	
	Mineral wool	0.260	0.040	6.500	
	Pantile	0.020	-	-	
	Surface resistances	-	-	0.170	
	Total	**0.293**	-	**6.732**	**0.15**
Roof 10%	Plaster board	0.013	0.211	0.062	
	Timber	0.260	0.131	1.985	
	Pantile	0.020	-	-	
	Surface resistances	-	-	0.170	
	Total	**0.293**	-	**2.217**	**0.45**
Floor	Parquet	0.020	0.200	0.100	
	Anhydrite	0.060	1.200	0.050	
	Polystyrol	0.100	0.035	2.857	
	Concrete	0.120	2.100	0.057	
	Surface resistances	-	-	0.170	
	Total	**0.300**	-	**3.234**	**0.31**
Window glazing	Glass (low-e)	0.004	-	-	
	Argon	0.016	-	-	
	Glass	0.004	-	-	
	Total	**0.024**	-	-	**1.1**

References

Gieseler, U. D. J., Bier, W. and Heidt, F. D. (2002) *Cost Efficiency of Ventilation Systems for Low-Energy Buildings with Earth-to-Air Heat Exchange and Heat Recovery*, Proceedings of the 19th International Conference on Passive and Low Energy Architecture (PLEA), Toulouse, France, pp577–583

Gieseler, U. D. J., Heidt, F. D. and Bier, W. (2004) 'Evaluation of the cost efficiency of an energy efficient building', *Renewable Energy Journal*, vol 29, pp369–376

Roth, S. (2000) 'Mitteilungsblatt des Wärmepumpentest und Ausbildungszentrums Winterthur-Töss', *WPZ Bulletin*, no 22, January 2000

TRNSYS (2005) *A Transient System Simulation Program*, Solar Energy Laboratory, University of Wisconsin, Madison, WI

9.5 Row house in the Temperate Climate Renewable Energy Strategy

Joachim Morhenne

Table 9.5.1 *Targets for the row house in the Temperate Climate Renewable Energy Strategy*

	Targets
Space heating	20 kWh/m^{2}a
Non-renewable primary energy: (space heating + water heating + electricity for mechanical systems)	60 kWh/m^{2}a

This section presents a conservation solution for the row houses in the temperate climate. As a reference for the temperate climate, the city of Zurich is used.

9.5.1 Solution 2: Solar domestic hot water and solar-assisted heating

The use of a solar combi-system and an efficient mechanical ventilation system with heat recovery are necessary measures to reach the space heating target of 20 kWh/m^{2}a (2400 kWh/a per unit). Both are given some freedom in reducing transmission losses. The space heating target in row houses with a typical area to volume ratio (A/V) can therefore be reached by construction being only slightly improved in insulation level compared to actual building codes. As a result, this strategy is also applicable to building renovation, since the building envelope then could be difficult to improve sufficiently without high costs.

Due to the low space heating demand, the primary energy target can be achieved by using a solar system only for the hot water production, with a 60 per cent solar fraction. Therefore, a solar combi-system will substantially reduce the primary energy demand, which will be shown by a solution called a high-performance case. A reduced efficiency of the ventilation heat exchanger and a larger solar collector area (base case) is also a possible solution. Even in the case of a building renovation (retrofit case), assuming that not all surfaces can be insulated well, the primary energy target can be fulfilled.

Why follow this strategy?

Solar gains make it possible to reach the energy targets without having to apply excessive conservation measures. Given that most high-performance houses today have a solar domestic hot water system, this strategy proposes to increase the solar system to also provide some space heating. The building envelope therefore only needs small improvements compared to the existing standards. This advantage also makes this strategy applicable to building retrofit with ambitious energy targets. Expensive high-performance windows can be avoided.

The heating power is not limited by the ventilation rate and the indoor comfort can be improved by radiant heating. Furthermore, this strategy provides an opportunity to compensate for low passive gains due to the orientation of the building or shading conditions and therefore gives more freedom for the design.

Main differences of the analysed solutions

- *High-performance*: envelope standard (basis), efficiency of ventilation heat recovery 90 per cent, collector area 11 m^2.
- *Base case*: envelope standard (basis), efficiency of ventilation heat recovery 80 per cent, collector area 12 m^2.
- *Retrofit case*: envelope with reduced insulation of the wall on the east–west side (10 cm instead of 12cm) and only 16 cm insulation in the roof instead of 20 cm. Efficiency of ventilation heat recovery 80 per cent, collector area 13 m^2. Instead of the reduced insulation of the wall, the higher heating demand could, as well, occur due to a less insulated roof or thermal bridges.
- Lightweight case: envelope standard U-values. The walls consist of wooden studs with mineral wool instead of massive layers. Efficiency of ventilation heat recovery 80 per cent. Collector area 12 m^2.

Space heating demand without solar system

- High performance: 16.9 kWh/m^2a.
- Base case: 19.1 kWh/m^2a.
- Retrofit: 21.2 kWh/m^2a.

Building envelope

The U-values are the same for the base case, the lightweight case and the high-performance case:

- *Opaque construction*: massive or lightweight walls with insulation on the outside. Thermal mass increases summer comfort and augments the solar gains slightly. In case of lightweight walls, the floor and ceiling have to be massive.
- *Windows*: frame ratio of 30 per cent, low-e coated glass with argon gas between panes.

Table 9.5.2 *Building envelope U-values*

Component	U-Value (W/m^2K)	
	Basis	**Retrofit**
Floor	0.37	0.37
Walls	0.30	0.30
Walls E/W (end units)	0.25	0.30
Roof	0.23	0.29
Window glass	1.2	1.2
Window frame	1.7	1.7

Mechanical systems

- *Ventilation*: mechanical ventilation with heat recovery. *Base case and retrofit*: η=80 per cent. *High-performance case*: η=90 per cent. *Ventilation rate*: 0.45 ach. *Infiltration rate*: 0.05 ach. *Electric consumption*: 0.3 W/m^2h (all cases).
- *Heat supply*: two-pipe heating grid for the row with biomass or condensing gas furnace. *Alternative*: heat pump (borehole).
- *Solar system*: individual solar combi-system for each row house or central systems.
- *Heat distribution*: hot water radiant heating, fresh air heating.

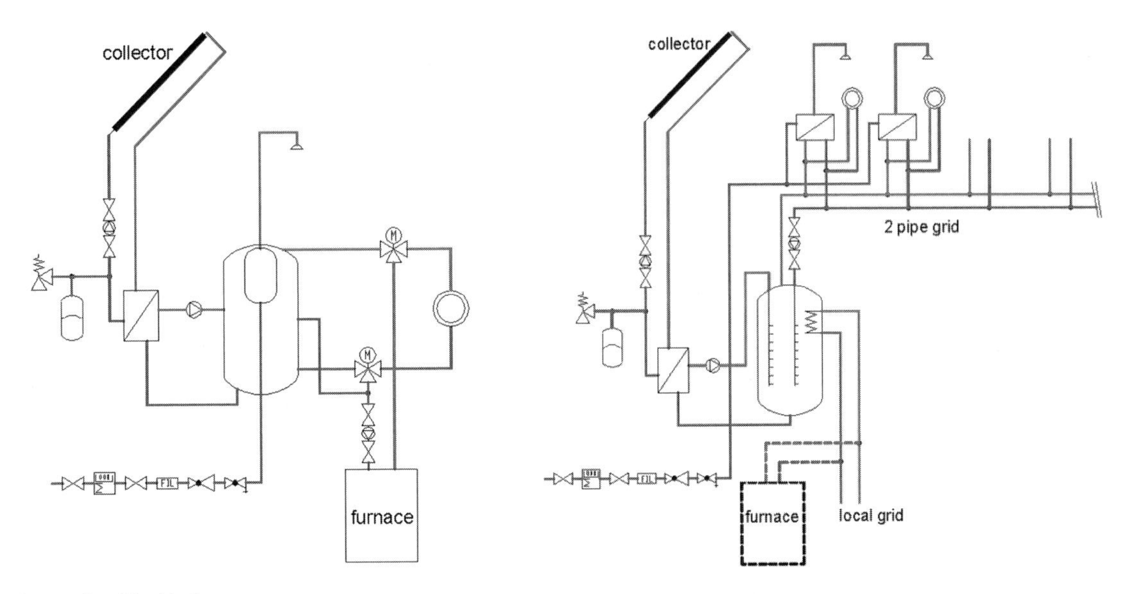

Source: Joachim Morhenne

Figure 9.5.1 *Scheme of the solar-assisted heating system: (a) individual solution; (b) central system for a row of houses*

Two-pipe grids with solar-assisted heating have the advantage of the lowest backup heating demand. Four-pipe grids might have cost benefits depending on grid length and transported energy. Important issues are the question of private or common property of the system and costs for investment and maintenance. Individual systems need more space for storage; the investment is higher but the grid losses are lower. Figure 9.5.1 explains the solar heating systems.

Energy performance

Space heating demand: simulation results from TRNSYS give the monthly space heating demand (total energy use) for a row house shown in Figure 9.5.2. The row consists of six units (two end and four mid houses). Results in Table 9.5.3 are mean values calculated from end and mid units. The heating season extends from October to May. Heating set point is 20°C for all cases.

Table 9.5.3 *Total energy use for space heating*

	High performance	Base case	Retrofit case	Base case lightweight
Space heating demand (mean) kWh/a	2140	2330	2550	2590
Space heating demand (mean) kWh/m²a	17.8	19.4	21.3	21.6
System losses kWh/m²a	2.0	1.9	1.7	1.8
Solar contribution	18%	19.5%	19.3%	19.4%

Peak load for space heating: the maximum peak loads for the different solutions are shown in Table 9.5.3. While the peak occurs in January, near peak demands also occur in February and December. Outside of these three months, the peaks fall off very sharply. The peak load occurs at an ambient temperature of –12.2°C.

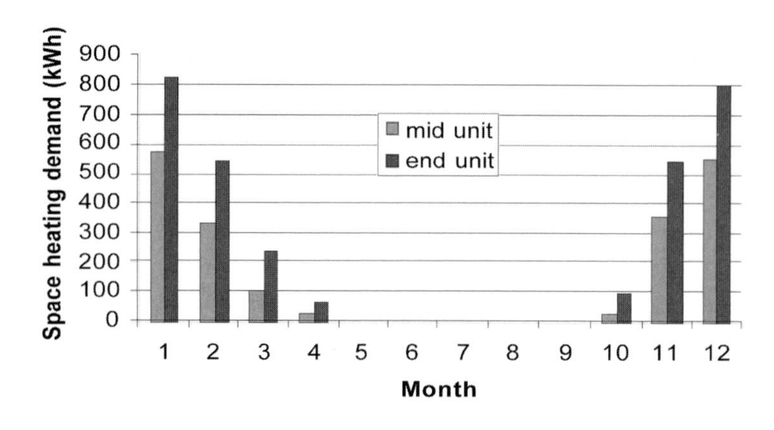

Source: Joachim Morhenne

Figure 9.5.2 *Space heating demand (base case)*

Table 9.5.4 *Maximum peak load at* $T_{ambient} = -12.2°C$

Case	Mid unit (W)	End unit (W)
High performance	1790	2400
Base case	1830	2490
Retrofit	1990	2680
Lightweight	1920	2550

Domestic hot water demand: the heat demand for DHW only differs slightly due to the different heat loads and the different collector sizes (see Table 9.5.5).

Table 9.5.5 *Delivered energy for DHW and solar contribution per unit (mean)*

Case/collector area	Delivered energy for DHW (kWh/a)	Delivered energy for DHW (kWh/m²a)	Solar contribution (%)
High performance/11m²	1470	12.2	58.6
Base/12 m²	1430	12.0	59.6
Retrofit/13 m²	1420	11.8	60.1
Base case lightweight/12 m²	1450	12.1	59.2

Total end energy use: the delivered energy for DHW and space heating sums up to 3990 kWh/a (base case). The electric consumption of fans, pumps and controls is 674 kWh/a. Figure 9.5.3 shows the energy balance for the base case and the reference case.

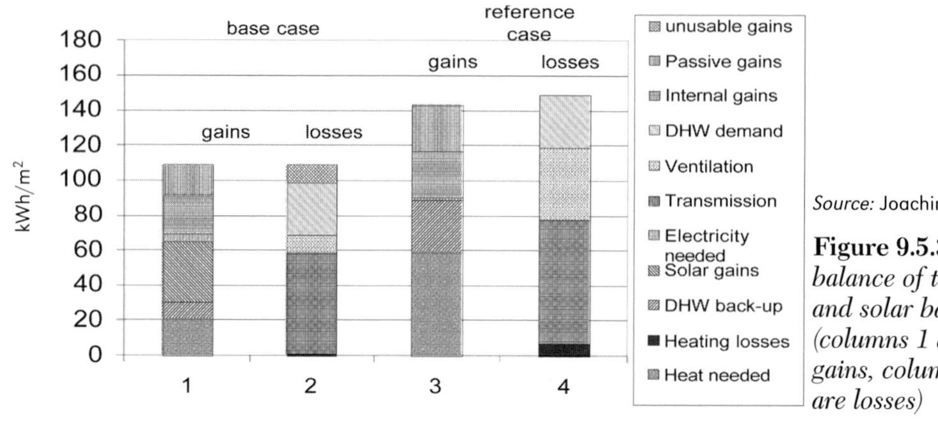

Source: Joachim Morhenne

Figure 9.5.3 *Energy balance of the reference and solar base case (columns 1 and 3 are gains, columns 2 and 4 are losses)*

Non-renewable primary energy demand and CO_2 emissions: the primary energy demand and the CO_2 emissions are shown in Table 9.5.6. Factors are taken from GEMIS (2004). All cases fulfil the given primary energy target. The primary energy factors used are:

- gas: 1.14; and
- electricity: 2.35.

The CO_2 emission factors are:

- gas: 0.247 kg/kWh; and
- electricity: 0.430 kg/kWh.

Table 9.5.6 *Delivered and non-renewable primary energy demand and CO_2 emissions*

Case	Delivered energy: gas (kWh/m^2a)	Delivered energy: electricity (kWh/m^2a)	Solar fraction (%)	Primary energy (non-renewable) (kWh/m^2a)	CO_2 (kg/m^2a)
High performance	32.1	5.8	39.4	49.7	10.3
Base	33.3	5.7	39.5	51.1	10.6
Retrofit	34.8	5.6	39.0	52.9	11.0
Lightweight construction	35.5	5.7	38.2	53.7	11.2

9.5.2 Summer comfort

The heating system is not active in summer; therefore, only internal and passive solar gains and the ventilation air contribute to overheating. Window size and orientation, as well as shading devices, influence the risk of overheating. The indoor temperature is shown in Figure 9.5.4. Due to the chosen shading and ventilation strategy, the indoor temperature in summer is comfortable. To further improve the thermal comfort, even better shading devices and increased night ventilation are recommended. The strategy used in this example is as follows:

- Mechanical ventilation is switched off, if possible, to save electric consumption. Note that in very air-tight houses, it is important to ventilate since the infiltration rate is extremely low.
- Night ventilation by windows is increased to 1 ach between 7 pm and 6 am during June to August. Night ventilation is very effective for cooling the building mass, either by opening windows or by using the mechanical ventilation system.
- External shading (windows facing west, south and east) is with 50 per cent transmittance and 10 per cent absorption.
- A summer bypass in the heat exchanger is recommended to reduce excessive indoor temperatures outside the heating season if the ventilation system is on.

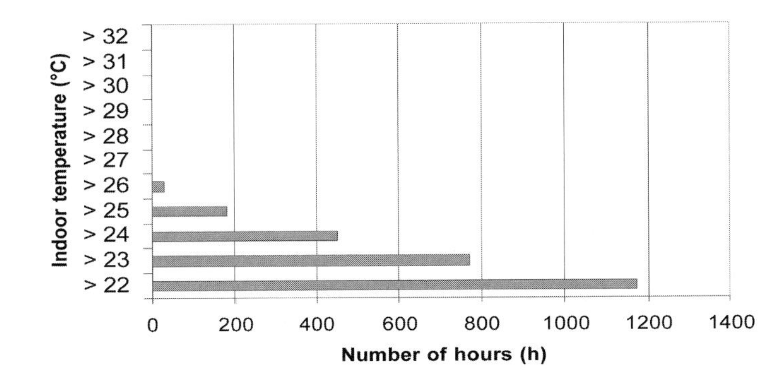

Source: Joachim Morhenne

Figure 9.5.4 *Indoor temperature of the end house (independent of cases except lightweight construction)*

9.5.3 Sensitivity analysis

System design

Two system designs, shown in Figure 9.5.5, have been evaluated. The presented energy performance is related to system type 1. The energy performance of system type 2 is only slightly lower, but the risk of overheating is increased.

Type 1 is a typical solar combi-system. Different solar combi-systems have been evaluated in IEA SHC Task 26 (Weiss, 2004). Type 2 is a directly heated heavy floor with a connected standard DHW system. The main difference from type 1 is the renunciation of the buffer storage, which has been replaced by a floor heating system acting as storage and heating system. To have full capacity of floor and ceiling, no screed is assumed. The furnace is, in all cases, placed centrally for the row. Figure 9.5.5 shows an installation for a single unit (see also Figure 9.5.1).

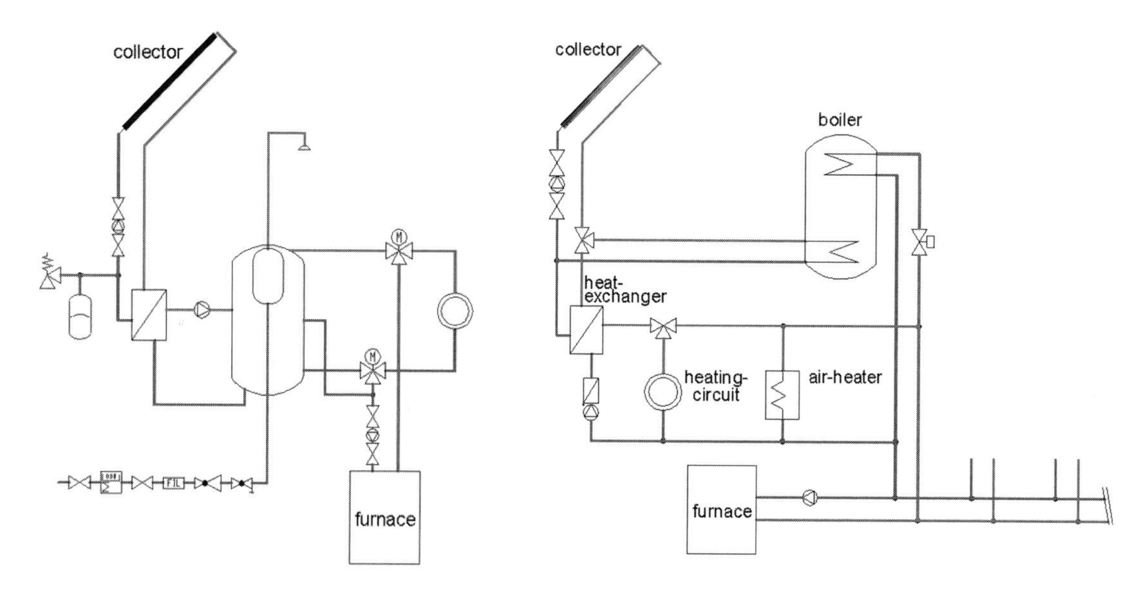

Source: Joachim Morhenne

Figure 9.5.5 *Schemes of the two systems: (a) type 1 is a typical individual solar combi-system; (b) type 2 uses the building mass as a heat storage*

The energy performance of system type 2 is 5 per cent lower and cannot be increased because of comfort limitations. The advantage of this system is its simplicity. The disadvantage is the risk of overheating and, at some times, increased indoor temperature. Compared to system type 1, the number of hours with increased indoor temperatures is 10 per cent. The most important parameters of the heating systems are summarized in Table 9.5.7.

Table 9.5.7 *Important parameters for the two systems*

	Temperate climate; solar strategy	
	System 1	**System 2**
Design temperature	45°C/40°C	35°C/30°C
Heated surface/heating power	3000 W	40 m² south, 20 m² north Pipe 0.3 m, tube 0.02 m
Collector area:		
base	12 m²	
high performance	11 m²	10 m²
retrofit	13 m²	
Collector type	Flat plate	Flat plate
Collector slope, south	48°	48°
Flow rate	12 l/m²	30 l/m² 12 l/m² DHW
Control	Maximum efficiency	Maximum efficiency
Heat exchanger	92%	92%
Storage	75 l/m²	30 l/m²
Main façade	South	South
Shading coefficient	0.5	0.5
Construction	Heavyweight	Heavyweight

Collector area

The size, orientation and slope of the collector affect the energy output. The influence of collector slope and azimuth angle can be taken from standard tables (Duffie and Beckman, 1991). The optimum value (azimuth south, slope 48°) has been used here. The influence of the collector area is minimal in the range of the chosen values. Figures 9.5.6 and 9.5.7 show the dependency.

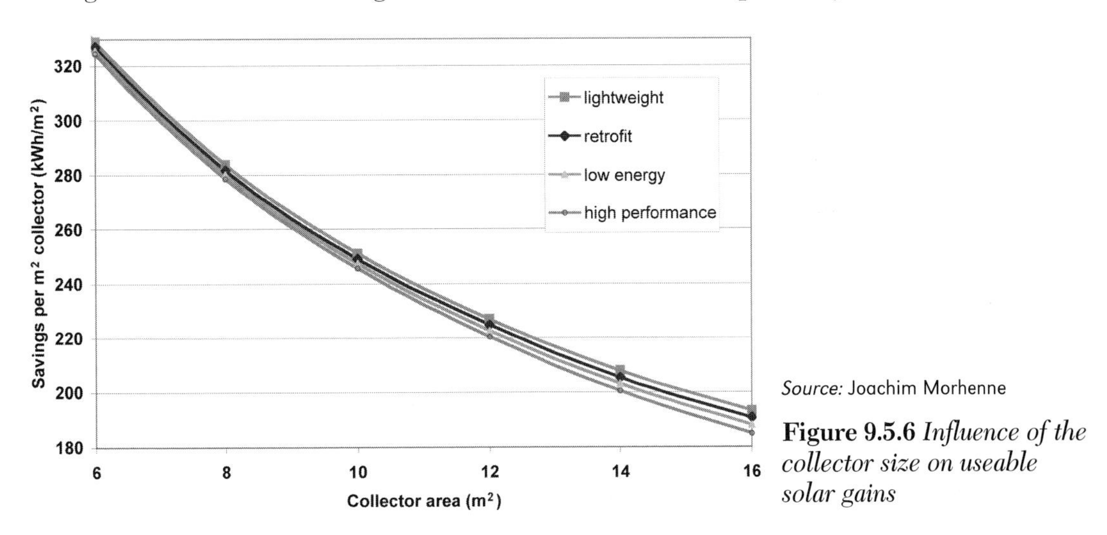

Source: Joachim Morhenne

Figure 9.5.6 *Influence of the collector size on useable solar gains*

The collector gains for a reduction of the space heating demand are limited by the heating demand. The better the building, the lower the useable gains for space heating purposes (see Figure 9.5.7). If the collector area exceeds the given figures, only a further reduction in the energy demand for DHW will be achieved (see Figure 9.5.6).

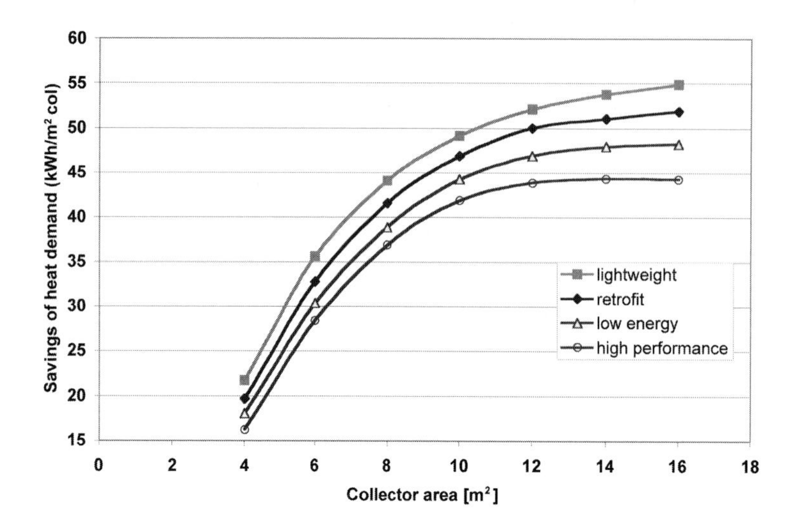

Source: Joachim Morhenne

Figure 9.5.7 *Savings in delivered energy (gas) due to collector gains*

Influence of the building's construction weight

The thermal mass of the building influences the heating demand by storing excess heat coming from passive and internal gains, as well as from an inexact control system. Due to limitations by overheating, this effect amounts to 9 per cent. The most important influence of the thermal mass is an increased summer comfort; but the LCA is influenced negatively.

An interesting issue is that the higher heating demand for lightweight construction can partly be compensated for by excess collector gains (see Figure 9.5.7). In addition, it is possible to use larger collector areas for further compensation.

The analysed lightweight building differs in the mass of walls to the ambient and the walls' dividing units. Lightweight constructions have wooden frames; the massive walls are made of limestone. The floors are massive in all cases. Simulations to optimize the building mass and performance are recommended.

Primary energy demand depending on collector area

Figure 9.5.8 shows the primary energy demand depending on the collector area. Depending on the space heating demand, the primary energy target is met using between 5 m^2 and 8 m^2 of collector area.

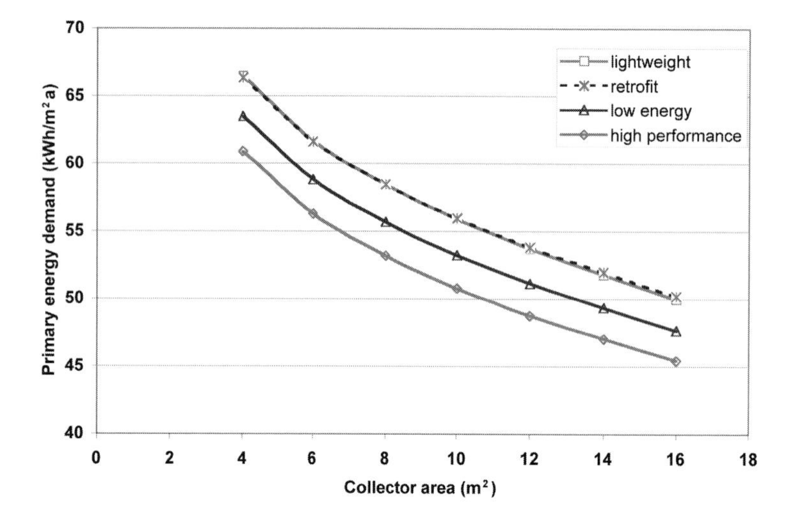

Source: Joachim Morhenne

Figure 9.5.8 *Primary energy demand depending upon collector area*

Influence of the supply temperature of the heating system

The operation temperature primarily affects the efficiency of solar systems. The temperature of the supply and return flow of the heating system therefore has a significant influence because the solar system is always connected to the return flow of the system. The influence of this parameter is shown in Figure 9.5.9.

Because of the reduced specific heating power of low temperature systems (per m^2 of surface), the reduced heating capability has to be compensated for by a larger surface area. Figure 9.5.9 shows the influence. Compared to the design temperature of radiators, the surface has to be increased by a factor of 2.5 to 5 to achieve acceptable solar gains. A return temperature of less than 40°C is recommended (design temperature, floating control by ambient temperature), which is a compromise between costs and performance.

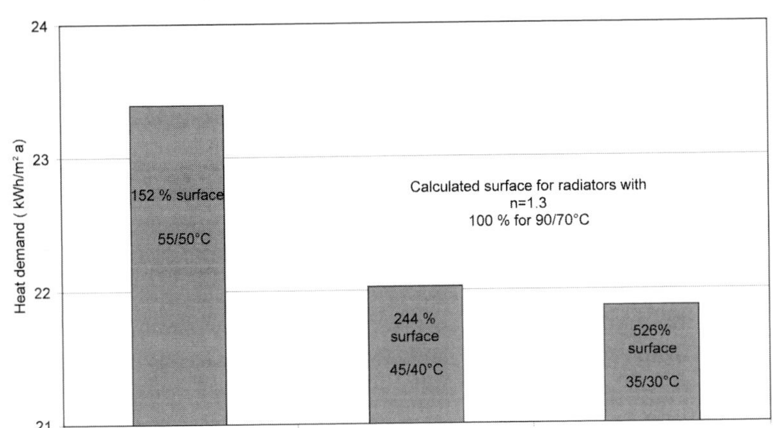

Source: Joachim Morhenne

Figure 9.5.9 *Influence of the heating system's design temperature on solar gains and the surface area of radiators*

Influence of the storage size

The influence of the storage volume of the solar system is low if the storage size exceeds a critical value. In the range of 45 to 85 l/m^2 of collector, the influence of the storage is less than 3 per cent.

Influence of the flow rate in the collector circuit

Simulation results show that for heating purposes, the temperature of the return flow of the heating system has to be exceeded by the collector outlet temperature to transfer any energy to the system. Therefore, the flow rate has to be lowered until the necessary outlet of the collector is reached independently of the efficiency. A dynamic control of the flow rate gives optimum performance. The collector operation for heating is a typical low flow application. In the simulations, 12 l/m^2h have been used.

In case of system type 2, the flow rate in the collector is not independent of the flow rate of the heating system. The flow rate in the floor heating system is limiting the minimum flow rate (value in simulations is 30 l/m^2h). In case of an installation of solar collectors with vacuum tubes, the minimum flow rate is limited by the tube array.

9.5.4 Design advice

In system type 1, a typical solar combi-system is used. The main parameters have been shown; for further information about solar combi-systems, see the results from IEA SHC Task 26 (Weiss, 2003). For system type 2, a simulation is recommended, especially if higher solar passive gains can be achieved.

In all cases (if not connected to a district heating system), a central heating furnace is recommended to reduce investment costs and to be able to install a solar system that is not oversized.

Central collector systems have the advantage of lower costs and have reduced losses compared to six individual systems; the performance thus increases and can be compensated for by a smaller collector area. Individual solar systems have the advantage that the heating grid can be switched off in summer, resulting in less grid losses. An additional advantage is the higher comfort of a hot water storage compared to solutions using a heat exchanger to heat the DHW.

To reduce the losses in the heating grid, it is recommended to install an electric backup heating in the storage for the summer backup if individual solar systems are chosen. This will, in most cases, use less than 5 per cent of the remaining energy for heating the DHW.

A further reduction of primary energy is possible when using a biomass furnace. A central furnace for a small group of buildings requires a co-operative ownership of the heating system or an operating firm.

Table 9.5.8 *Construction of different cases (building envelope target)*

	Material	Thickness			Conductivity	Per cent	Studs	Studs	Resistance			U-value		
		m			λ (W/mK)	%	λ (W/mK)	%	(m²K/W)			(W/m²K)		
		Temp ref	Temp basis	Temp retrofit					Temp ref	Temp basis	Temp retrofit	Temp ref	Temp basis	Temp retrofit
wall	exterior surface								0.04	0.04	0.04			
	plaster	0.02	0.02	0.02	0.9	100%			0.02	0.02	0.02			
	polystyrol	0.059	0.1	0.1	0.035	100%			1.69	2.86	2.86			
	limestone	0.175	0.175	0.175	0.560	100%			0.31	0.31	0.31			
	plaster	0.015	0.015	0.015	0.7	100%			0.02	0.02	0.02			
									0.13	0.13	0.13			
		0.269	**0.31**	**0.31**					**2.21**	**3.38**	**3.38**	**0.45**	**0.30**	**0.30**
wall E/W	exterior surface								0.04	0.04	0.04			
	plaster	0.02	0.02	0.02	0.9	100%			0.02	0.02	0.02			
	polystyrol	0.08	0.12	0.1	0.035	100%			2.29	3.43	2.86			
	limestone	0.175	0.175	0.175	0.560	100%			0.31	0.31	0.31			
	plaster	0.015	0.015	0.015	0.7	100%			0.02	0.02	0.02			
									0.13	0.13	0.13			
		0.29	**0.33**	**0.31**					**2.81**	**3.95**	**3.38**	**0.36**	**0.25**	**0.30**
roof	exterior surface								0.04	0.04	0.04			
	roof tiles	0.050	0.050	0.050		100%								
	air gap	0.045	0.045	0.045		100%								
	protection foil	0.0025	0.0025	0.0025		100%								
	mineral wool	0.146	0.200	0.160	0.039	85%			3.74	4.59	3.67			
	wood	0.200	0.220	0.160			0.13	15%	1.54	1.69	1.23			
	PE- foil					100%								
	gypsum board	0.013	0.013	0.013	0.210	100%			0.06	0.06	0.06			
	interior surface								0.10	0.10	0.10			
		0.4565	**0.5305**	**0.4305**					**3.61**	**4.36**	**3.51**	**0.28**	**0.23**	**0.29**
floor														
	concrete	0.012	0.012	0.012	2.1	100%			0.01	0.01	0.01			
	mineral wool	0.076	0.080	0.080	0.035	100%			2.17	2.29	2.29			
	pavement	0.060	0.060	0.060	0.800	100%			0.08	0.08	0.08			
	wood	0.020	0.020	0.020	0.130	100%			0.15	0.15	0.15			
	interior surface								0.17	0.17	0.17			
		0.156	**0.160**	**0.160**					**2.58**	**2.69**	**2.69**	**0.39**	**0.37**	**0.37**
window					**glass/LE/gas**									
	pane	0.004	0.004	0.004	low emissivity									
	gas	0.016	0.016	0.016	Argon									
	pane	0.004	0.004	0.004	clear									
		0.024	**0.024**	**0.024**								**1.2**	**1.2**	**1.2**
	frame				wood							**1.70**	**1.70**	**1.70**

References

Duffie, J. A. and Beckman, W. A. (1991) *Solar Engineering of Thermal Processes*, John Wiley and Sons, New York

TRNSYS (2005) *A Transient System Simulation Program*, Solar Energy Laboratory, University of Wisconsin, Madison, WI

Weiss, W. (ed) (2003) *Solar Heating Systems for Houses: A Design Handbook for Solar Combisystems*, James and James Ltd, London

9.6 Life-cycle analysis for row houses in a temperate climate

Alex Primas and Annick Lalive d'Epinay

For the ecological assessment with life-cycle analysis (LCA), row housing was selected since it represents a frequent building type within low energy buildings. The results presented in this section refer to a row house with six units in a temperate climate (reference location: Zurich).

9.6.1 Description of the studied building and system variations

Apart from the reference building (building code 2001), two strategies resulting in low energy demand were included:

1 *Strategy 1 (conservation)*: all six units have their own condensing gas boiler for space heating and DHW. Each unit has 4 m^2 of solar flat-plate collectors for DHW production.

2 *Strategy 2 (renewable energy)*: besides a central condensing gas boiler for heat production, each unit has 12 m^2 of solar flat-plate collectors and a small condensing gas boiler for hot water production.

In the reference house, each of the six units has its own gas boiler for heat and hot water production. The following building construction types were investigated (see Figure 9.6.1):

- *Type A*: massive construction with brick walls and compact insulation made of expanded polystyrene (EPS) for the external walls. The main floor insulation (foam glass) lies below the concrete floor. Above the concrete floor, only a thin acoustic insulation is placed (in both floors).
- *Type B*: massive construction with limestone walls and a ventilated wood façade insulated with rock wool. The main floor insulation (rock wool) lies above the concrete floor. Above the floor a thin acoustic insulation is also placed (in both floors).
- *Type C*: hybrid building construction with a wood frame construction for the wall elements insulated with cellulose flakes and a ventilated wood façade. The floors are made of reinforced concrete. The main floor insulation (extruded polystyrene, XPS) lies below the concrete floor. Above the floor only a thin acoustic insulation is placed (in both floors).

Further details of the assessed building construction are found in Lalive d'Epinay et al (2004).

The approach

Within the system boundaries, the production, renewal and disposal of all materials that have an influence on the total energy demand of the house were considered. As electricity from the grid, the Union for the Coordination of Transmission of Electricity (UCTE) electricity mix on a low voltage level was used. This leads to larger differences compared to the cumulative energy demand (CED) factor for EU 17 electricity taken from GEMIS (see www.oeko.de/service/gemis/en/index.htm), used within the energy analyses of the typical solutions in this book (see also Appendix 2). The UCTE conversion factor for electricity (non-renewable) used in this chapter was 3.56, while the CED

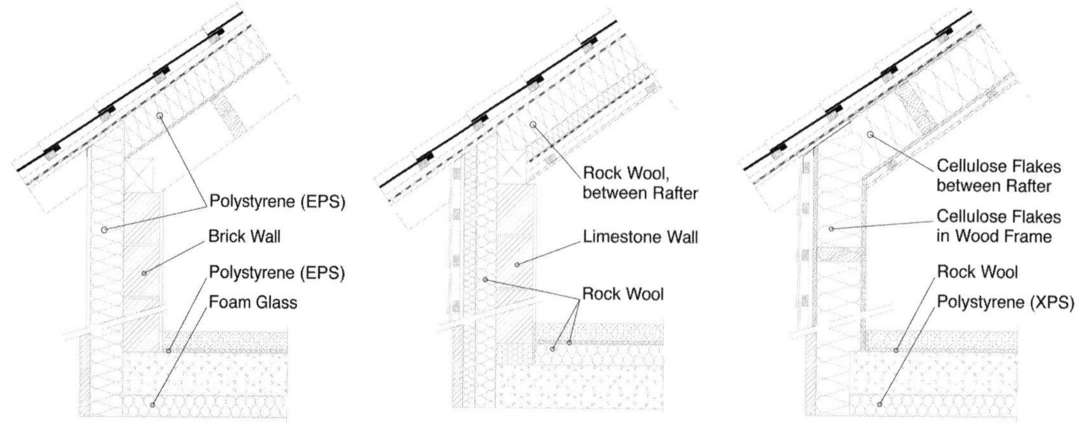

Source: Alex Primas and Annick Lalive d'Epinay

Figure 9.6.1 *Construction types: (a) construction type A; (b) construction type B; (c) construction type C*

conversion factor from GEMIS was 2.35. The EU 17 electricity mix was not available within the LCA data used (Frischknecht et al, 1996). The electricity demand of the household appliances was excluded in the discussion of the results.

9.6.2 Results from the analyses

For all the studies, the two strategies were each calculated with the three construction types. In Figures 9.6.2 to 9.6.6, the different bars show the strategies used (1 = conservation; 2 = solar) and the different construction types (A, B, C) for strategies 1 and 2. The reference building 'R' (construction type A) is always showed as a comparison. All results refer to the loads per m^2 net carpet area and year (building life span: 80 years).

Influence of the life-cycle phases

In Figure 9.6.2 the share of construction, renewal, disposal, transportation and operation on the total impact over the entire life cycle is shown. The analysis was carried out with the weighting method Eco-indicator 99 (hierarchist). The most important results are as follows:

- The total impact over the life cycle of the high-performance building with strategy 1 is 54 per cent to 58 per cent of the total impact of the reference building. For strategy 2, the impact is 60 per cent to 64 per cent, compared to the reference building.
- The operation phase makes up only 47 per cent to 54 per cent of the total impact of the high-performance building.
- The impact from the material used for renewal over the whole building lifetime is in the same order as the impact from the initial building construction.
- The destruction of the building and disposal of the materials from renovation and destruction are not negligible and amount to 11 per cent to 14 per cent of the total impact.
- The impact of the transport of materials to the construction site is very low, with about 2 per cent to 3 per cent of the total impact of the high-performance building.

In the calculation of the CED (non-renewable energy), the share of the operation phase of the typical solution shows with 65 per cent to 72 per cent of the total impact, a clearly higher value compared to the calculation with Eco-indicator 99. This is, to a large extent, caused by the low impact of the disposal (5 per cent to 7 per cent of the total impact) in the calculation with CED.

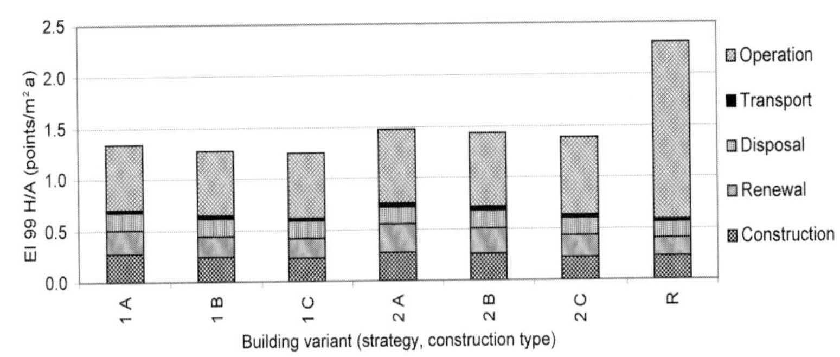

Source: Alex Primas and Annick Lalive d'Epinay

Figure 9.6.2 *Life-cycle phases, Eco-indicator 99 H/A*

If the renewable energy part of the CED is included, there is a difference between the construction types due to wood used for construction, especially in construction type C. The non-renewable CED has a share of 90 per cent (type A) to 82 per cent to 84 per cent (type C) of the total CED (renewable + non-renewable).

Influence of the building components

In Figure 9.6.3, the share of the impact of different material groups on the total impact over the entire life cycle was analysed with the Eco-indicator 99 method. The most important results are as follows:

- The impact of the insulation materials is only 2 per cent of the total impact of the high-performance building for construction type B and C, and 5 per cent to 7 per cent for construction type A.
- The massive building materials amount to 17 per cent to 26 per cent of the total impact over the life cycle of the high-performance building. This impact is caused mainly by the cement and concrete in the floor constructions. The smallest shares are found in construction type C.
- The impact of the heating, DHW and ventilation system makes up only 4 per cent to 6 per cent of the total impact over the life cycle of the high-performance building.
- In strategy 2 (solar), the thermal solar collector (12 m^2 per unit) amounts to about 6 per cent to 7 per cent of the total impact of the high-performance building. In strategy 1 (conservation), the impact of the small thermal solar collector for DHW (4 m^2 per unit) amounts to 2 per cent to 3 per cent of the total impact of the building.

The calculation with CED (non-renewable) shows similar results here with, the difference that the shares of the components are smaller due to the higher importance of the operation energy (see Figure 9.6.4). The largest difference lies within the impact of the massive building materials, where the calculation of the CED shows an impact share of 8 per cent to 12 per cent of the total impact. The share of the thermal solar collectors in the solution for strategy 2 is also, with 3 per cent to 4 per cent of the total impact, clearly lower.

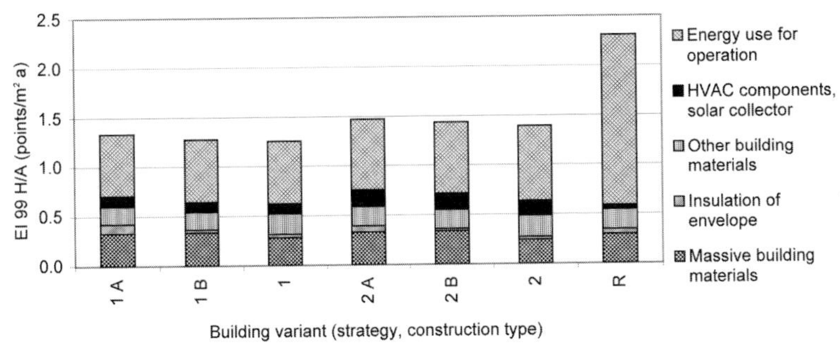

Source: Alex Primas and Annick Lalive d'Epinay

Figure 9.6.3 *Building components, Eco-indicator 99 H/A*

Source: Alex Primas and Annick Lalive d'Epinay

Figure 9.6.4 *Building components, cumulative energy demand (non-renewable)*

Influence of the heating system

In Table 9.6.1, the basic data for the investigated heating systems are shown. For all solutions, the construction type B (massive walls, rock wool insulation) was used. Data used for the reference building refer to a gas boiler with an efficiency of 94 per cent (lower heating value (LHV)).

Table 9.6.1 *Basic parameters of the investigated heating systems; the collector area is per row house unit*

System	Efficiency	Strategy 1	Strategy 2
Gas boiler	Efficiency 100% (LHV)	+ 4 m^2 solar collector	+ 12 m^2 solar collector
Oil boiler	Efficiency 98% (LHV)	+ 4 m^2 solar collector	+ 12 m^2 solar collector
Heat pump	COP 2.52 (heat + DHW)	No solar collector	Not investigated
Pellet boiler	Efficiency 85% (LHV)	Not investigated	+ 12 m^2 solar collector

In Figure 9.6.5, the influence of different heating systems on the total impact over the entire life cycle was analysed with the Eco-indicator 99 method (hierarchist). This figure shows the three damage categories that are differentiated within the Eco-indicator 99 methodology.

Figure 9.6.5 shows clearly the advantages of the heat pump in strategy 1 (conservation) and the use of wood in strategy 2 (renewable energy). These two variants show, with 47 per cent and 55 per cent, respectively, the highest impact reduction over the whole life cycle compared to the reference building. This reduction is achieved only within the damage category 'depletion of resources'. Within the damage categories 'human health' and 'ecosystem quality', the heat pump shows the highest impacts of all variants (including the reference house) due to the electricity consumption. The heat pump would also have an advantage in these categories if an electricity mix with a high amount of renewable energy would be used. The comparison between oil and gas shows a slight advantage for gas.

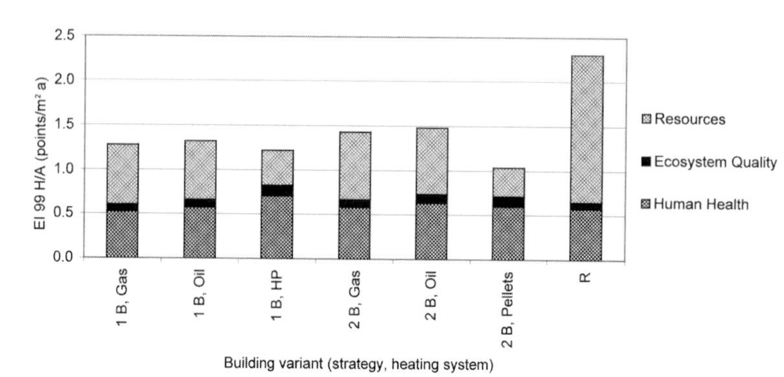

Source: Alex Primas and Annick Lalive d'Epinay

Figure 9.6.5 *Influence of the heating system, Eco-indicator 99 H/A*

The calculation with CED gives different results for the heat pump. Here, the low efficiency of the electricity generation and the low COP of the air–water heat pump is responsible for the poor result. The heat pump shows only an impact reduction of 36 per cent compared to the reference building, while the variant with the pellet furnace in strategy 2 shows, with 63 per cent, the highest impact reduction over the whole life cycle compared to the reference building.

Influence of the solar collector area

The influence of the solar collector area on the total impact over the entire life cycle was analysed for the typical solutions with strategy 2 (construction type B) using the Eco-indicator 99 method. Figure 9.6.6 shows that the total impact does not get much lower for collector areas over 10 m^2 to 12 m^2 per unit (120 m^2 net carpet area). For the assessment with CED (non-renewable), the collector area, for which the total impact does not get much lower, is slightly higher, with about 16 m^2 collector area per unit.

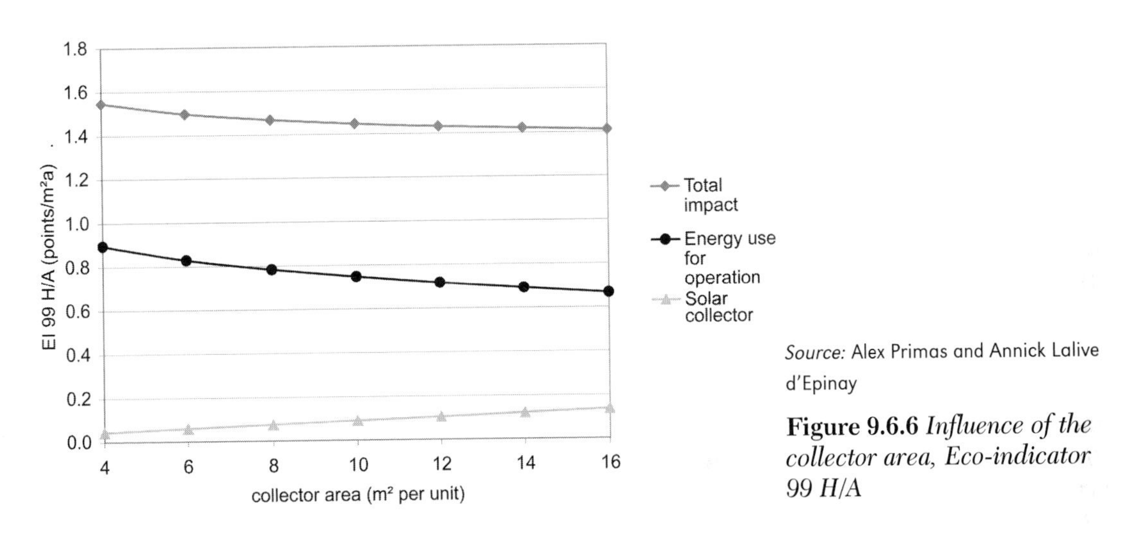

Source: Alex Primas and Annick Lalive d'Epinay

Figure 9.6.6 *Influence of the collector area, Eco-indicator 99 H/A*

9.6.3 Conclusions

For the energy optimized buildings, the impact from the operation phase makes up between only one third to two thirds of the total impact over the life cycle. The actual share is dependent on the building type and assessment method. Therefore, the choice of material becomes more important. It is clear that an ecological optimization considering the impact over the life cycle is needed for building parts that affect the energy demand (for example, insulation, solar collectors and ventilation unit).

The results show that the share of the insulation materials on the total impact is small. If attention is given to the choice of the insulation materials, the share of the insulation materials was below 3 per cent of the total impact. Therefore, for ecologically optimal buildings, the amount of insulation material could be far higher. On the other hand, a larger solar collector as used in strategy 2 (0.1 m^2 per m^2 net heated floor area) does not give a clear ecological advantage. For the ventilation system, the use of ventilators with a low power consumption is important (direct currant motors with total 0.3 W per m^2/h or less for the ventilation unit). Here also, the balance between energy demand for electricity and the heat recovery rate has to be considered (pressure losses).

Generally, it can be stated that the renewal phase causes just as high impacts over the entire building lifespan as the construction of the building. Therefore, the use of long-lasting and recyclable building materials (especially for the interior) is important. A large part of the impact also results from the materials used for the primary structure of the house (floor, walls and roof). This impact depends on the basic concept of the building and can therefore only be influenced in the early planning stages. The results showed clearly that in all cases transportation to the building site has a small impact if the massive construction materials are not transported over far distances.

References

Frischknecht, R., Bollens, U., Bosshart, S., Ciot, M., Ciseri, L., Doka, G., Hischier, R., Martin, A., Dones, R. and Gantner, U. (1996) *Ökoinventare von Energiesystemen, Grundlagen für den ökologischen Vergleich von Energiesystemen und den Einbezug von Energiesystemen in Ökobilanzen für die Schweiz*, Bundesamt für Energie, (BfE), Bern, Switzerland

Lalive d'Epinay, A., Primas, A. and Wille, B. (2004) *Ökologische Optimierung von Solargebäuden über deren Lebenszyklus*, Schlussbericht, IEA SHC Task 28/ECBCS Annex 38, Sustainable Solar Housing, Bundesamt für Energie (BFE), Basler & Hofmann AG, Bern, Switzerland

9.7 Apartment building in the Temperate Climate Conservation Strategy

D.I. Sture Larsen

Table 9.7.1 *Targets for apartment building in the Temperate Climate Conservation Strategy*

	Targets
Space heating	15 kWh/m²a
Primary energy (space heating + water heating + electricity for mechanical systems)	60 kWh/m²a

This section presents a solution for the apartment building in the temperate climate. As a reference for the temperate climate, the city of Zurich is used. The solution is based on energy conservation, minimizing the heat losses of the building. The apartment building itself is compact and has a favourable ratio of envelope area to building volume, making a relatively moderate insulation level of the envelope sufficient to meet the target (assuming thermal bridges are eliminated). A mechanical ventilation system with heat recovery is used to reduce the ventilation losses.

9.7.1 Solution 1: Conservation with condensing gas boiler and solar DHW

Key elements

- Mechanical ventilation with heat recovery (80 per cent).
- Auxiliary heat distribution by a central heating system.
- Condensing gas boiler.
- Solar domestic hot water (75 m^2 collector area).

Building envelope

The building construction consists of concrete floors and masonry insulated externally with polystyrene insulation. The external insulation has the advantages of both avoiding thermal bridges and providing thermal mass inside the envelope, which helps to store passive solar gain. The thermal mass also improves summer comfort.

The windows are double glazed with one low-e coating and are filled with argon. The frame area is only 20 per cent of the window area. This choice is made to increase gains and reduce losses since the frame has a slightly higher U-value than the glazing and cannot transmit solar radiation. This solution requires special attention, such as letting the insulation of the wall overlap a large part of the window frame.

U-values are shown in Table 9.7.2. Another useful parameter for the insulation level of the total building envelope is the specific heat load, which in this case is 0.78 W/m²K (floor area) for the envelope without ventilation heat recovery and 0.50 W/m²K with heat recovery.

Table 9.7.2 *U-values of the building components*

Component	U-Value (W/m²K)
Walls	0.28 (25 cm masonry +12 cm polystyrene)
Roof	0.21
Floor (excluding ground)	0.33
Windows (frame and glass)	1.32
Window frame	1.42
Window glass	1.30
Average for building envelope	0.40

Mechanical systems

Ventilation: the balanced mechanical ventilation system with heat recovery includes a winter and a summer mode. The summer mode has a bypass of the heat exchanger and is used for cooling with night air. If neither heat recovery nor cooling is needed, the mechanical ventilation system can be shut off, assuming occupants will open windows to ventilate.

Space and DHW heating: the space heating is supplied by a hydronic central heating system using a condensing gas boiler. The DHW heating is based on solar collectors and, when needed, energy from the gas boiler. There is no backup with electricity.

Energy performance

Space heating demand and peak load: the space heating demand is 16,200 kWh/a (10 kWh/m²a) according to DEROB-LTH. Assumptions are shown in Table 9.7.3. Hourly loads of the heating system are calculated from the DEROB simulation results without direct solar radiation in order to simulate a shaded building. The annual peak load is then 18.0 kW or 11.3 W/m² at –13.1°C.

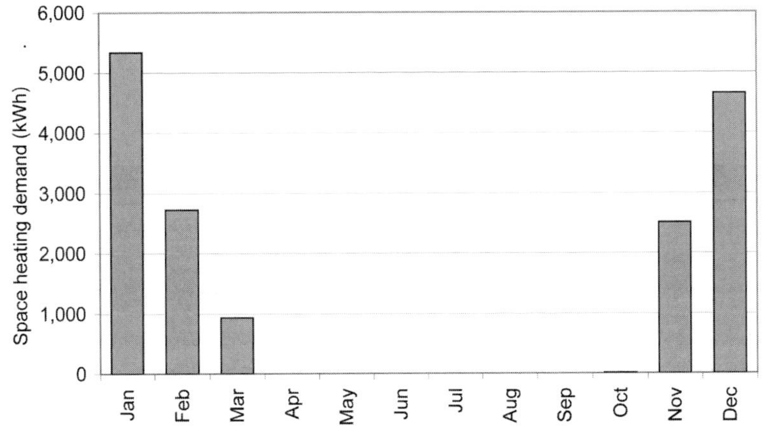

Source: D. I. Sture Larsen

Figure 9.7.1 *Monthly space heating demand*

Table 9.7.3 *Assumptions for the simulations*

Heating set point	20°C
Maximum room temperatures	23°C (assumes shading and window ventilation) 26°C for June, July and August
Ventilation rate	0.45 ach
Infiltration rate	0.05 ach
Heat recovery efficiency	80%

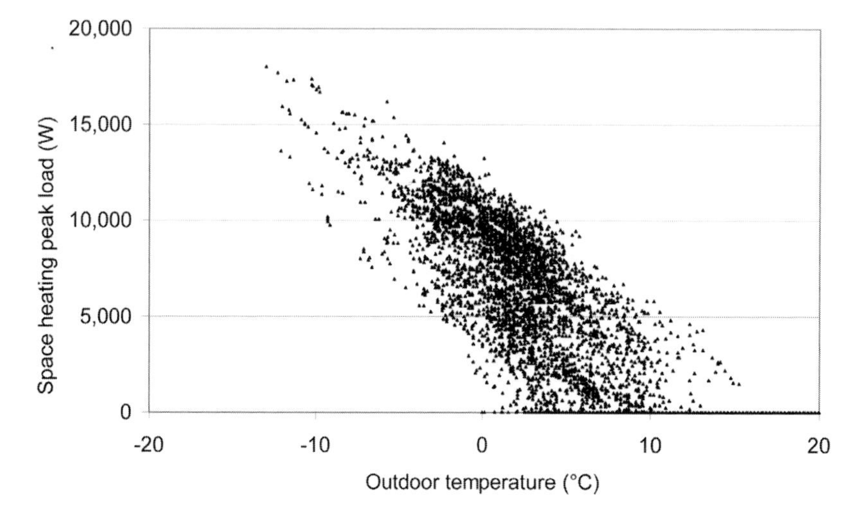

Source: D. I. Sture Larsen

Figure 9.7.2 *Space heating peak load; results from simulations without direct solar radiation*

Domestic hot water demand: the net DHW heat demand is approximately 36,800 kWh (23 kWh/m^{2}a). The DHW demand is larger than the space heating demand. This is typical for a compact building with a low specific heat load and a low ratio envelope-to-volume area.

Household electricity: the use of household electricity for each apartment is 22 kWh/m^{2}a for two adults and one child. The primary energy target does not include household electricity since this factor can vary considerably depending on the occupants' behaviour and if efficient household appliances are installed or not.

Non-renewable primary energy demand and CO_2 emissions: the total end energy use for space heating, DHW, system losses and mechanical systems is approximately 40 kWh/m^{2}a. After taking into account the solar contribution for DHW, the delivered energy is approximately 21 kWh/m^{2}a (see Table 9.7.4). The non-renewable primary energy demand is approximately 30 kWh/m^{2}a (48600 kWh/a) and CO_2 equivalent emissions are 6.2 kg/m^{2}a.

The results are clearly below the targets of 15 kWh/m^{2}a and 60 kWh/m^{2}a, respectively. However, it is important not to reduce the insulation level too much since the building envelope should be robust for many years. The use of active solar for DHW is an important means of reducing the demand for primary energy.

Table 9.7.4 *Total energy use, non-renewable primary energy demand and CO_2 equivalent emissions for the apartment building with condensing gas boiler and solar DHW*

Net Energy (kWh/m²a)		Total Energy Use (kWh/m²a)				Delivered energy (kWh/m²a)		Non renewable primary energy		CO_2 equivalent emissions	
		Energy use		Energy source				factor	(kWh/m²a)	factor (kg/kWh)	(kg/m²a)
Mechanical systems	5.0	Mechanical systems	5.0	Electricity	5.0	Electricity	5.0	2.35	11.8	0.43	2.2
Space heating	10.1	Space heating	10.1	Gas	16.4	Gas	16.4	1.14	18.7	0.25	4.0
DHW	23.1	DHW	23.1								
		Tank and circulation losses	1.8								
		Conversion losses	0.0	Solar	18.6						
Total	38.2		40.0		40.0		21.4		30.5		6.2

9.7.2 Sensitivity analysis

Glazing area

Three different glazing areas (A, B and C) were used in the simulations (see Table 9.7.5). The results showed that an increased south glazing area (22 per cent to 47 per cent glazing to façade area ratio) had no significant influence on the space heating demand (see Figure 9.7.3). The window type is double glazing with one low-e coating and argon.

Increased losses caused by larger windows towards the south were largely compensated for by higher passive solar gains. Note that this conclusion is valid for a heavy construction. A very high window to façade area ratio combined with a lightweight construction will lead to higher space heating demand and more problems with overheating. Therefore, the glazing area B was used in this solution. It is important to study indoor temperatures and comfort and not only heating demand when choosing the glass area.

Table 9.7.5 *Variations A, B and C of the glazing area*

Window/façade area ratio for the north and south façades				
North	Glazing		17%	
	Windows		21%	
		Glazing area A	Glazing area B	Glazing area C
South	Glazing	22%	30%	47%
	Windows	28%	38%	58%

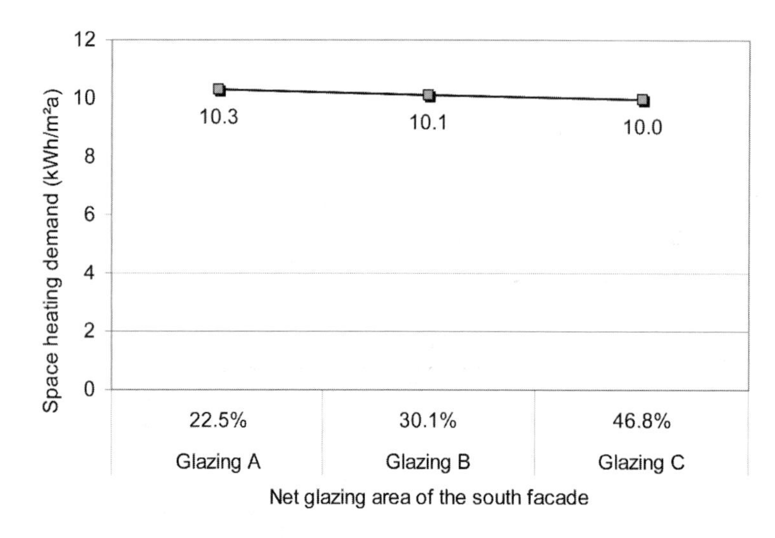

Source: D. I. Sture Larsen

Figure 9.7.3 *Space heating demand with different glazing areas (double glazing, one low-e coating and argon)*

Building envelope and heat recovery

Figure 9.7.4 shows the space heating demand for different insulation levels for the envelope with glazing area B. The insulation level with the mean U-value 0.40 W/m^2K and with ventilation heat recovery was found to meet the target for the apartment building with conservation strategy.

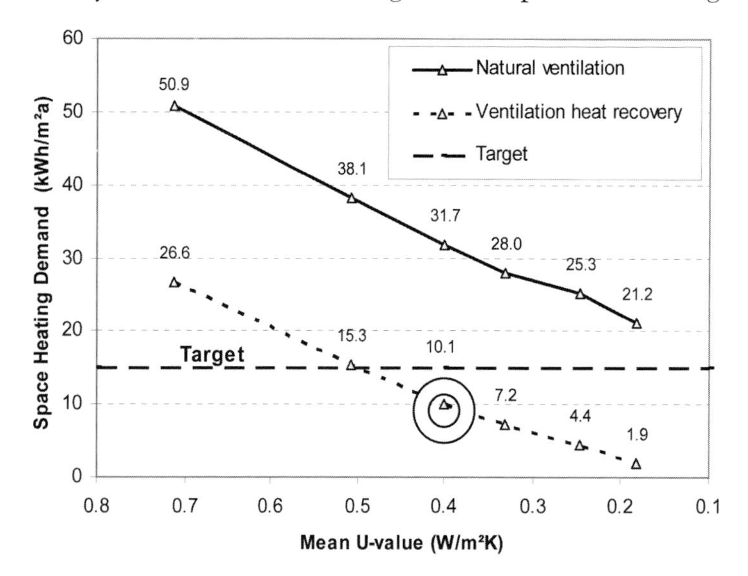

Note: The points represent different simulated cases. All cases are with the glazing area B. The solution is marked with a double circle. The horizontal line shows the space heating target of 15 kWh/m^2a.

Source: D. I. Sture Larsen

Figure 9.7.4 *Space heating demand for different insulation levels with and without ventilation heat recovery*

The efficient ventilation heat recovery of 80 per cent has a large influence on the space heating demand, which decreases with approximately 22 kWh/m^2a for all of the insulation levels. The apartment building without ventilation heat recovery does not meet the space heating target of 15 kWh/m^2a.

The slope of the lower end of the curves (for both natural ventilation and mechanical ventilation with heat recovery) is slightly different than the upper parts due to the fact that in the two last simulation cases, the window type was changed from double glazed to triple glazed with different g-values, which alters the useful passive solar gains compared to transmission losses.

9.7.3 Conclusions

A moderate insulation level of this compact building is sufficient to reach the target as long as a mechanical ventilation system with efficient heat recovery is used. The amount of glazing is not critical within a reasonable range in a building with sufficient constructional storage mass.

Active solar for DHW should be considered to be a normal and very useful part of the energy concept. The primary energy demand can also be reduced by an increased use of renewable energy, such as biomass.

Table 9.7.6 *Building envelope constructions*

Element		Layer	Thickness [m]	Conductivity λ [W/mK]	Resistance R [m² K/W]	U-Value [W/m² K]
Wall		Plaster	0.015	1.000	0.015	
		Brick	0.250	0.580	0.431	
		Polystyrol	0.120	0.041	2.927	
		Plaster	0.005	0.700	0.007	
		Surface resistances			0.170	
		Total	0.390		3.550	**0.28**
Roof		Earth	0.100	1.160	0.086	
		Drain layer (air)	0.030		0.160	
		Plastic foil	0.002	0.190	0.011	
		Protection felt	0.002	0.220	0.009	
		Polystyrol	0.160	0.038	4.211	
		Protection felt	0.002	0.220	0.009	
		Vapour barrier (PE)	0.001	0.200	0.005	
		Screed	0.030	1.400	0.021	
		Reinforced concrete	0.200	2.330	0.086	
		Plaster	0.015	1.000	0.015	
		Surface resistances			0.140	
		Total	0.542		4.753	**0.21**
Floor over basement		Parquet	0.015	0.200	0.075	
		Screed	0.060	1.400	0.043	
		Acoustic boards (min.fibre)	0.030	0.035	0.857	
		Leveling granulate	0.040	0.700	0.057	
		Reinforced concrete	0.200	2.330	0.086	
		Cork, expanded (R=140)	0.070	0.040	1.750	
		Plaster	0.015	1.000	0.015	
		Surface resistances			0.170	
		Total	0.430		3.053	**0.33**
Window		Glass	0.004			
		Argon	0.015			
		Glass, low-e 10%	0.004			
		Total glazing	0.023			1.30
		Frame (20% of window)				1.42
		Window average				**1.32**

References

Kvist, H. (2005) *DEROB-LTH for MS Windows, User Manual Version 1.0–20050813*, Energy and Building Design, Lund University, Lund, Sweden, www.derob.se

Meteotest (2004) *Meteonorm 5.0 – Global Meteorological Database for Solar Energy and Applied Meteorology*, Bern, Switzerland, www.meteotest.ch

Polysun (2005) *Polysun 3.3.5j (Larsen Version)*, Swiss Federal Office of Energy (Bundesamt für Energie), SPF Rapperswil, Solar Energy Laboratory SPF (Institut für Solartechnik), Bern, Switzerland, www.solarenergy.ch

9.8 Apartment building in the Temperate Climate Renewable Energy Strategy

D.I. Sture Larsen

Table 9.8.1 *Targets for apartment building in the Temperate Climate Renewable Energy Strategy*

	Targets
Space heating	20 kWh/m^{2}a
Primary energy (space heating + water heating + electricity for mechanical systems)	60 kWh/m^{2}a

This section presents a solution for the apartment building in the temperate climate. As a reference for the temperate climate, the city of Zurich is used. The solution is based on renewable energy in order to minimize the consumption of non-renewable primary energy and, consequently, the emission of CO_2. The apartment building itself is compact and has a favourable ratio of envelope area to building volume, making a relatively moderate insulation level of the envelope sufficient. A mechanical ventilation system with heat recovery is used to reduce the ventilation losses.

9.8.1 Solution 2: Renewable energy with a biomass boiler and solar DHW

Key elements

- Mechanical ventilation with heat recovery (80 per cent).
- Auxiliary heat distribution by a central heating system.
- Biomass boiler (pellets).
- An active solar combi-system for DHW and space heating (100 m^2 collector area).

Building envelope

The building construction consists of concrete floors and masonry insulated externally with polystyrene insulation. The external insulation has the advantages of both avoiding thermal bridges and providing thermal mass inside the envelope, which helps to store passive solar gain. The thermal mass also improves summer comfort.

The windows are double glazed with one low-e coating and are filled with argon. The frame area is only 20 per cent of the window area. This choice is made to increase gains and reduce losses since the frame has slightly higher U-value than the glazing and cannot transmit solar radiation. This solution requires special attention, such as letting the insulation of the wall overlap a large part of the window frame.

U-values are shown in Table 9.8.2. Another useful parameter for the insulation level of the total building envelope is the specific heat load, which, in this case, is 0.78 W/m^{2}K (floor area) for the envelope without ventilation heat recovery and 0.50 W/m^{2}K with heat recovery.

Table 9.8.2 *U-values of the building components*

Component	U-Value (W/m^{2}K)
Walls (25 cm masonry +12 cm polystyrene)	0.28
Roof	0.21
Floor (excluding ground)	0.33
Windows (frame and glass)	1.32
Window frame	1.42
Window glass	1.30
Average for building envelope	0.40

Mechanical systems

Ventilation: the balanced mechanical ventilation system with heat recovery includes a winter and a summer mode. The summer mode has a bypass of the heat exchanger and is used for cooling with night air. If neither heat recovery nor cooling is needed, the mechanical ventilation system can be shut off. However, care should be taken to ensure high indoor air quality. If the ventilation system is shut off part of the time, it is important to ventilate by opening windows.

Domestic Hot Water and space heating systems: the space heating is supplied by a hydronic central heating system using a condensing biomass boiler. The DHW heating is based on solar collectors and, when needed, energy from the biomass boiler. There is no backup with electricity.

Energy performance

Space heating demand and peak load: the space heating demand is 16,200 kWh/a (10 kWh/m^{2}a) according to DEROB-LTH. Assumptions are shown in Table 9.8.3. Hourly loads of the heating system are calculated from the DEROB simulation results without direct solar radiation in order to simulate a shaded building. The annual peak load is then 18.0 kW or 11.3 W/m^2 at –13.1°C.

Table 9.8.3 *Assumptions for the simulations*

Heating set point	20°C
Maximum room temperatures	23°C (assumes shading and window ventilation) 26°C for June, July and August
Ventilation rate	0.45 ach
Infiltration rate	0.05 ach
Heat recovery efficiency	80%

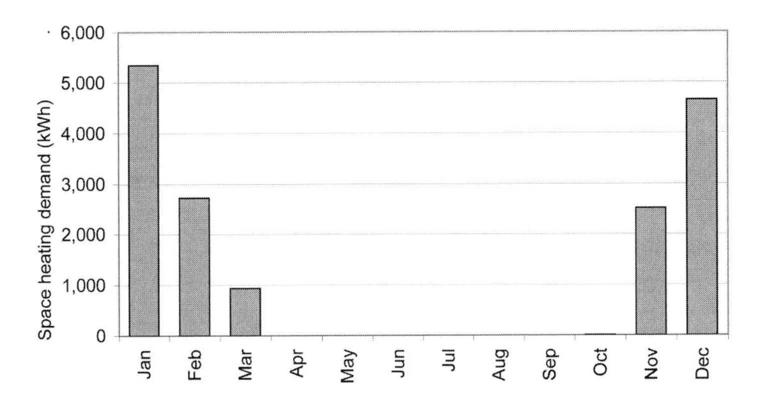

Source: D. I. Sture Larsen

Figure 9.8.1 *Monthly space heating demand*

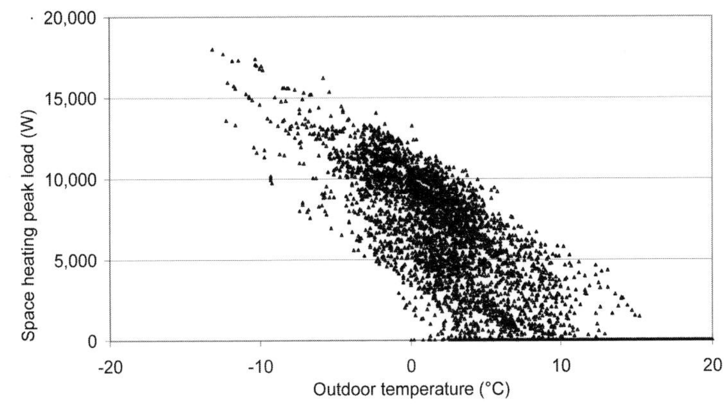

Source: D. I. Sture Larsen

Figure 9.8.2 *Space heating peak load; results from simulations without direct solar radiation*

Domestic hot water demand: the net DHW heat demand is approximately 36,800 kWh (23 kWh/m^{2}a). The DHW demand is larger than the space heating demand. This is typical for a compact building with a low specific heat load and a low ratio envelope to volume area.

Household electricity: the use of household electricity for each apartment is 22 kWh/m^{2}a for two adults and one child. The primary energy target does not include household electricity since this factor can vary very considerably depending on the occupants' behaviour and if efficient household appliances are installed or not.

Non-renewable primary energy demand and CO_2 emissions: the total end energy use for space heating, DHW, system losses and mechanical systems is approximately 43 kWh/m^{2}a. After taking into account the solar contribution for DHW and space heating, the delivered energy is approximately 23 kWh/m^{2}a (see Table 9.8.4). The non-renewable primary energy demand is approximately 14 kWh/m^{2}a (22,800 kWh/a) and CO_2 equivalent emissions are 3.0 kg/m^{2}a.

The results are clearly below the targets of 15 kWh/m^{2}a and 60 kWh/m^{2}a respectively. However, it is important not to reduce the insulation level too much since the building envelope should be robust for many years. The use of active solar is an important means of reducing the demand for primary energy.

Table 9.8.4 *Total energy use, non-renewable primary energy demand and CO_2 equivalent emissions for the apartment building with biomass boiler and solar DHW and space heating*

Net Energy (kWh/m² a)		Total Energy Use (kWh/m² a)				Delivered energy (kWh/m² a)		Non renewable primary energy		CO_2 equivalent emissions	
		Energy use		Energy source				factor	(kWh/m² a)	factor (kg/kWh)	(kg/m² a)
Mechanical systems	5.0	Mechanical systems	5.0	Electricity	5.0	Electricity	5.0	2.35	11.8	0.43	2.2
Space heating	10.1	Space heating	10.1	Bio-pellets	18.0	Bio-pellets	18.0	0.14	2.5	0.04	0.8
DHW	23.1	DHW	23.1								
		Tank and circulation losses	2.2								
		Conversion losses	2.7	Solar	20.1						
Total	38.2		43.1		43.1		23.0		14.3		3.0

9.8.2 Sensitivity analysis

Collector area

Six different collector areas (0 m^2, 50 m^2, 75 m^2, 100 m^2, 150 m^2 and 200 m^2) were used in the simulations. The results showed that the most significant influence on the demand for primary energy and on the CO_2 emission came from the decision to use an active solar system. The area of the collector was less important. The improvement achieved by increasing the collector area decreases with the size of the collector.

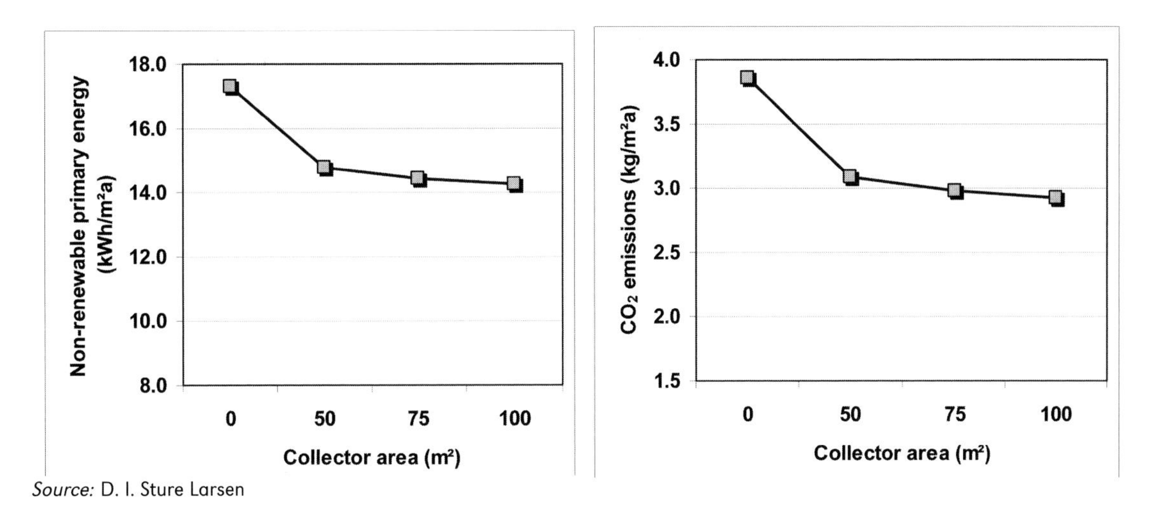

Source: D. I. Sture Larsen

Figure 9.8.3 *Influence of the collector area on the primary energy demand* (left) *and CO_2 emissions* (right)*; solution with biomass boiler and solar combi-system*

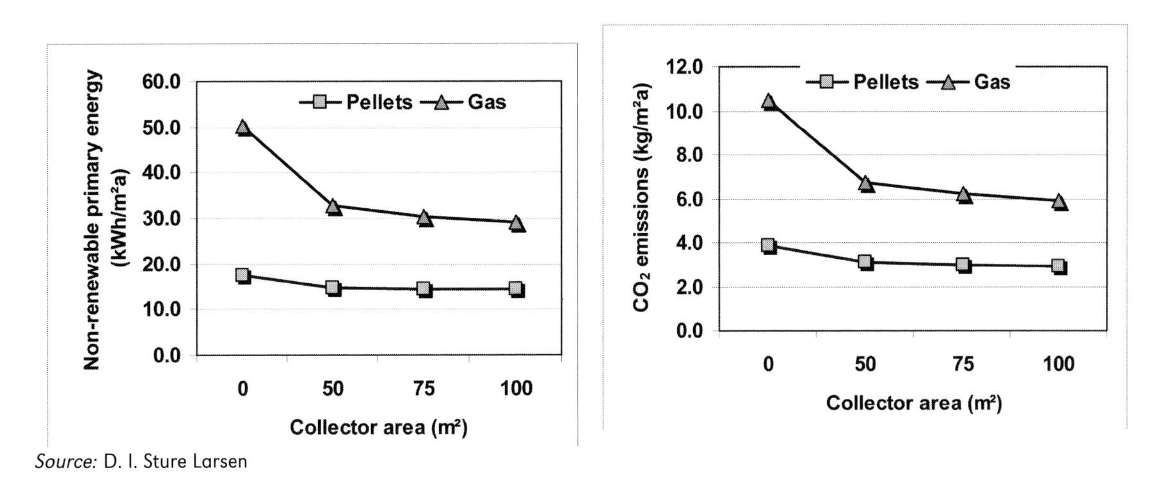

Source: D. I. Sture Larsen

Figure 9.8.4 *Influence of the choice of energy source on primary energy demand* (left) *and CO_2 emissions* (right)*; a comparison between biomass fuel and gas*

9.8.3 Design advice

Given that the heat loss of a building is defined by transmission and ventilation losses, ventilation heat recovery is important. For the non-renewable primary energy demand, both a solar system and the choice of fuel have an important impact on the results. A solar system should always be considered.

In this compact apartment building, solar-heated DHW is very important because the demand for DHW is even larger than the demand for space heating. The additional cost of coupling the solar system to a central heating system is minimal and worthwhile. Biomass or wood pellets are an interesting alternative to fossil fuels. To reduce the primary energy demand and CO_2 emissions, biomass is an important factor. The CO_2 emissions caused by burning wood are the same as when the waste wood is allowed to rot in the forest.

Table 9.8.5 *Building envelope constructions*

Element		Layer	Thickness (m)	Conductivity λ (W/mK)	Resistance R (m^2 K/W)	U-Value (W/m^2 K)
Wall		Plaster	0.015	1.000	0.015	
		Brick	0.250	0.580	0.431	
		Polystyrol	0.120	0.041	2.927	
		Plaster	0.005	0.700	0.007	
		Surface resistances			0.170	
		Total	0.390		3.550	**0.28**
Roof		Earth	0.100	1.160	0.086	
		Drain layer (air)	0.030		0.160	
		Plastic foil	0.002	0.190	0.011	
		Protection felt	0.002	0.220	0.009	
		Polystyrol	0.160	0.038	4.211	
		Protection felt	0.002	0.220	0.009	
		Vapour barrier (PE)	0.001	0.200	0.005	
		Screed	0.030	1.400	0.021	
		Reinforced concrete	0.200	2.330	0.086	
		Plaster	0.015	1.000	0.015	
		Surface resistances			0.140	
		Total	0.542		4.753	**0.21**
Floor over basement		Parquet	0.015	0.200	0.075	
		Screed	0.060	1.400	0.043	
		Acoustic boards (min.fibre)	0.030	0.035	0.857	
		Leveling granulate	0.040	0.700	0.057	
		Reinforced concrete	0.200	2.330	0.086	
		Cork, expanded (R=140)	0.070	0.040	1.750	
		Plaster	0.015	1.000	0.015	
		Surface resistances			0.170	
		Total	0.430		3.053	**0.33**
Window		Glass	0.004			
		Argon	0.015			
		Glass, low-e 10%	0.004			
		Total glazing	0.023			1.30
		Frame (20% of window)				1.42
		Window average				**1.32**

References

Kvist, H. (2005) *DEROB-LTH for MS Windows, User Manual Version 1.0–20050813*, Energy and Building Design, Lund University, Lund, Sweden, www.derob.se

Meteotest (2004) *Meteonorm 5.0 – Global Meteorological Database for Solar Energy and Applied Meteorology*, Bern, Switzerland, www.meteotest.ch

Polysun (2005) *Polysun 3.3.5j (Larsen Version)*, Swiss Federal Office of Energy (Bundesamt für Energie), SPF Rapperswil, Solar Energy Laboratory SPF (Institut für Solartechnik), Bern, Switzerland, www.solarenergy.ch

10

Mild Climates

10.1 Mild climate design

Maria Wall and Johan Smeds

The example solutions for conservation and renewables in this chapter are compared to reference buildings that fulfil the building code from the year 2001 in northern Italy and other similar climates south of the Alps that still have a heating demand. The example solutions have been designed to fulfil the energy targets of this book while achieving superior comfort.

10.1.1 Mild climate characteristics

The climate of Milan was used to present the mild climate region. Milan, located at 45.3° N latitude, has an average yearly temperature of 11.7°C (Meteotest, 2004). A comparison with the other two climate regions studied in this book is shown in Figure 10.1.1.

The mild climate is characterized by a heating season with a low number of degree days, higher solar gains than the other two climate regions and higher outdoor temperatures. Figure 10.1.2 shows the monthly average outdoor temperature and global solar radiation for Milan. The monthly global solar radiation on a horizontal plane varies between 28 kWh/m^2 (December) and 188 kWh/m^2 (July).

This mild climate has, until now, been an excuse to allow the construction of poorly insulated buildings. Instead, we should now see it as a great opportunity to achieve high-performance houses by means of standard technologies largely tested across other parts of Europe.

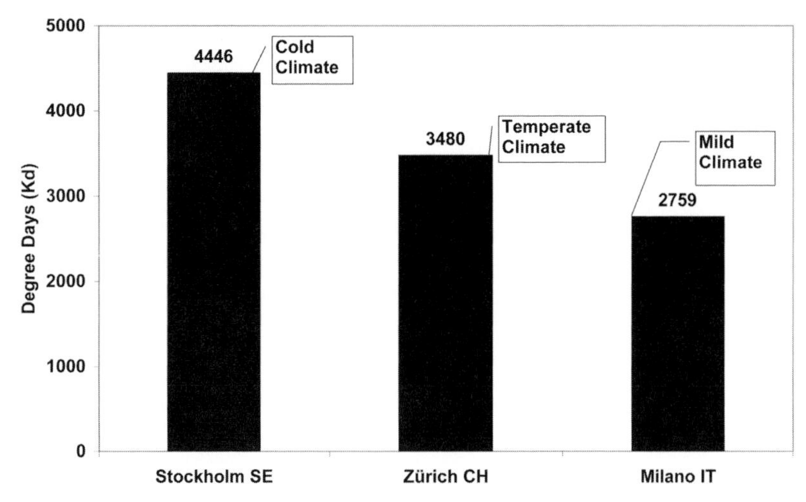

Source: Maria Wall and Johan Smeds

Figure 10.1.1 *Degree days (20/12) in cold, temperate and mild climate cities*

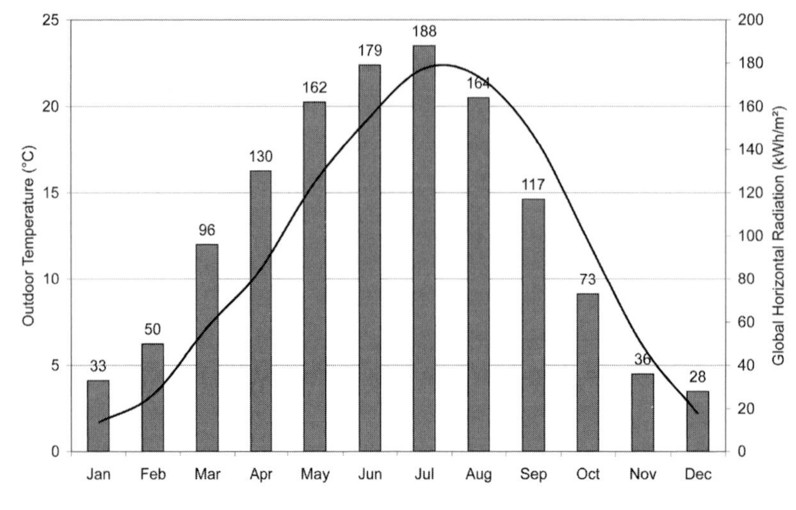

Source: Maria Wall and Johan Smeds

Figure 10.1.2 *Monthly average outdoor temperature and solar radiation (global horizontal) for Milano*

The energy targets set out in this book are relatively easily achieved in this climate. Nevertheless, a high-performance solution demands significantly higher insulation standards than is common practice. A mild climate encourages many different solutions; but it is still important to reduce heat losses before designing the supply system, which preferably should be based on renewable energy sources.

High performance in this climate also requires the envelope to be air tight and the use of mechanical ventilation with heat recovery. This is important to ensure an adequate fresh air supply. Natural ventilation could theoretically be sufficient in the mild climate; but in a tightly constructed building it is too strongly user dependent. On the other hand, some of the added costs of these measures can be offset by the smaller capacity heating system which then becomes possible.

In a mild climate, passive solar gains even in mid winter very frequently exceed heat losses through windows. Window size must be limited, however, by the overheating risk and resulting increase in needed heating capacity. Thermal mass in the building can dampen temperature fluctuations and thus reduce peak loads and increase thermal comfort. Window shading is essential.

Solar energy is, by comparison, plentiful, whereas biomass is more geographically limited (for example, in mountain regions). District heating is increasingly common in urban areas.

10.1.2 Single family house

Single family house reference design

The reference single family house with 150 m^2 floor area has an average envelope U-value of 0.74 W/m^2K. This reflects the average insulation standard according to the building code in 2001 for northern Italy. A condensing gas boiler for DHW and space heating is assumed. Ventilation and infiltration rate is in total 0.6 ach and no heat recovery is assumed. The space heating demand for this reference house is 100 kWh/m^2a according to simulations with TRNSYS. The DHW heating is 18.7 kWh/m^2a. Household electricity is assumed to be 29 kWh/m^2a for the reference single family house.

The total energy use for DHW and space heating, mechanical systems and system losses is approximately 131 kWh/m^2a. Using electricity and gas as the energy sources results in a non-renewable primary energy demand of 155 kWh/m^2a and CO_2 equivalent emissions of 33 kg/m^2a.

Single family house solution examples

Different solutions are possible to easily meet the space heating target of 20 kWh/m^{2}a for strategy 1 (energy conservation) and 25 kWh/m^{2}a for strategy 2 (renewable energy). The example solutions have a mechanical ventilation with 80 per cent heat recovery.

The examples are as follows.

CONSERVATION: SOLUTION 1

U-value of the whole building:	0.28 W/m^{2}K
Space heating demand:	18.7 kWh/m^{2}a
Heating distribution:	hot water radiant heating
Heating system:	condensing gas boiler
DHW heating system:	solar collectors and condensing gas boiler

RENEWABLE ENERGY: SOLUTION 2

U-value of the whole building:	0.32 W/m^{2}K
Space heating demand:	23.6 kWh/m^{2}a
Heating distribution:	hot water radiant heating
DHW and space heating system:	solar combi-system and condensing gas boiler

Figure 10.1.3 shows the total energy demand, delivered energy and non-renewable primary energy demand for the two solutions and the reference single family house. Compared to the reference house, the two solutions have less than 40 per cent of the total energy demand and delivered energy demand, and only 33 per cent of the non-renewable primary energy demand. Solution 2 has almost the same demand of delivered energy and non-renewable primary energy as solution 1. This is caused by the fact that the higher contribution from the solar collectors in solution 2 is approximately the same as the increased transmission losses caused by the higher U-value of the building envelope in solution 2 compared to solution 1.

The CO_2 equivalent emissions are shown in Figure 10.1.4. The CO_2 emissions are only 10 kg/m^{2}a for the two solutions, which is 32 per cent of the emissions from the reference house, which is solely heated by gas.

The results show that different strategies and system solutions can give rise to similar energy performance and CO_2 emissions.

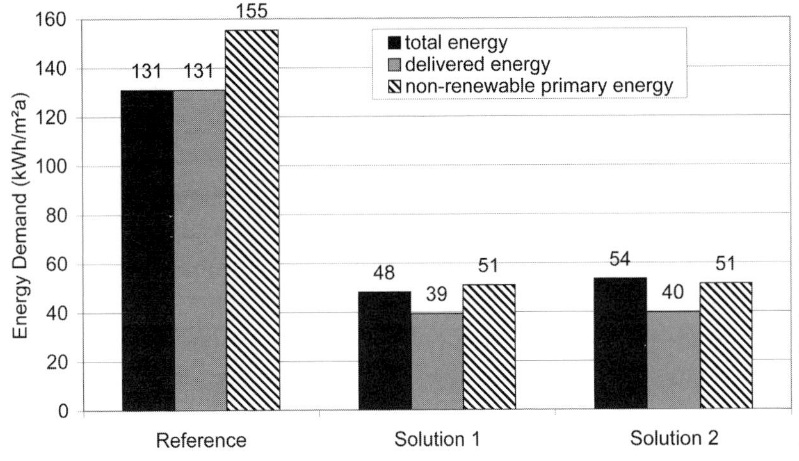

Source: Maria Wall and Johan Smeds

Figure 10.1.3 *Overview of the total energy use, the delivered energy and the non-renewable primary energy demand for the single family houses; the reference house has a condensing gas boiler*

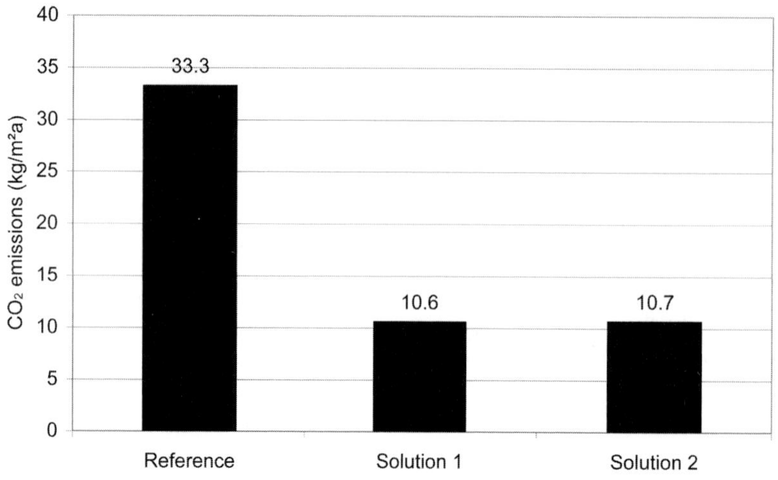

Source: Maria Wall and Johan Smeds

Figure 10.1.4 *Overview of the CO_2 equivalent emissions for the single family houses; the reference building has a condensing gas boiler*

10.1.3 Row houses

Row house reference design

A row of six houses is assumed, each 120 m^2. The reference row house has an average U-value of the building envelope of 0.86 W/m^2K. A condensing gas boiler for DHW and space heating is assumed. The combined natural ventilation and infiltration rate totals 0.6 ach and no heat recovery is assumed. The space heating demand for this reference house is 64.7 kWh/m^2 according to TRNSYS simulations. The DHW heating is 23.4 kWh/m^2a. Household electricity is assumed to be 36 kWh/m^2a for the reference house.

The total energy use for DHW and space heating, mechanical systems and system losses are approximately 102 kWh/m^2a for the reference. Using electricity and gas as the energy sources results in a non-renewable primary energy demand of 122 kWh/m^2a and CO_2 equivalent emissions of 26 kg/m^2a.

Row house solution examples

Example solutions to achieve the space heating target of 15 kWh/m^2a for strategy 1 (energy conservation) and 20 kWh/m^2a for strategy 2 (renewable energy) are shown in sections 10.4 and 10.5. A good option for a row of houses is a common heating system, and this is, indeed, acceptable in some regions. The solutions shown have mechanical ventilation with 80 per cent heat recovery.

The main examples are as follows.

CONSERVATION: SOLUTION 1

U-value of the whole building:	0.38 W/m^2K
Space heating demand:	13.2 kWh/m^2a
Heating distribution:	hot water radiant heating
DHW and space heating system:	outdoor air to water heat pump

RENEWABLE ENERGY: SOLUTION 2

U-value of the whole building:	0.46 W/m^2K
Space heating demand:	16.1 kWh/m^2a
Heating distribution:	hot water radiant heating
DHW and space heating system:	borehole heat pump

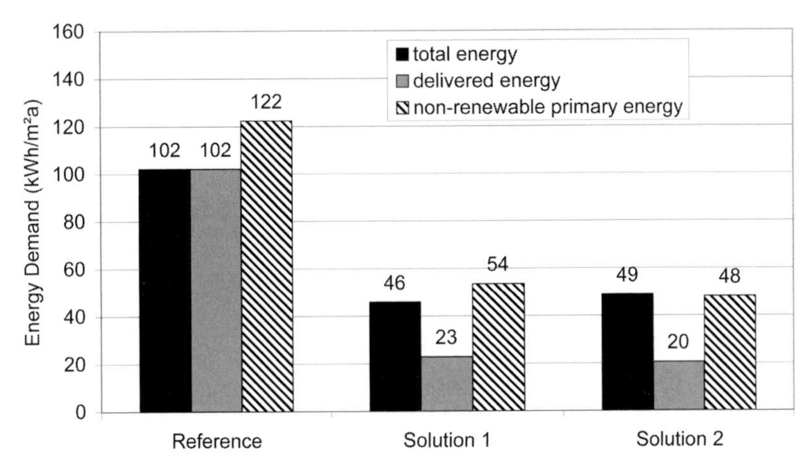

Source: Maria Wall and Johan Smeds

Figure 10.1.5 *Overview of the total energy use, the delivered energy and the use of non-renewable primary energy for the row houses; the reference house has a condensing gas boiler*

Figure 10.1.5 shows the total energy use, delivered energy and the use of non-renewable primary energy for the reference row house and the example solutions. Solution 2 with a borehole heat pump, yields slightly lower delivered energy and primary energy demands than Solution 1, with an outdoor air heat pump, even if the building envelope is less insulated in Solution 2. This is caused by a higher coefficient of performance (COP) for the borehole heat pump (3.1) than for the outdoor heat pump (2.3). The total energy demand is only 50 per cent of the energy demand for the reference house and the delivered energy demand is only around 20 per cent of the reference. The non-renewable primary energy demand is reduced to only 40 per cent. The CO_2 equivalent emissions are only 33 per cent to 37 per cent of the emissions from the reference house (see Figure 10.1.6). Other solutions (for example, with solar systems) are, of course, possible.

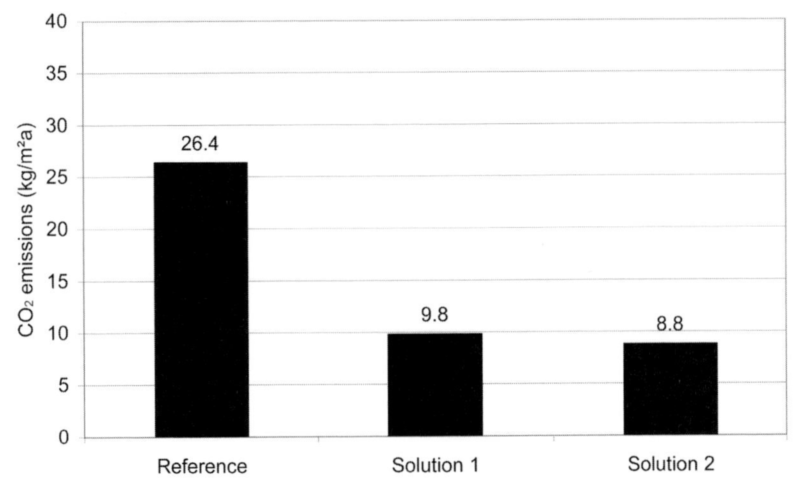

Source: Maria Wall and Johan Smeds

Figure 10.1.6 *Overview of the CO_2 equivalent emissions for the row houses; the reference house has a condensing gas boiler*

References

Meteotest (2004) *Meteonorm 5.0 – Global Meteorological Database for Solar Energy and Applied Meteorology*, Bern, Switzerland, www.meteotest.ch

TRNSYS (2005) *A Transient System Simulation Program*, Solar Energy Laboratory, University of Wisconsin, Madison, WI

10.2 Single family house in the Mild Climate Conservation Strategy

Luca Pietro Gattoni

Table 10.2.1 *Targets for single family house in the Mild Climate Conservation Strategy*

	Targets
Space heating	20 kWh/m²a
Primary energy (space heating + water heating + electricity for mechanical systems)	60 kWh/m²a

This section presents an example solution for a high-performance single family house in a mild climate, according to a conservation strategy. As a reference for the mild climate, the city of Milan is considered. This climate requires considering heating demand as well as summer comfort.

10.2.1 Solution 1: Conservation with condensing gas boiler and solar DHW

Building envelope

A single family house has an unfavourable ratio between the area of the building envelope and the heated volume. This leads to large energy losses if the quality of the envelope is poor. It is more difficult to reach the space heating target for a single family house than for other building types (row houses or apartments).

Compared to the reference building, the proposed solution means a large increase in insulation thickness. For economical and technical reasons, the roof insulation can be thicker than for the walls. In addition, for low-rise buildings, the influence of the roof insulation is of the same magnitude or higher than for walls. The increased insulation level of the opaque envelope leads to an average U-value of 0.18 W/m^2K. The conductivity of the insulation is assumed as $\lambda = 0.035$ W/mK. The U-values are:

- $U_{wall} = 0.20$ W/m^2K, corresponding to 16 cm of insulation; and
- $U_{roof} = 0.16$ W/m^2K, corresponding to 20 cm of insulation.

The window area, which is the part of the envelope with a higher U-value, is kept as large as needed for daylight purposes and large enough to the south to benefit from the small solar gains needed to reach the target. The glazed area is smaller for the proposed solution than for the reference building. Nevertheless, solar gains are of the same magnitude and, due to the much higher insulation level, they are also more efficient in reducing the heat demand.

The proposed window type has a double glazing with a low-e coating; the gap is filled with argon gas ($U_{glass} = 1.10$ W/m^2K). This window type is not unusual. Even if it is uncommon today at mild latitudes, it is quite standard in Central Europe. The U-value of the chosen glass unit is lower than for the frame. Therefore, it is good to keep the frame area as small as possible. For this reason, the frame area ratio for the south windows is assumed to be 25 per cent instead of 30 per cent as for the reference case.

Mechanical systems

Ventilation: to reach the space heating target, ventilation losses have to be reduced. Therefore, a mechanical ventilation system with 80 per cent heat recovery is chosen. The ventilation rate is 0.45 ach and the infiltration rate is 0.05 ach.

Domestic hot water and space heating systems: for the proposed solution, the availability of natural gas is assumed. A condensing gas boiler with a high efficiency is then affordable. However, it is not possible to rely entirely on this system in order to meet the primary energy target. The DHW demand needs partly to be covered by renewable energy. A solar collector area of 4 m^2 that supplies at least 40 per cent of the DHW heating is sufficient. The energy demand is covered according to the following:

- The space heating demand is covered: 100 per cent by a gas boiler (primary energy factor, or PEF = 1.14; efficiency of the plant = 1.00).
- The DHW demand is covered: 60 per cent by a gas boiler (PEF = 1.14; efficiency of the plant = 0.85) and 40 per cent by solar collectors (PEF = 0).
- The electricity demand for mechanical systems is covered: 100 per cent by electricity (PEF = 2.35).

The electricity demand for mechanical systems is estimated as 5 kWh/m^2a. The occupants require 160 litres per day of DHW. The incoming cold water is 13.5°C and is heated to 55°C. The system losses are estimated as 1.45 W/K, and the efficiency of the DHW heating is 85 per cent.

Different supply systems for DHW and space heating are possible to use in order to meet the primary energy target of 60 kWh/m^2a. If natural gas is not available, or if a solar system is not possible or desired by the user, the target can be reached by means of a heat pump covering all of the heating needs. A similar solution, leading to similar figures in terms of primary energy, is proposed for row houses in section 10.4 (see also the sensitivity analysis in section 10.3).

Energy performance

Space heating demand: these basic principles, using a heavyweight construction, result in a space heating demand of 18.7 kWh/m^2a, which fulfils the target. For a lightweight structure with identical U-values, the space heating demand is 20.0 kWh/m^2a.

The effect on the ventilation losses is shown in Figure 10.2.1 and Table 10.2.2. The ventilation losses are 8.6 kWh/m^2a for the high-performance solution and 34.5 kWh/m^2a for the reference house. In order to improve savings through a better ventilation system, the air tightness of the envelope also has to be improved (Table 10.2.2 infiltration rate changing from 0.10 ach to 0.05 ach).

Table 10.2.2 *Comparison of key numbers for the construction and energy performance of the single family house; for the proposed solution the percentage of the south window frame is 25% instead of 30%*

			Reference building	Proposed solution
Envelope				
Wall	U_{wall}	W/m^2K	0.60	0.20
Roof	U_{roof}	W/m^2K	0.48	0.16
Floor	U_{floor}	W/m^2K	0.80	0.34
Window				
Glazing 70%	$U_{glazing}$	W/m^2K	2.80	1.10
	g	-	0.75	0.60
Frame 30%	U_{frame}	W/m^2K	2.40	1.80
Window area (window/façade area ratio)				
South	A_S	m^2 (%)	9.0 (28%)	12.8 (40%)
East	A_E	m^2 (%)	3.0 (8%)	3.0 (8%)
West	A_W	m^2 (%)	9.0 (25%)	3.0 (8%)
North	A_N	m^2 (%)	1.6 (5%)	1.6 (5%)
Air change rate				
Infiltration	n_{inf}	ach	0.10	0.05
Ventilation	n_{vent}	ach	0.50	0.45
Heat recovery	η	–	–	0.80
Space heating demand	Q_h	kWh/m^2a	100.1	18.7

Thanks to the conservation strategy, the heating season goes from November to March (see Figure 10.2.2), while for the reference case it extends from October to April. The peak load is approximately 2800 W, reached only for a short time, when the outside temperature is around –10°C, which is uncommon for the city of Milan (see Figure 10.2.3). The coldest hours are more representative for some rural regions at higher altitudes in the north of the mild climate area.

Table 10.2.3 *Simulation results for the energy balance in the heating period (1 October–30 April)*

			Reference building	Proposed solution
Losses				
Transmission	Q_{TRA}	kWh/m²a	98.4	38.0
Ventilation	Q_{VENT}	kWh/m²a	34.5	8.6
Gains				
Solar	Q_{SOL}	kWh/m²a	17.6	16.1
Internal	Q_{INT}	kWh/m²a	16.1	11.8
Space heating demand	Q_h	kWh/m²a	100.1	18.7

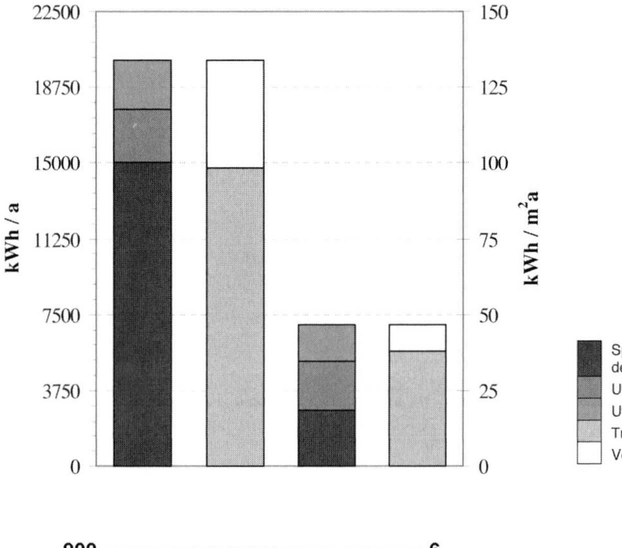

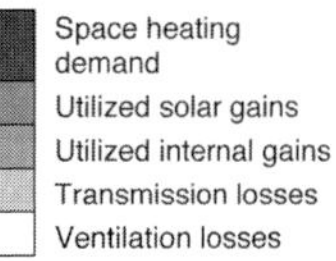

Note: This figure shows the comparison between the reference building (first two bars) and the proposed solution (third and fourth bars). The second and fourth bars represent the losses, whereas the first and third bars indicate gains and heat demand.

Source: Luca Pietro Gattoni

Figure 10.2.1 *Simulation results for the energy balance of the single family house according to Table 10.2.3*

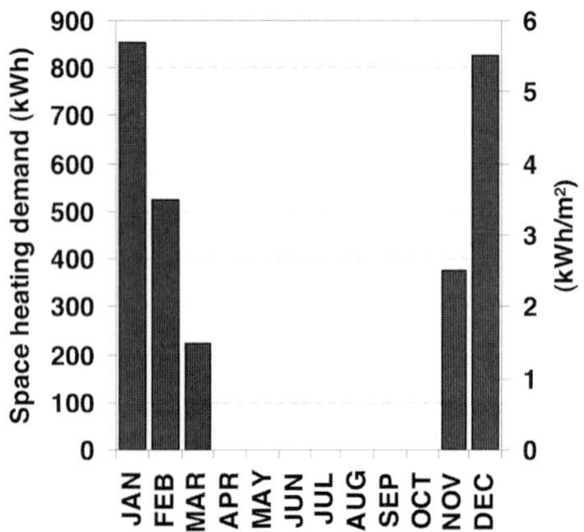

Source: Luca Pietro Gattoni

Figure 10.2.2 *Monthly space heating demand for the single family house*

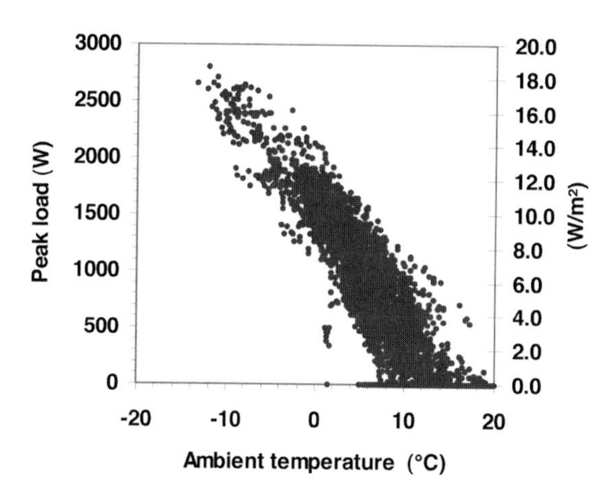

Source: Luca Pietro Gattoni

Figure 10.2.3 *Simulation results for the hourly peak load without direct solar radiation*

Household electricity: the use of household electricity is approximately 2500 kWh (16.6 kWh/m^{2}a) for the two adults and two children. The primary energy calculation shown in Table 10.2.4 does not include household electricity since the use is very much dependent on the occupants.

Non-renewable primary energy demand and CO_2 emissions: the system presented is evaluated in terms of non-renewable primary energy demand and CO_2 emissions. The results are shown in Table 10.2.4 and the proposed solution fulfils the primary energy target with the following result:

- non-renewable primary energy demand: 50.9 kWh/m^{2}a; and
- CO_2 equivalent emissions: 10.6 kg/m^{2}a.

Table 10.2.4 *Total energy demand, primary energy demand and CO_2 equivalent emissions*

Net Energy (kWh/m² a)		Total Energy Use (kWh/m² a) Energy use		Total Energy Use (kWh/m² a) Energy source		Delivered energy (kWh/m² a)		Non renewable primary energy factor (-)	Non renewable primary energy (kWh/m² a)	CO_2 factor (kg/kWh)	CO_2 equivalent emissions (kg/m² a)
Mechanical systems	5.0	Mechanical systems	5.0	Electricity	5.0	Electricity	5.0	2.35	11.8	0.43	2.2
Space heating	18.7	Space heating	18.7	Natural gas	34.3	Natural gas	34.3	1.14	39.1	0.25	8.5
DHW	18.7	DHW	18.7								
		Tank and circulation losses	3.4	Solar	8.9						
		Conversion losses	2.3								
Total	42.4		48.2		48.2		39.3		50.9		10.6

10.2.2 Summer comfort

In order to evaluate the summer conditions, the number of hours with a certain indoor temperature has been studied. This evaluation has been carried out according to two basic strategies:

1. Night cooling by natural ventilation: the air change rate during night time is 1 ach.
2. Shading of the windows: the solar shading factor is 0.5 or 0.8.

The results presented in Figure 10.2.4 clearly show that buildings in the mild climate are sensitive to overheating and therefore need to be designed to avoid poor comfort or a mechanical cooling system.

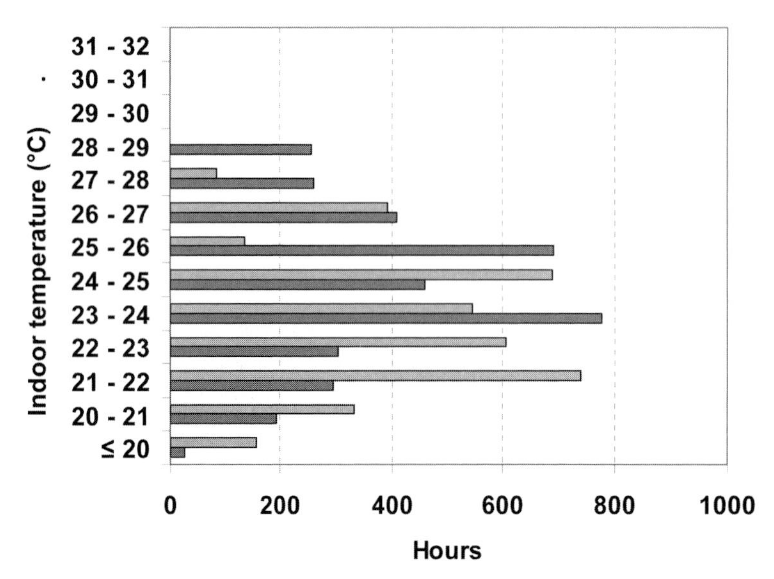

Source: Luca Pietro Gattoni

Figure 10.2.4 *Number of hours with a certain indoor temperature: The simulation period is 1 May to 30 September; night ventilation and shading devices during daytime are used*

Shading and night ventilation are efficient at reducing overheating.

Night-time ventilation should be evaluated taking into account the local climatic conditions that can vary considerably from place to place (rural or urban context) in terms of air flow rate (here assumed equal to 1 ach) and temperature.

The solutions found to lower the heating demand also need to be evaluated regarding the risk of overheating. Some of the measures are beneficial for both purposes (insulation), while others (large glazed surfaces) could lead to poor indoor conditions.

10.2.3 Sensitivity analysis

The influence of window and glazing area, glazing type and wall insulation thickness on the space heating demand are shown in this section. Results are presented for the actual high-performance house, but also represent general trends useful to study during an early design stage.

Figure 10.2.5 shows the space heating demand as a function of the glazing-to-floor area ratio, in combination with variation of percentage of window area. The mesh in Figure 10.2.5 is approximately

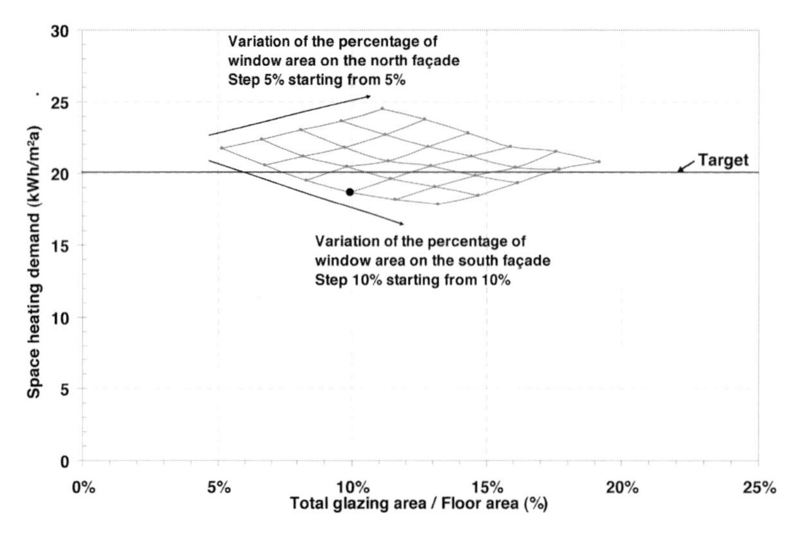

Source: Luca Pietro Gattoni

Figure 10.2.5 *Space heating demand as a function of the glazing-to-floor area ratio, in combination with variation of percentage of window area on the south façade (dots on descending lines) and on the north façade (dots on rising lines); the frame area is always 30% of the window area*

a rectangle. Two of the four sides represent the increase of the south and north window area, respectively. Every single dot represents a case with a well-defined area for south and north windows, which results in a unique ratio of total glazing to floor area (x-axis). The bold dot represents the solution described in Table 10.2.2. The dots on the descending lines represent the variation of the percentage of window area on the south façade (with steps of 10% starting from 10%), while the points on the rising lines show the variation of the percentage of window area on the north façade (with steps of 5% starting from 5%).

Figure 10.2.5 shows the variation of the glazing area that is responsible both for solar gains and for major thermal losses. The relation between gains and losses will influence the space heating demand. On the x-axis, the ratio between the total glazed area and the floor area is shown. This value is in many building regulations a measure of the availability of daylight inside the rooms. As one can see from Figure 10.2.5, similar ratios could result in different space heating demands. This is due to the varying proportion of south to north fenestration area while the total glazing-to-floor area ratio remains the same.

From Figure 10.2.5 we can learn that:

- The space heating target of 20 kWh/m^2a is possible to achieve with many combinations of south and north window area.
- Increasing the window area facing south will reduce the space heating demand, while increasing the north window area will increase the space heating demand.

Starting from the proposed solution, as described in Table 10.2.2, a parametric study is presented in Figures 10.2.6 and 10.2.7. These figures show the variation of the space heating demand as a function of the glazing-to-floor area ratio when an element of the envelope (wall or glass) changes its thickness or thermal property (essentially U- and g-values). Here, only the south glazing area is changed, while the north glazed area is kept at a minimum. The influence of different insulation levels of the walls are shown in Figure 10.2.6. In Figure 10.2.7, similar curves are shown for various glazing systems. The U-value = 0.7 W/m^2K represents a triple glazing with two low-e coatings and argon gas filling. The U-value 1.1 represents double glazing with low emissivity coating and argon gas filling, and U = 1.4 W/m^2K represents the same double glazing but with air filling.

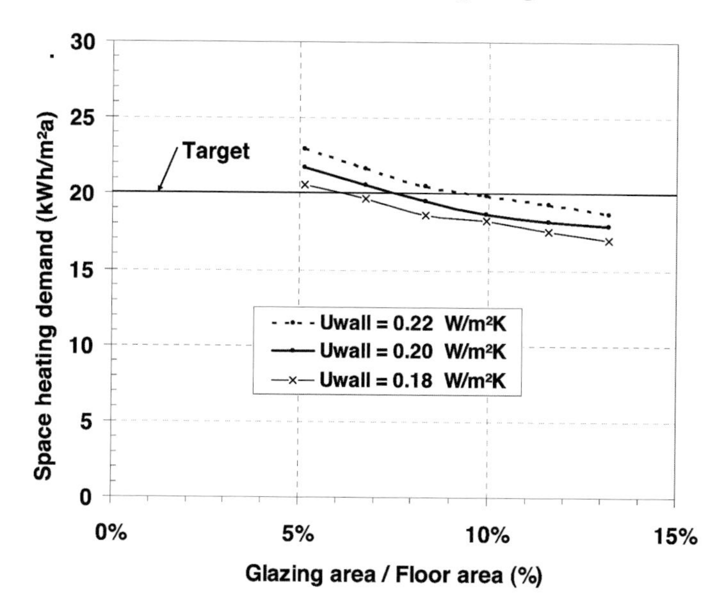

Source: Luca Pietro Gattoni

Figure 10.2.6 *Space heating demand as a function of glazing-to-floor area ratio: All numbers are kept constant as indicated in Table 10.2.2 apart from the south glazing area that varies in order to reach the total glazing area/floor area shown on the x-axis; the different lines represent different wall insulation levels*

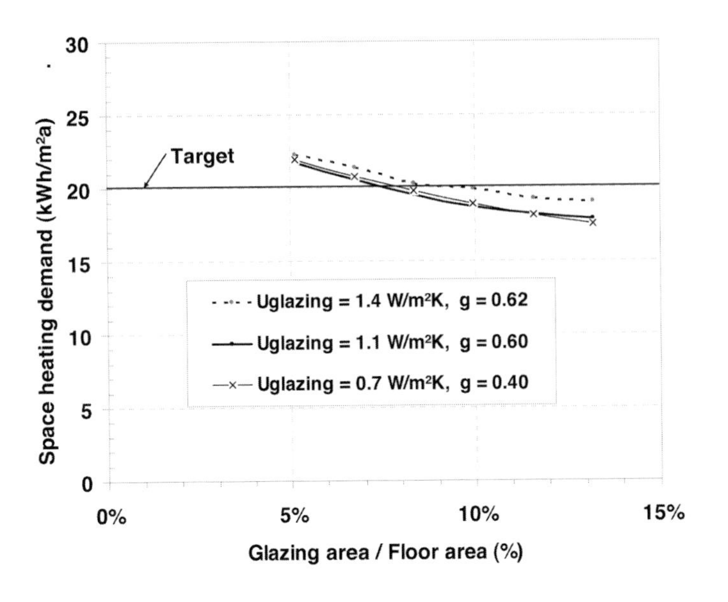

Source: Luca Pietro Gattoni

Figure 10.2.7 *Space heating demand as a function of glazing-to-floor area ratio: All numbers are kept constant as indicated in Table 10.2.2 apart from the south glazing area that varies in order to reach the total glazing area/floor area shown on the x-axis; the different lines represent different types of glazing*

Both Figures 10.2.6 and 10.2.7 show that when moving towards a higher percentage of the glazed area for the south façade, the space heating demand will be reduced. This is because both the gains and the losses increase; but, considering the solar availability in the mild climate, the gains exceed the losses. However, a high increase in the glazed area could have negative consequences on the total energy performance of the building and on the thermal comfort. Even if the space heating demand would be reduced, too large a window area would give rise to higher peak loads during nights and cloudy days.

References

TRNSYS (2005) *A Transient System Simulation Program*, Solar Energy Laboratory, University of Wisconsin, Madison, WI

10.3 Single family house in the Mild Climate Renewable Energy Strategy

Luca Pietro Gattoni

Table 10.3.1 *Targets for single family house in the Mild Climate Renewable Energy Strategy*

	Targets
Space heating	25 kWh/m²a
Primary energy (space heating + water heating + electricity for mechanical systems)	60 kWh/m²a

This section presents an example solution for a high-performance single family house in a mild climate, according to a renewable energy strategy. The primary energy target is the same as for the conservation strategy. However, a somewhat higher space heating demand is accepted in order to allow the shifting of investment costs from the conservation side (envelope) to the renewable energy supply. As a reference for the mild climate, the city of Milan is considered. This takes into account both heating demand as well as summer comfort.

10.3.1 Solution 2: Renewable energy with solar combi-system and condensing gas boiler

Building envelope

Compared to the reference house, the insulation level of the opaque envelope is increased and leads to an average U-value of 0.23 W/m^{2}K. The conductivity of the insulation is assumed as λ = 0.035 W/mK. The transmission losses are reduced by 55 per cent. U-values for all external walls and roof are:

- U_{wall} = 0.25 W/m^{2}K, corresponding to 12 cm of insulation; and
- U_{roof} = 0.22 W/m^{2}K, corresponding to 15 cm of insulation.

In comparison with the conservation example in section 10.2, the difference only concerns a somewhat less insulated opaque building envelope. The windows are the same for the renewable and the conservation solution. The chosen window type with Uwindow ~ 1.30 W/m^{2}K is intended to be a basis for high-performance houses in a mild climate to achieve high energy performance and thermal comfort.

Mechanical systems

Ventilation: to reach the space heating target the ventilation losses also have to be reduced. Therefore, a mechanical ventilation system with 80% heat recovery is chosen. The ventilation rate is 0.45 ach and the infiltration rate is 0.05 ach.

Domestic hot water and space heating systems: a solar combi-system with a gas boiler is chosen. The energy demand is covered according to the following:

- The space heating demand is covered: 70 per cent by a gas burner (primary energy factor, or PEF = 1.14; efficiency of the plant = 1.00), 30 per cent by a solar combi-system (PEF = 0).
- The DHW demand is covered: 70 per cent by a gas burner (PEF = 1.14; efficiency of the plant = 0.85), 30 per cent by a solar combi-system (PEF = 0).
- The electricity demand for mechanical systems is covered: 100 per cent by electricity (PEF = 2.35).

The electricity demand for mechanical systems is estimated as 5 kWh/m^{2}a. Furthermore, the occupants require 160 litres per day of DHW. The incoming cold water is 13.5°C and is heated to 55°C. The system losses are estimated as 1.45 W/K, and the efficiency of the DHW heating is 85 per cent.

Energy performance

Space heating demand: using a heavyweight construction, a space heating demand of 23.6 kWh/m^{2}a is obtained (see Table 10.3.3). This meets the space heating target and represents less than one quarter of the demand for the reference case.

Figure 10.3.1 shows the energy balance for the example solution, as well as the reference building. The monthly space heating demand is shown in Figure 10.3.2 and the peak load in Figure 10.3.3. The results are similar to the example for the conservation strategy in section 10.2.

Table 10.3.2 *Comparison of key numbers for the construction and the energy performance of the single family house; for the proposed solution the percentage of the south window frame is 25% instead of 30%*

			Reference building	Proposed solution
Envelope				
Wall	U_{wall}	W/m^2K	0.60	0.25
Roof	U_{roof}	W/m^2K	0.48	0.22
Floor	U_{floor}	W/m^2K	0.80	0.34
Window				
Glazing 70%	$U_{glazing}$	W/m^2K	2.80	1.10
	g	-	0.75	0.60
Frame 30%	U_{frame}	W/m^2K	2.40	1.80
Window area (window/façade area ratio)				
South	A_S	m2 (%)	9.0 (28%)	12.8 (40%)
East	A_E	m2 (%)	3.0 (8%)	3.0 (8%)
West	A_W	m2 (%)	9.0 (25%)	3.0 (8%)
North	A_N	m2 (%)	1.6 (5%)	1.6 (5%)
Air change rate				
Infiltration	n_{inf}	ach	0.10	0.05
Ventilation	n_{vent}	ach	0.50	0.45
Heat recovery	η	–	–	0.80
Space heating demand	Q_h	kWh/m^2a	100.1	23.6

Table 10.3.3 *Simulation results for the energy balance in the heating period (1 October–30 April)*

			Reference building	Proposed solution
Losses				
Transmission	Q_{TRA}	kWh/m^2a	98.4	43.9
Ventilation	Q_{VENT}	kWh/m^2a	34.5	8.5
Gains				
Solar	Q_{SOL}	kWh/m^2a	17.6	16.7
Internal	Q_{INT}	kWh/m^2a	16.1	12.1
Space heating demand	Q_h	kWh/m^2a	100.1	23.6

Household electricity: the use of household electricity is approximately 2500 kWh (16.6 kWh/m^2a) for the two adults and two children. The primary energy calculation shown in Table 10.3.4 does not include household electricity since the use is very much dependent on the occupants.

Non-renewable primary energy demand and CO_2 emissions: the example solution is evaluated in terms of non-renewable primary energy demand and CO_2 emissions. The results are shown in Table 10.3.4. The solution fulfils the primary energy target with the following results:

- non-renewable primary energy: 51.4 kWh/m^2a; and
- CO_2 equivalent emissions: 10.7 kg/m^2a.

Compared to the conservation strategy example, there is an increase of about 15 per cent of the delivered energy (the energy paid by the user). Investment and running costs can be further optimized by means of a detailed analysis of the fraction covered by the combi-system in relation to the insulation of the envelope.

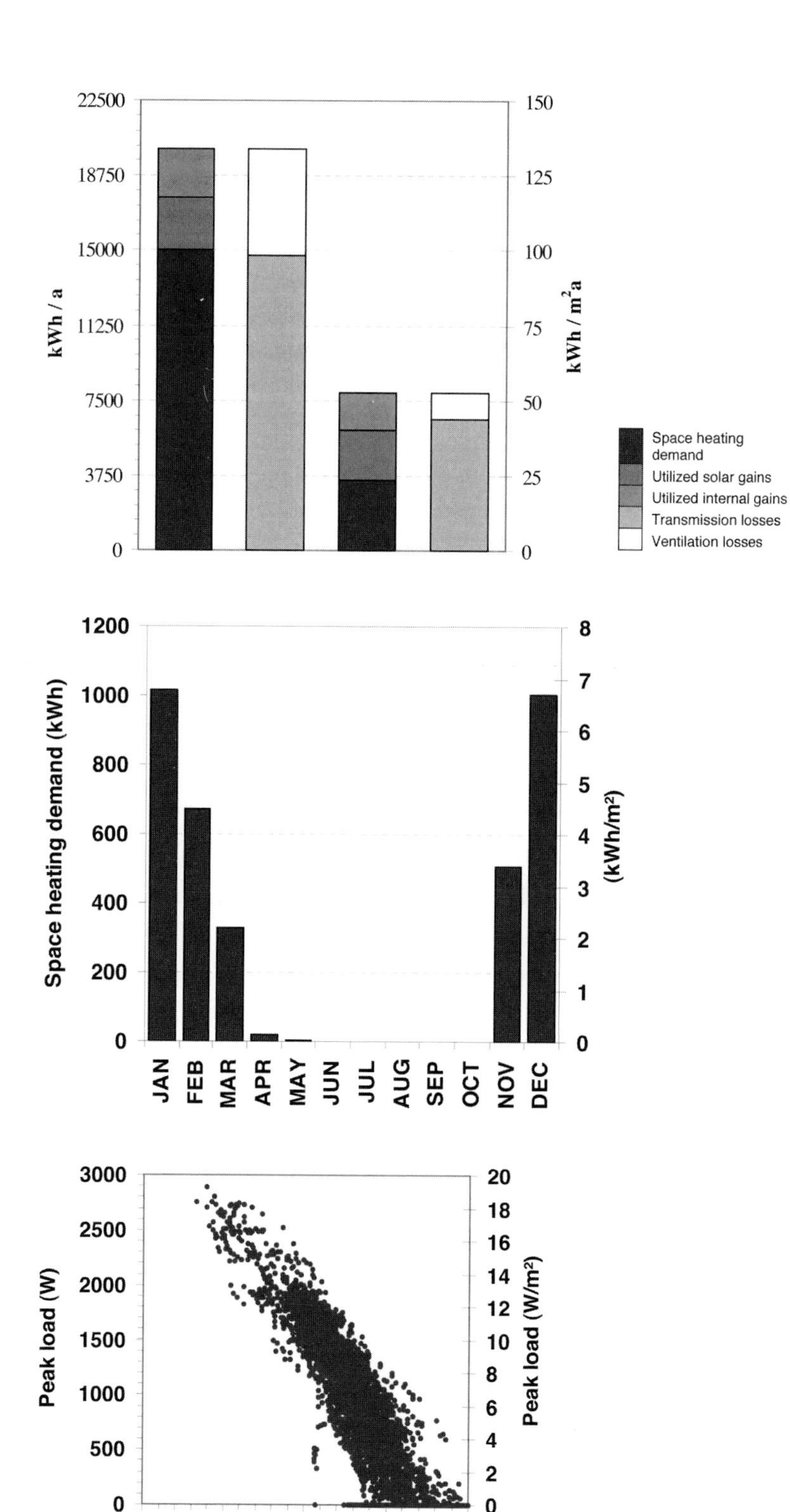

Note: This figure shows the comparison between the reference building (first two bars) and the proposed solution (third and fourth bars). The second and fourth bars represent the losses, whereas the first and third bars indicate gains and heat demand.

Source: Luca Pietro Gattoni

Figure 10.3.1 *Simulation results for the energy balance of the single family house according to Table 10.3.3*

Source: Luca Pietro Gattoni

Figure 10.3.2 *Monthly space heating demand for the single family house*

Source: Luca Pietro Gattoni

Figure 10.3.3 *Simulation results for the hourly peak load without direct solar radiation*

Table 10.3.4 *Total energy demand, primary energy demand and CO_2 equivalent emissions*

Net Energy (kWh/m² a)		Total Energy Use (kWh/m² a): Energy use		Total Energy Use (kWh/m² a): Energy source		Delivered energy (kWh/m² a)		Non renewable primary energy factor (-)	Non renewable primary energy (kWh/m² a)	CO_2 factor (kg/kWh)	CO_2 equivalent emissions (kg/m² a)
Mechanical systems	5.0	Mechanical systems	5.0	Electricity	5.0	Electricity	5.0	2.35	11.8	0.43	2.2
Space heating	23.6	Space heating	23.6	Natural gas	34.7	Natural gas	34.7	1.14	39.6	0.25	8.6
DHW	18.7	DHW	18.7								
		Tank and circulation losses	3.4	Solar	13.7						
		Conversion losses	2.7								
Total	47.3		53.5		53.5		39.7		51.4		10.7

10.3.2 Summer comfort

In order to evaluate the summer conditions, the number of hours with a certain indoor temperature has been studied. This evaluation has been carried out according to two basic strategies:

1 Night cooling by natural ventilation: the air change rate during night time is 1 ach.
2 Shading of the windows: the solar shading factor is 0.5 or 0.8.

The number of hours with certain indoor temperatures is shown in Figure 10.3.4. The number of hours above 26°C is an indicator of the risk of overheating and the following can be concluded from the parametric studies:

- Shading is a priority measure since a decrease of 37 per cent of overheating hours for the renewable strategy is accomplished when the shading coefficient is changed from 0.5 to 0.8. For the conservation strategy example in section 10.2, the overheating hours are reduced even more (62 per cent).
- To insulate the most exposed surfaces (for example, the roof) results in increased thermal comfort. Using shading coefficient 0.8 and the conservation solution with U_{roof} =0.18 W/m²K compared to the renewable solution with the U_{roof} =0.34 W/m²K gives a reduction of approximately 45 per cent of the number of hours above 26°C.
- The increase in insulation level gives good results in terms of overheating protection due to low heat transfer from outside to inside compared to a poorly insulated envelope, but only if solar protection is very good.

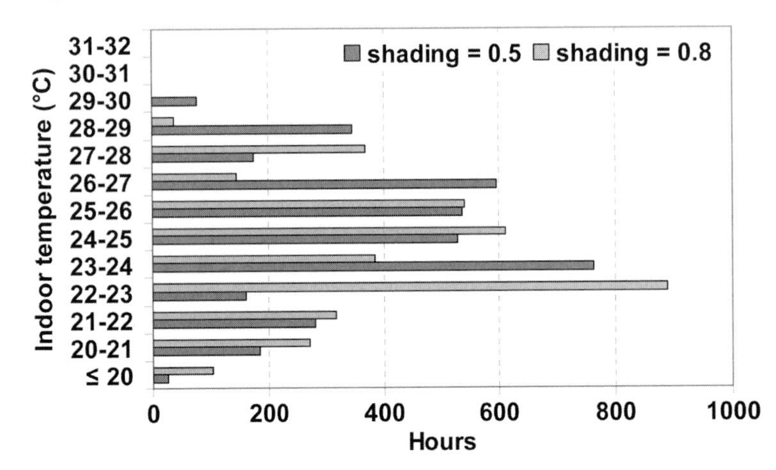

Source: Luca Pietro Gattoni

Figure 10.3.4 *Number of hours with a certain indoor temperature: The simulation period is 1 May to 30 September; night ventilation and shading devices during daytime are used*

10.3.3 Sensitivity analysis

To find realistic solutions for the renewable energy strategy in a mild climate is neither a technical problem on the building side nor a problem on the mechanical system side. The final selection of a good combination of technologies depends on many factors, such as installation and running costs; local availability of renewable energy; the client's wishes; and the competence within the design and construction team.

The proposed solution relies on a combination of the two largest energy sources in the mild climate area: natural gas and, from the renewable point of view, the sun. Figure 10.3.5 shows clearly that if only fossil fuels would be used (gas, line A), it is very difficult to reach the primary energy target. Increasing the use of solar energy will give rise to a non-renewable primary energy demand lower than the target of 60 kWh/m^{2}a. This is possible because of the low space heating demand. The proposed solution (Q_h = 23.6 kWh/m^{2}a) results in the following primary energy demand with different energy supply systems:

- Line B: gas burner + DHW with 40 per cent solar – non-renewable primary energy 56.5 kWh/m^{2}a, CO_2 equivalent emissions 11.8 kg/m^{2}a.
- Line C: outdoor air to water heat pump with COP = 2.3 – non-renewable primary energy 58.0 kWh/m^{2}a, CO_2 equivalent emissions 10.7 kg/m^{2}a.
- Line D: gas boiler + solar combi-system – non-renewable primary energy 51.4 kWh/m^{2}a, CO_2 equivalent emissions 10.7 kg/m^{2}a.
- Line E: gas boiler + solar combi-system + photovoltaics for the mechanical systems – non-renewable primary energy 39.6 kWh/m^{2}a, CO_2 equivalent emissions 8.6 kg/m^{2}a. An alternative to fossil fuels could be biomass (see line F in Figure 10.3.5).
- Line F: biomass + DHW with 40 per cent solar – non-renewable primary energy 20.4 kWh/m^{2}a, CO_2 equivalent emissions 4.8 kWh/m^{2}a.

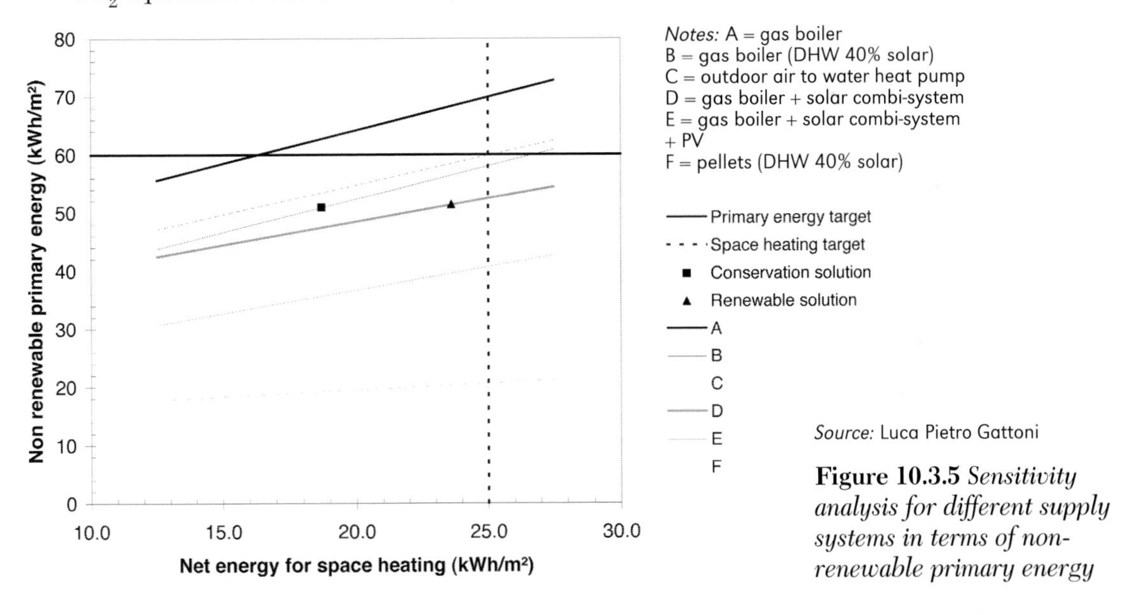

Notes: A = gas boiler
B = gas boiler (DHW 40% solar)
C = outdoor air to water heat pump
D = gas boiler + solar combi-system
E = gas boiler + solar combi-system + PV
F = pellets (DHW 40% solar)

Source: Luca Pietro Gattoni

Figure 10.3.5 *Sensitivity analysis for different supply systems in terms of non-renewable primary energy*

References

TRNSYS (2005) *A Transient System Simulation Program*, Solar Energy Laboratory, University of Wisconsin, Madison, WI

10.4 Row house in the Mild Climate Conservation Strategy

Luca Pietro Gattoni

Table 10.4.1 *Targets for row house in the Mild Climate Conservation Strategy*

	Targets
Space heating	15 kWh/m^{2}a
Primary energy (space heating + water heating + electricity for mechanical systems)	60 kWh/m^{2}a

This section presents a solution for a high-performance row house in a mild climate, according to a conservation strategy. As the reference for a mild climate, the city of Milan is considered. This takes into account both heating demand as well as summer comfort.

10.4.1 Solution 1: Conservation with outdoor air to water heat pump

Building envelope

A row with four mid units and two end units is studied. The opaque building envelope, excluding the ground floor, consists of three main parts:

1 roof (common for all units);
2 walls facing north and south (the only façades for the mid units); and
3 walls facing east and west (end units).

The roof represents a considerable part of the external surfaces. The main south and north façades include a major part of the window area and therefore it is of interest to study the influence of this window area.

The end units have a large façade area that will influence the peak load and space heating demand. The concept is to use more insulation in the roof and the end walls. The average U-value of the opaque envelope is 0.25 W/m^{2}K. Table 10.4.2 shows the U-values for the reference case and the proposed solution. The conductivity of the insulation is assumed as λ = 0.035 W/mK. The U-values for the solution are:

- Walls: U_{wall} = 0.30 W/m^{2}K, corresponding to 10 cm of insulation; $U_{wall-end\ unit}$ = 0.25 W/m^{2}K, corresponding to 12 cm of insulation.
- Roof: U_{roof} = 0.22 W/m^{2}K, corresponding to 15 cm of insulation.

The windows are concentrated on the south façade in order to profit from solar gains; the other openings are kept at a minimum to control losses (north) and to avoid overheating in summer (west and east). Concerning the glass type, both insulation properties and solar transmission are important. A glass with a U-value of 1.10 W/m^{2}K and a g-value of 0.60 is a good compromise even in a mild climate, also for comfort reasons. Compared to, for example, the Italian building regulations, a common double glazing (U=2.80 W/m^{2}K) would be sufficient.

Mechanical systems

Ventilation: to reach the space heating target, ventilation losses also have to be reduced. Therefore, a mechanical ventilation system with 80 per cent heat recovery is chosen. The ventilation rate is 0.45 ach and the infiltration rate is 0.05 ach.

Domestic hot water and space heating systems: the proposed solution is based on using an alternative to natural gas. For the conservation strategy, an air to water heat pump is considered. This solution relies fully on electricity:

- The space heating demand is covered: 100 per cent by a heat pump (PEF = 2.35; COP = 2.30).
- The DHW demand is covered: 100 per cent by a heat pump (PEF = 2.35; COP = 2.30).
- The electricity demand for mechanical systems is covered: 100 per cent by electricity (PEF = 2.35).

The electricity demand for mechanical systems is estimated as 5 kWh/m^{2}a. The occupants require 160 litres per day of DHW. The incoming cold water is 13.5°C and is heated to 55°C. The system losses are estimated as 1.45 W/K and the efficiency of the DHW heating is 85 per cent.

Energy performance

Space heating demand: the average space heating demand for the row of six units is Q_h = 13.2 kWh/m^{2}a for a heavy construction (see Tables 10.4.2 and 10.4.3). In detail:

- mid unit: Q_h = 10.9 kWh/m^{2}a, which is a low demand due to the small external surfaces; and
- end unit: Q_h = 18.0 kWh/m^{2}a; this does not alone fulfil the target for the row of six units, but is sufficient for the end units.

The heating season extends from November to March (see Figure 10.4.2), while for the reference case it extends from October to April. The space heating peak load is approximately 2400 W (see Figure 10.4.3).

Table 10.4.2 *Comparison of key numbers for the construction and energy performance of the row house; for the proposed solution the percentage of the south window frame is 25% instead of 30%*

			Reference case	Proposed solution
Envelope				
Wall	U_{wall}	W/m^{2}K	0.60	0.30
	U_{wall}	W/m^{2}K	0.60	0.25
Roof	U_{roof}	W/m^{2}K	0.55	0.22
Floor	U_{floor}	W/m^{2}K	0.80	0.34
Window				
Glazing 70%	$U_{glazing}$	W/m^{2}K	2.80	1.10
	g	-	0.75	0.60
Frame 30%	U_{frame}	W/m^{2}K	2.40	1.80
Window area (window/façade area ratio)				
South	A_S	m^2 (%)	11.7 (40%)	13.1 (45%)
East (end unit)	A_E	m^2 (%)	4.0 (10%)	3.2 (8%)
West (end unit)	A_W	m^2 (%)	4.0 (10%)	3.2 (8%)
North	A_N	m^2 (%)	5.8 (20%)	4.4 (15%)
Air change rate				
Infiltration	n_{inf}	ach	0.10	0.05
Ventilation	nvent	ach	0.50	0.45
Heat recovery	η	-	-	0.80
Space heating demand	Q_h	kWh/m^{2}a	64.7	13.2
Mid unit	Q_h	kWh/m^{2}a	58.1	10.9
End unit	Q_h	kWh/m^{2}a	77.9	18.0

Table 10.4.3 *Simulation results for the energy balance in the heating period (1 October–30 April)*

			Reference case	Proposed solution
Losses				
Transmission	Q_{TRA}	kWh/m^{2}a	76.8	39.2
Ventilation	Q_{VENT}	kWh/m^{2}a	27.8	8.5
Gains				
Solar	Q_{SOL}	kWh/m^{2}a	23.7	15.3
Internal	Q_{INT}	kWh/m^{2}a	16.1	12.4
Space heating demand	Q_h	kWh/m^{2}a	64.7	13.2

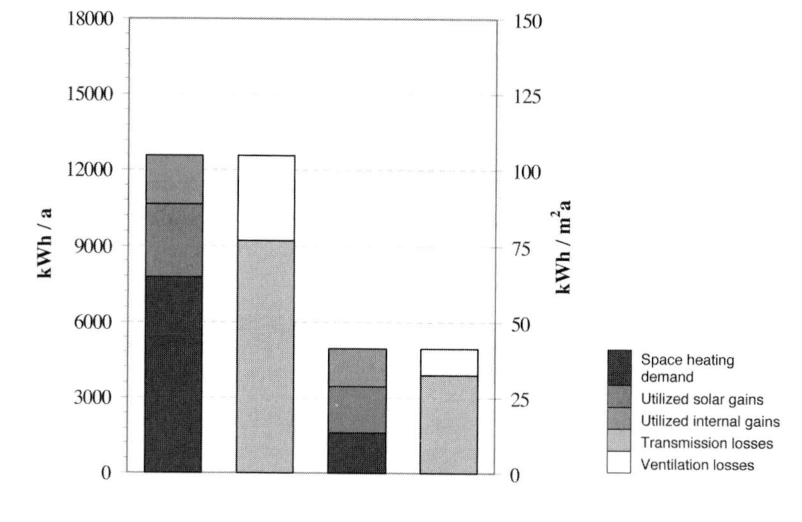

Source: Luca Pietro Galloni

Figure 10.4.1 *Simulation results for the energy balance of the row house according to Table 10.4.3: Average for the row with two end units and four mid units*

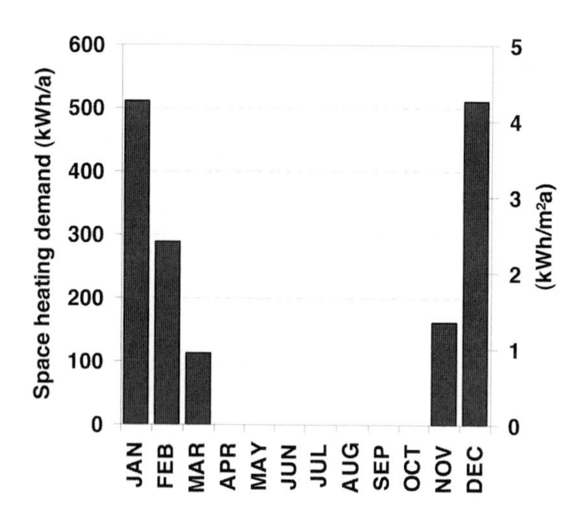

Source: Luca Pietro Galloni

Figure 10.4.2 *Monthly space heating demand for the row house: Average for the row with two end units and four mid units*

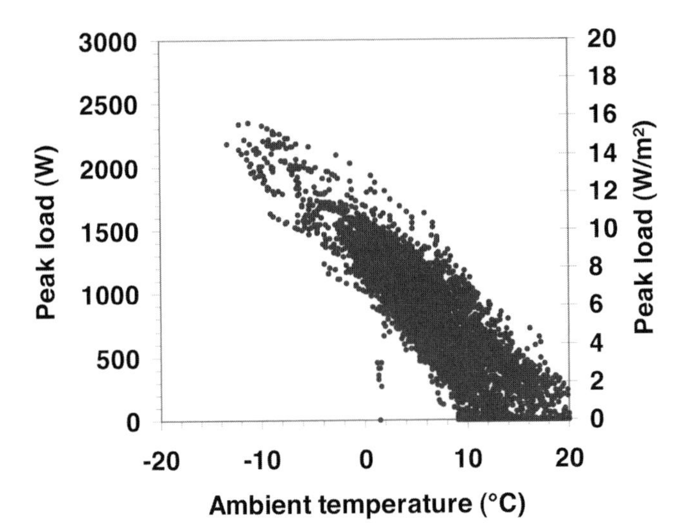

Source: Luca Pietro Galloni

Figure 10.4.3 *Simulation results for the hourly peak load for an end unit without direct solar radiation*

Household electricity: the use of household electricity is approximately 2500 kWh (20.8 kWh/m^{2}a) for the two adults and two children. The primary energy calculation shown in Table 10.4.4 does not include household electricity since the use is very much dependent on the occupants.

Non-renewable primary energy demand and CO_2 emissions: the system presented is evaluated in terms of non-renewable primary energy demand and CO_2 emissions. The results are shown in Table 10.4.4. The proposed solution reaches the primary energy target with the following results:

- non-renewable primary energy demand: 53.5 kWh/m^{2}a;
- CO_2 equivalent emissions: 9.8 kg/m^{2}a.

Table 10.4.4 *Total energy demand, primary energy demand and CO_2 equivalent emissions*

Net Energy (kWh/m² a)		Total Energy Use (kWh/m² a): Energy use		Total Energy Use (kWh/m² a): Energy source		Delivered energy (kWh/m² a)		Non renewable primary energy factor (-)	Non renewable primary energy (kWh/m² a)	CO_2 factor (kg/kWh)	CO_2 equivalent emissions (kg/m² a)
Mechanical systems	5.0	Mechanical systems	5.0	Electricity	5.0	Electricity	5.0	2.35	11.8	0.43	2.2
Space heating	13.2	Space heating	13.2	Electricity Heat pump	17.8	Electricity Heat pump	17.8	2.35	41.8	0.43	7.6
DHW	23.4	DHW	23.4								
		Tank and circulation losses	4.2	Outdoor air	23.1						
		Conversion losses	0.0								
Total	41.6		45.9		45.9		22.8		53.5		9.8

10.4.2 Summer comfort

In order to evaluate the summer conditions, the number of hours with a certain indoor temperature has been studied. This evaluation has been carried out according to two basic strategies:

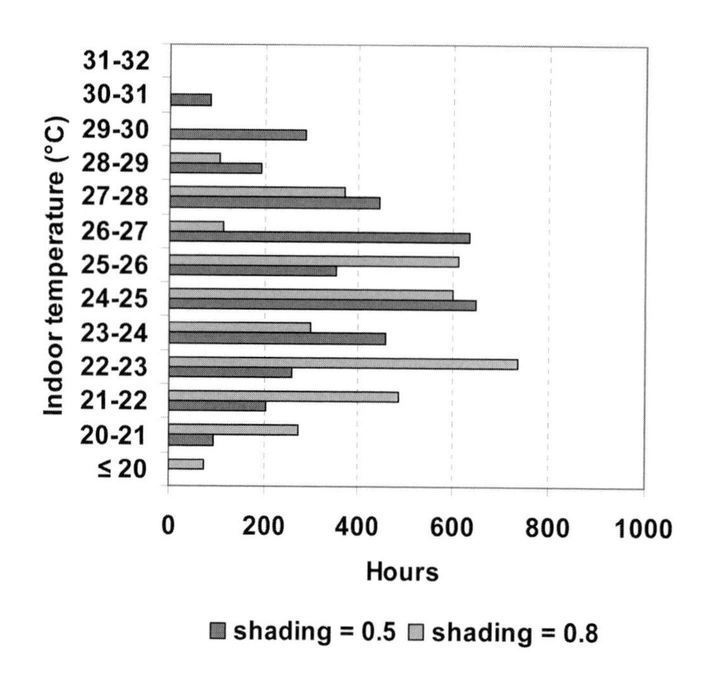

Source: Luca Pietro Galloni

Figure 10.4.4 *Number of hours with a certain indoor temperature: The simulation period is 1 May to 30 September; night ventilation and shading devices during daytime are used*

1 Night cooling by natural ventilation: the air change rate during night time is 1 ach.
2 Shading of the windows: the solar shading factor is 0.5 or 0.8.

The results presented in Figure 10.4.4 clearly show that the row houses are sensitive to overheating problems in the mild climate and therefore need to be designed to avoid poor comfort or a mechanical cooling system. Shading and night ventilation are efficient to reduce overheating.

10.4.3 Sensitivity analysis

The influence of window and glazing area, glazing type and wall insulation thickness on the space heating demand are shown in this section. Results are presented for the actual high-performance house, but also represent general trends useful to study during an early design stage.

Figure 10.4.5 shows the space heating demand as a function of the glazing-to-floor area ratio, in combination with variation in percentage of window area. The mesh in Figure 10.4.5 is approximately

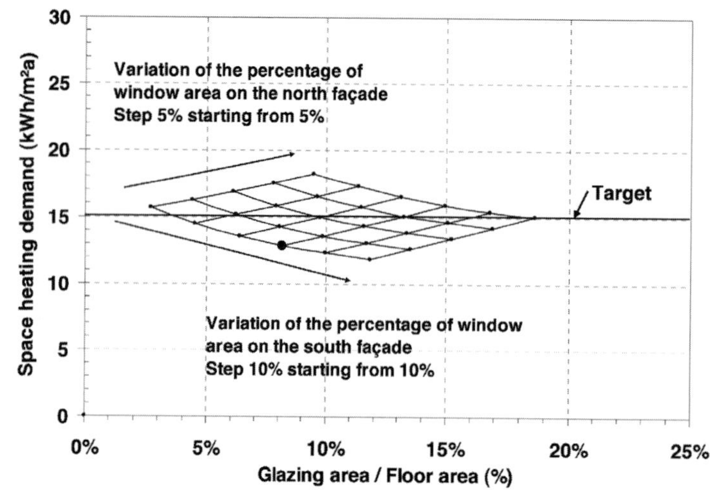

Source: Luca Pietro Galloni

Figure 10.4.5 *Space heating demand as a function of the glazing-to-floor area ratio, in combination with variation of percentage of window area on the south façade (dots on descending lines) and on the north façade (dots on rising lines); the frame area is always 30% of the window area*

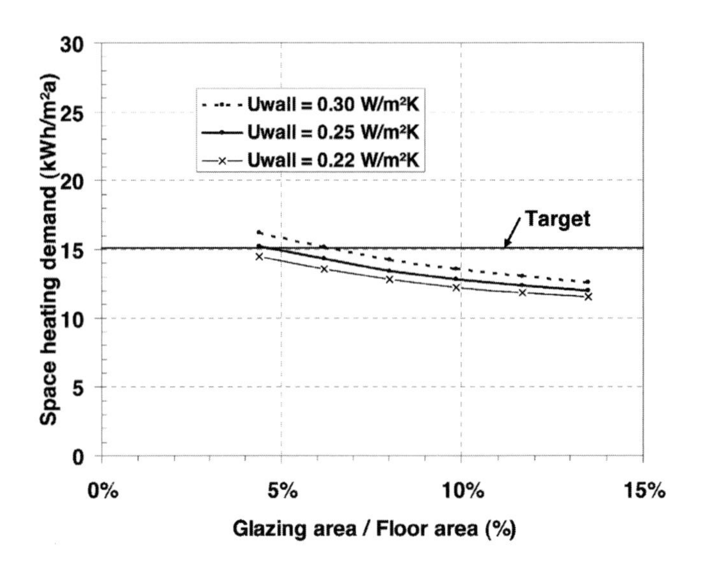

Source: Luca Pietro Galloni

Figure 10.4.6 *Space heating demand as a function of glazing-to-floor area ratio: All numbers are kept constant as shown in Table 10.4.2 apart from the south glazing area that varies in order to reach the total glazing area/floor area shown on the x-axis; the different lines represent different wall insulation levels*

a rectangle. Two of the four sides represent the increase of the south and north window areas, respectively. Every single dot represents a case with a well-defined area for south and north windows, which results in a unique ratio of total glazing-to-floor area (x-axis). The bold dot represents the solution described in Table 10.4.2. The dots on the descending lines represent the variation in the percentage of window area on the south façade (with steps of 10 per cent, starting from 10 per cent), while the points on the rising lines show the variation in the percentage of window area on the north façade (with steps of 5 per cent, starting from 5 per cent).

Figure 10.4.5 shows that an increased south window area will decrease the space heating demand. In addition, the influence of the window size is less with a higher insulation level of the building envelope, compared to the single family house. This demonstrates that a robust envelope is better in tolerating low solar gains in case of a shaded south façade and is also less sensitive to having windows on the north side of the building.

Figure 10.4.6 shows the effect of the insulation level in reduced space heating demand without influencing the usability of solar gains and the internal mass.

Figure 10.4.7 shows that the difference between the glazing types in terms of space heating demand becomes more significant for large surfaces (starting with 40 per cent of south façade). This

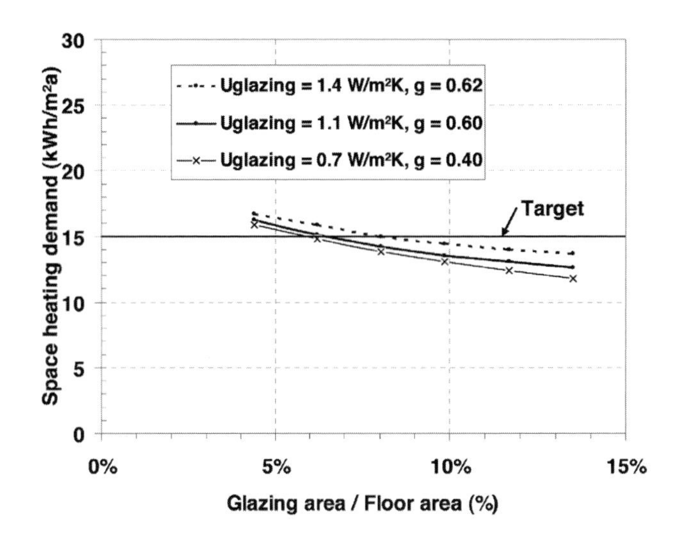

Source: Luca Pietro Galloni

Figure 10.4.7 *Space heating demand as a function of glazing-to-floor area ratio: All numbers are kept constant as shown in Table 10.4.2 apart from the south glazing area that varies in order to reach the total glazing area/floor area shown on the x-axis; the different lines represent different types of glazing*

trend is larger for a less insulated building envelope. The U-value = 0.7 W/m^{2}K represents a triple glazing with two low-e coatings and argon gas filling. The U-value 1.1 represents double glazing with low emissivity coating and argon gas filling, and U = 1.4 W/m^{2}K represents the same double glazing but with air filling.

In addition, when moving towards large glazed areas, triple glazing is a better choice (U = 0.7 W/m^{2}K) regarding surface temperatures and thermal comfort. The lower g-value reduces the solar gains in summertime. Nevertheless, large glazed surfaces mean a higher risk of overheating.

References

TRNSYS (2005) *A Transient System Simulation Program*, Solar Energy Laboratory, University of Wisconsin, Madison, WI

10.5 Row house in the Mild Climate Renewable Energy Strategy

Luca Pietro Gattoni

Table 10.5.1 *Targets for row house in the Mild Climate Renewable Energy Strategy*

	Targets
Space heating	20 kWh/m^{2}a
Primary energy (space heating + water heating + electricity for mechanical systems)	60 kWh/m^{2}a

This section presents a solution for a high-performance row house in a mild climate, according to a renewable strategy. As a reference for the mild climate, the city of Milan is considered. This takes into account both heating demand as well as summer comfort.

10.5.1 Solution 2: Renewable energy with borehole heat pump

Building envelope

The building envelope is more insulated than the reference building, which results in transmission losses of 40.9 kWh/m^{2}a instead of 76.8 kWh/m^{2}a as for the reference (see Tables 10.5.2 and 10.5.3). The conductivity of the insulation is assumed as λ = 0.035 W/mK. The mean U-value of the opaque envelope is 0.30 W/m^{2}K, where:

- Walls: U_{wall} = 0.35 W/m^{2}K, corresponding to 8 cm of insulation; $U_{wall - end\ unit}$ = 0.30 W/m^{2}K, corresponding to 10 cm of insulation.
- Roof: U_{roof} = 0.28 W/m^{2}K, corresponding to 12 cm of insulation.

The south façade has a large glazed area (60 per cent), while the north façade has minimum-sized windows. The higher space heating target for the renewable strategy and the favourable area to volume ratio for the mid unit allow the use of double glazing with low-e coating and air gas filling; $U_{glazing}$ = 1.40 W/m^{2}K (g-value 0.62). This window type represents the basic solution. A similar double glazing but with argon filling, with the U-value 1.10 W/m^{2}K (g=0.60), offers a better performance very close to a triple-glazed window and represents an alternative solution.

Mechanical systems

Ventilation: to reach the space heating target, ventilation losses also have to be reduced. Therefore, a mechanical ventilation system with 80 per cent heat recovery is chosen. The ventilation rate is 0.45 ach and the infiltration rate is 0.05 ach.

Domestic hot water and space heating systems: a system with a borehole heat pump is chosen instead of the traditional use of natural gas. This solution is not usual nowadays; but it is very promising and the installation costs are expected to decrease with the increase in realized projects. In addition, the running costs are very low due to the high efficiency:

- The space heating demand is covered: 100 per cent by a borehole heat pump (PEF = 2.35; COP = 3.1).
- The DHW demand is covered: 100 per cent by a borehole heat pump (PEF = 2.35; COP = 2.7).
- The electricity for mechanical systems is covered: 100 per cent by electricity (PEF = 2.35).

The electricity demand for mechanical systems is estimated as 5 kWh/m²a. The occupants require 160 litres per day of DHW. The incoming cold water is 13.5°C and is heated to 55°C. The system losses are estimated as 1.45 W/K and the efficiency of the DHW heating is 85 per cent.

Energy performance

Space heating demand: for a heavy construction, the average space heating demand is 16.1 kWh/m²a for the row with four mid units and two end units. The demand for each type of unit is:

- mid unit: Q_h = 13.2 kWh/m²a; and
- end unit: Q_h = 21.9 kWh/m²a.

Table 10.5.2 *Comparison of key numbers for the construction and energy performance of the single family house; for the proposed solution the percentage of the south window frame is 25% instead of 30%*

			Reference building	Proposed solution
Envelope				
Wall	U_{wall}	W/m²K	0.60	0.35
	U_{wall}	W/m²K	0.60	0.30
Roof	U_{roof}	W/m²K	0.55	0.28
Floor	U_{floor}	W/m²K	0.80	0.34
Window				
Glazing 70%	$U_{glazing}$	W/m²K	2.80	1.40
	g	-	0.75	0.62
Frame 30%	U_{frame}	W/m²K	2.40	1.80
Window area (window/façade area ratio)				
South	A_S	m² (%)	11.7 (40%)	17.5 (60%)
East (end unit)	A_E	m² (%)	4.0 (10%)	4.8 (12%)
West (end unit)	A_W	m² (%)	4.0 (10%)	4.8 (12%)
North	A_N	m² (%)	5.8 (20%)	4.4 (15%)
Air change rate				
Infiltration	n_{inf}	ach	0.10	0.05
Ventilation	n_{vent}	ach	0.50	0.45
Heat recovery	η	–	–	0.80
Space heating demand	Q_h	kWh/m²a	64.7	16.1
Mid unit	Q_h	kWh/m²a	58.1	13.2
End unit	Q_h	kWh/m²a	77.9	21.9

Table 10.5.3 *Simulation results for the energy balance in the heating period (1 October–30 April)*

			Reference building	Proposed solution
Losses				
Transmission	Q_{TRA}	kWh/m²a	76.8	40.9
Ventilation	Q_{VENT}	kWh/m²a	27.8	8.8
Gains				
Solar	Q_{SOL}	kWh/m²a	23.7	21.1
Internal	Q_{INT}	kWh/m²a	16.1	12.4
Space heating demand	Q_h	kWh/m²a	64.7	16.1

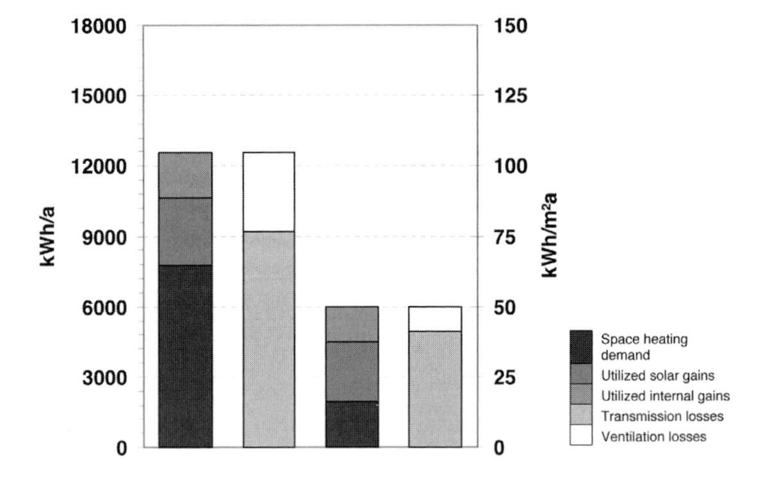

Source: Luca Pietro Gattoni

Figure 10.5.1 *Simulation results for the energy balance of the row house according to Table 10.5.3: Average for the row with two end units and four mid units*

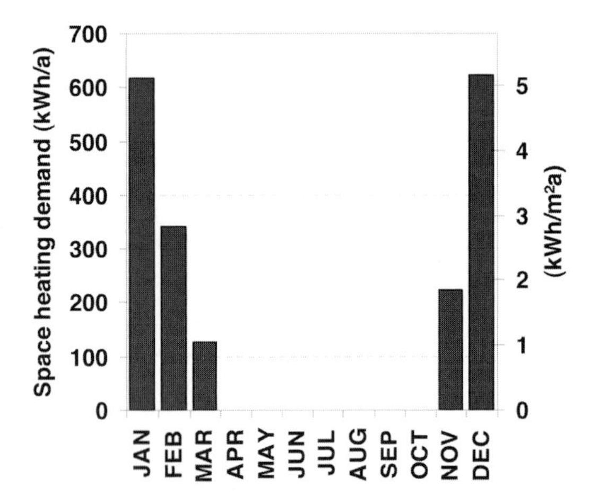

Source: Luca Pietro Gattoni

Figure 10.5.2 *Monthly space heating demand for the row house: Average for the row with two end units and four mid units*

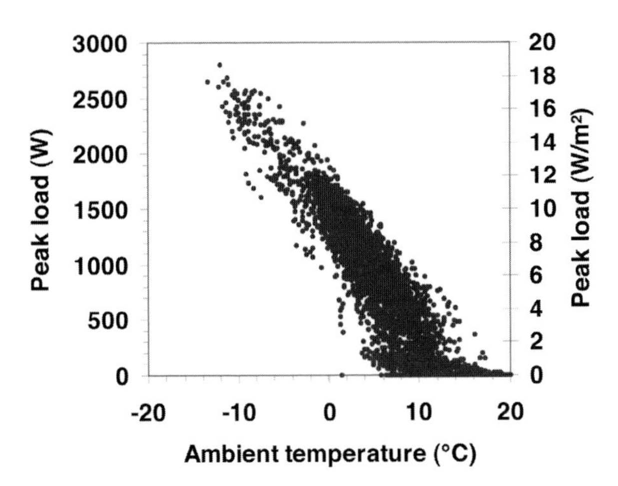

Source: Luca Pietro Gattoni

Figure 10.5.3 *Simulation results for the hourly peak load for an end unit without direct solar radiation*

Household electricity: the use of household electricity is approximately 2500 kWh (20.8 kWh/m^{2}a) for the two adults and two children. The primary energy calculation shown in Table 10.5.4 does not include household electricity since the use is very much dependent on the occupants.

Non-renewable primary energy demand and CO_2 emissions: the solution is evaluated in terms of non-renewable primary energy demand and CO_2 emissions. The results are shown in Table 10.5.4. The primary energy demand is 48 kWh/m^{2}a, thus fulfilling the target of 60 kWh/m^{2}a. The CO_2 equivalent emissions are 8.8 kg/m^{2}a.

Table 10.5.4 *Total energy demand, primary energy demand and CO_2 equivalent emissions*

Net Energy (kWh/m² a)		Total Energy Use (kWh/m² a)				Delivered energy (kWh/m² a)		Non renewable primary energy factor (-)	Non renewable primary energy (kWh/m² a)	CO_2 factor (kg/kWh)	CO_2 equivalent emissions (kg/m² a)
		Energy use		Energy source							
Mechanical systems	5.0	Mechanical systems	5.0	Electricity	5.0	Electricity	5.0	2.35	11.8	0.43	2.2
Space heating	16.1	Space heating	16.1	Electricity Heat pump	15.4	Electricity Heat pump	15.4	2.35	36.2	0.43	6.6
DHW	23.4	DHW	23.4	Ground	28.4						
		Tank and Circulation losses	4.2								
		Conversion losses	0.0								
Total	44.5		48.8		48.8		20.4		48.0		8.8

10.5.2 Summer comfort

In order to evaluate the summer conditions, the number of hours with a certain indoor temperature has been studied. This evaluation has been carried out according to two basic strategies:

1 Night cooling by natural ventilation: the air change rate during night time is 1 ach.
2 Shading of the windows: the solar shading factor is 0.5 or 0.8.

The results presented in Figure 10.5.4 show clearly that the row houses are sensitive to overheating in the mild climate and therefore need to be designed to avoid poor comfort or a mechanical cooling system. As can be seen, due to larger window area to the south compared to the conservation strategy, the number of overheating hours is larger.

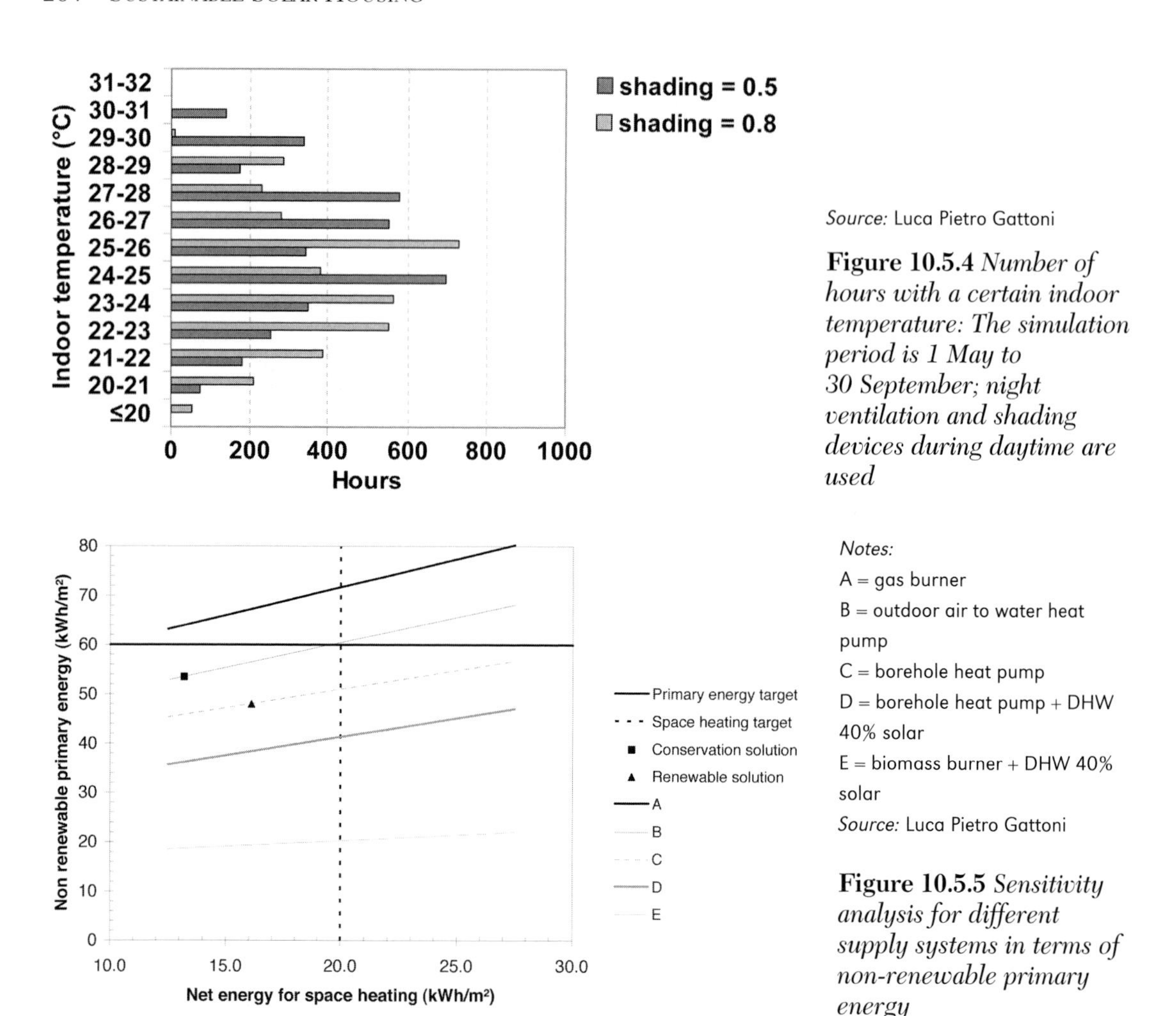

Source: Luca Pietro Gattoni

Figure 10.5.4 *Number of hours with a certain indoor temperature: The simulation period is 1 May to 30 September; night ventilation and shading devices during daytime are used*

Notes:
A = gas burner
B = outdoor air to water heat pump
C = borehole heat pump
D = borehole heat pump + DHW 40% solar
E = biomass burner + DHW 40% solar
Source: Luca Pietro Gattoni

Figure 10.5.5 *Sensitivity analysis for different supply systems in terms of non-renewable primary energy*

10.5.3 Sensitivity analysis

Figure 10.5.5 shows the primary energy demand for different space heating demand, using different energy sources. The proposed solution with the space heating demand Q_h = 16.1 kWh/m²a relies on a borehole heat pump (line C, marked point). Figure 10.5.5 shows clearly that if only fossil fuels would be used (gas, line A), it is not possible to fulfil the primary energy target. Increasing the use of solar energy will give rise to a non-renewable primary energy demand lower than the target of 60 kWh/m²a. This is possible because of the low space heating demand.

In the mild climate, a ground source heat pump is a promising technology despite higher installation costs compared to other solutions. For row houses, the installation costs can be shared among a group of houses. In addition, the seasonal efficiency is high due to the small ground temperature fluctuation, which leads to lower running and maintenance costs.

It is also possible to design the system in order to cool the building in an efficient way in terms of energy use and thermal comfort. This is a big advantage in a mild climate where the cooling demand may be significant.

References

TRNSYS (2005) *A Transient System Simulation Program*, Solar Energy Laboratory, University of Wisconsin, Madison, WI

APPENDIX 1

Reference Buildings: Constructions and Assumptions

Johan Smeds

A1.1 Introduction

Building codes for heating demand calculations in residential buildings usually differ substantially from country to country. Despite the fact that physical laws are the same in all countries and differences in climate conditions can be rather small when comparing two neighbouring countries, every country has its own thermal building regulation and its own way of calculating energy balances for dwellings.

There are four main methods applied in building codes. One of the methods is component specific, giving detailed information on maximum permissible U-values for construction components as, for example, walls, windows or roof constructions. Another method is to set an average U-value for the whole building, allowing a high degree of flexibility for each building component. Yet another method is the performance-based building regulation, setting a limit for energy use that should not be exceeded. Then there are also building codes that, besides being performance based, take heating systems into account. This allows higher U-values when specific heating systems are used.

Earlier research comparing building code requirements in the MURE database case study (Eichhammer and Schlomann, 1999) lacks descriptions of the model house and it is unclear what calculation methods have been used. In the case of MURE's report, internal and solar gains and ventilation losses were left out of the calculations since they were assumed to be equal. It is therefore necessary to define own reference buildings for International Energy Agency (IEA) Task 28/Annex 38. It is important for comparison with building simulation work of IEA Task 28 to have three different building geometries defined for an apartment building, a row house and a detached single family house. The calculations also have to be carried out with an internationally accepted method that includes all energy gains and losses.

Benchmarking of building codes enables one to see which countries need to improve their regulations. The procedure of benchmarking is made possible by calculating energy balances for dwellings built according to the national building codes of 2001. Thirteen countries participating in IEA Task 28 have therefore given input data to the energy balance calculations. Experts from each participating country have created constructions according to their national building regulation for the three different building types: an apartment building, a row house and a detached single family house. The geometry of each building type is fixed so that the internal dimensions are the same. The U-values of the building components, assumptions for heat recovery and climatic data vary. For each country a representative city is chosen for creating climatic data files. By using only one calculation method, the results of the energy calculations can easily be compared. The computer program Bilanz (Heidt, 1999), developed by the University of Siegen in Germany was chosen because it calculates the energy balance of dwellings according to the European standard EN 832.

The results from the calculations are used to create reference house constructions for mild, temperate and cold climate regions. The energy use of the reference houses indicates the current heating demand level for dwellings in each climate region. The geometries of the reference houses also function as a basis for calculations on well-insulated high-performance houses. The magnitude of energy saving potentials in dwellings can thereby be indicated.

A1.2 Reference apartment building

For energy balance calculation work, we have chosen a 4-storey apartment building with 16 apartments. All surface areas and dimensions noted are based on the inner surfaces of exterior walls. The dimensions to the outer edge of outer walls will depend on the insulation thickness.

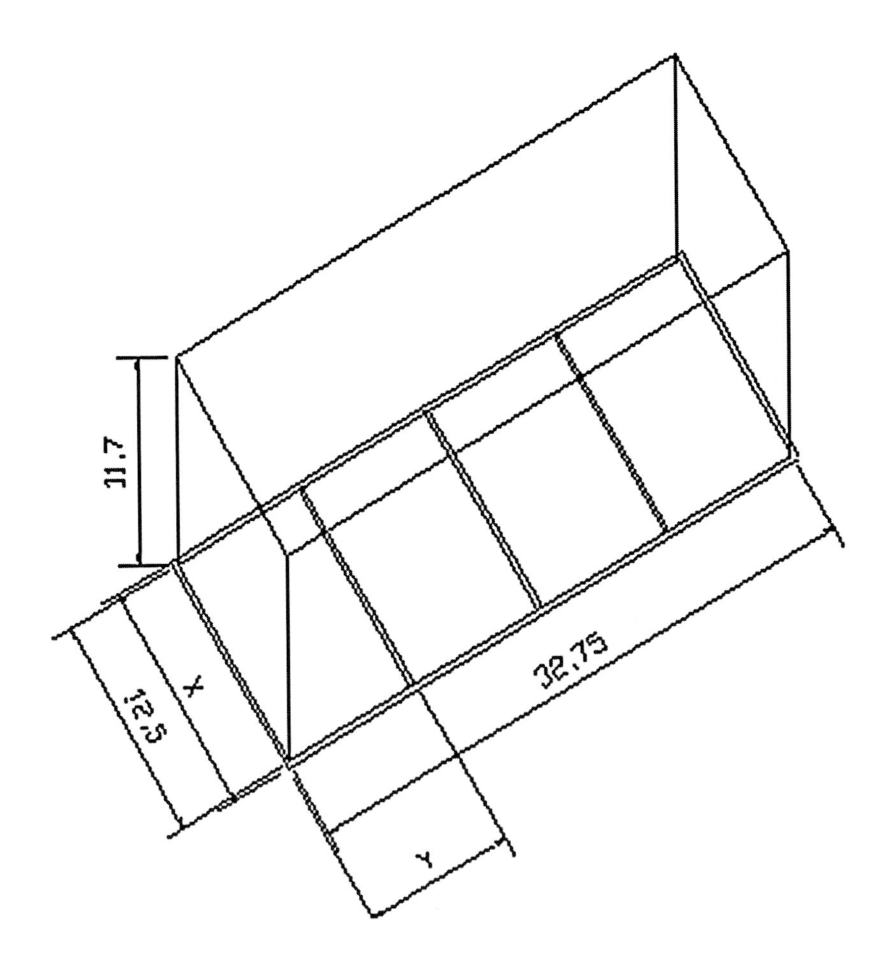

Source: AEU Ltd

Figure A1.1 *Geometry of the apartment building*

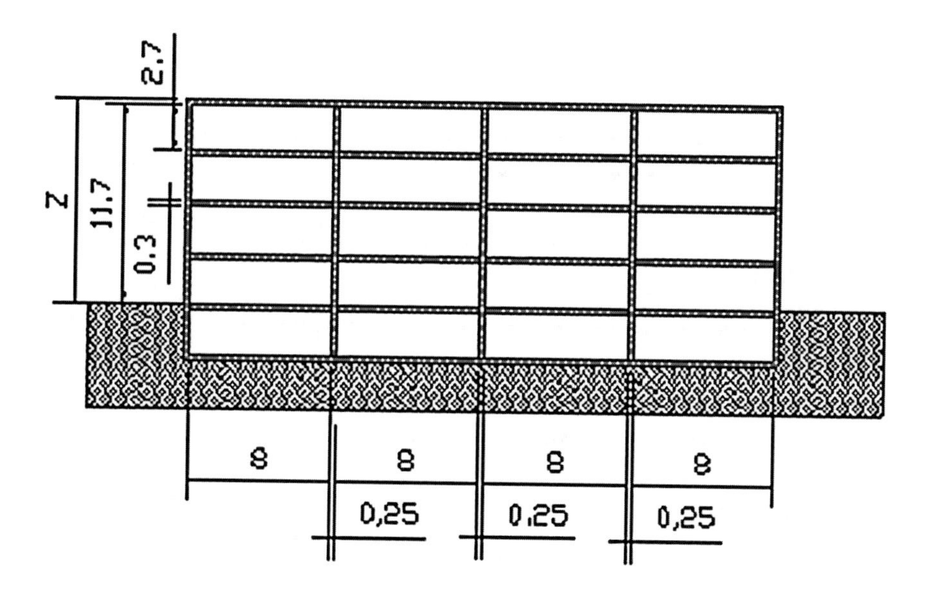

Source: AEU Ltd

Figure A1.2 *Section of the apartment building*

Table A1.1 *General information on the apartment building*

Building volume with internal dimensions	4789.7 m^3
Total envelope area with internal dimensions	1877.6 m^2
Net heated floor area (carpet area)	1600.0 m^2 ; 100 m^2 per apartment
Surface to volume ratio with internal dimensions	0.39 m^{-1}
Opaque wall area	834.8 m^2
Roof area	409.4 m^2
Floor area	409.4 m^2
Window area north	80.0 m^2
Window area south	144.0 m^2
Frame ratio	30.0%
Ventilation	0.2 ach infiltration + 0.4 ach ventilation = 0.6 ach
Occupants per apartment	Two adults, one child

A1.3 Reference row house

For energy balance calculation work, we have chosen an end unit and a mid unit. The end unit is the end house in a row and the mid unit is between other units. All dimensions given in Figure A1.3 and Tables A1.2 and A1.3 are measured from the interior surface of the outer walls. Figure A1.3 shows the geometry of a row house unit.

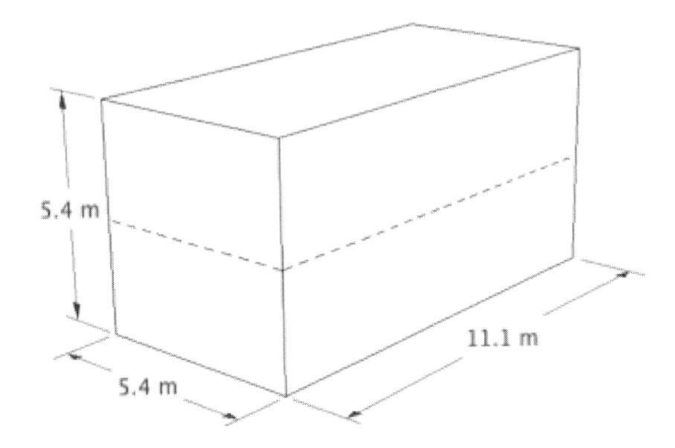

Source: Johan Smeds

Figure A1.3 *Geometry of a row house unit*

Table A1.2 *General information on the row house mid unit*

Building volume with internal dimensions	323.5 m^3
Total envelope area with internal dimensions	178.3 m^2
Net heated floor area (carpet area)	120.0 m^2
Surface to volume ratio with internal dimensions	0.55 m^{-1}
Opaque wall area	39.4 m^2
Roof area	60.0 m^2
Floor area	60.0 m^2
Window area north	3.0 m^2
Window area south	14.0 m^2
Frame ratio	30.0 %
Ventilation	0.2 ach infiltration + 0.4 ach ventilation = 0.6 ach
Occupants	Two adults and two children

Table A1.3 *General information on the row house end unit*

Building volume with internal dimensions	323.5 m^3
Total envelope area with internal dimensions	238.3 m^2
Net heated floor area (carpet area)	120.0 m^2
Surface to volume ratio with internal dimensions	0.74 m^{-1}
Opaque wall area	96.4 m^2
Roof area	60.0 m^2
Floor area	60.0 m^2
Window area north	3.0 m^2
Window area south	14.0 m^2
Window area east or west	3.0 m^2
Frame ratio	30.0 %
Ventilation	0.2 ach infiltration + 0.4 ach ventilation = 0.6 ach
Occupants	Two adults and two children

A1.4 Reference detached house

For energy balance calculation work, we have chosen a simplified model of a two-storey (1.5 storey) single family house. All dimensions given in Figure A1.4 and Table A1.4 are measured from the interior surface of the outer walls.

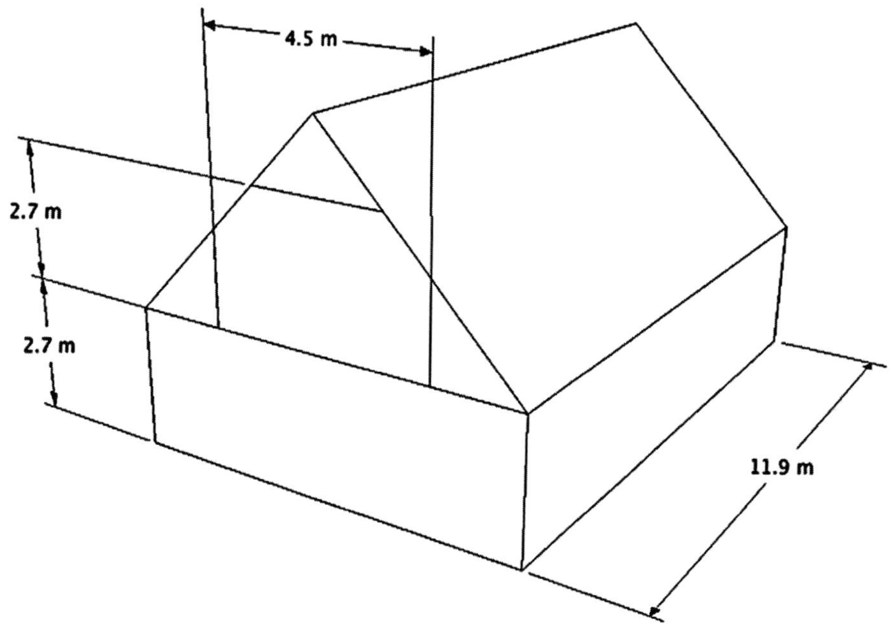

Source: Johan Smeds

Figure A1.4 *Geometry of the single family house*

Table A1.4 *General information on the detached house*

Building volume with internal dimensions	436.7 m^3
Total envelope area with internal dimensions	363.8 m^2
Surface to volume ratio with internal dimensions	0.93 m^{-1}
Ventilation volume	389.4 m^3
Net heated floor area (carpet area)	150.0 m^2
Opaque wall area	113.6 m^2
Roof area	129.7 m^2
Floor area	96.4 m^2
Window area north	1.0 m^2
Window area south	9.0 m^2
Window area east	3.0 m^2
Window area west	9.0 m^2
Frame ratio	30.0 %
Ventilation	0.2 ach infiltration + 0.4 ach ventilation = 0.6 ach
Occupants	Two adults and two children

A1.5 National and regional reference U-values

Tables A1.5 to A1.12 summarize the input for energy balance calculations according to EN 832. National U-values for constructions are given and result in average U-values for the mild, temperate and cold climate regions. Since heat exchangers are used in Sweden and The Netherlands, the reduction in ventilation losses is compensated for by reducing the U-value of the building shell and having natural ventilation with 0.4 (ach) and 0.2 (ach) infiltration. Constructions for other climates that cannot be referred to as mild, temperate or cold (Rome, Sendai and Toronto) are also presented; but they do not affect the regional U-values.

A1.5.1 U-values of the reference buildings

Table A1.5 *U-values of the reference buildings*

Climate Region	Cold climate			Regional climates			Other climates			
Country	Sweden	Norway	finland	Mild	Temperate	Cold	Italy	Japan	Canada1*	Canada2†
Reference climate	Stockholm	Oslo	Helsinki	Milan	Zurich	Stockholm	Rome	Sendai	Toronto	Toronto
Degree Days	4446	4566	5137	2759	3480	4446	1586	2747	4286	4286
Detached single family house (SFH)										
Ventilation (ach)	0.5	0.4	0.4	0.4	0.4	0.4	0.4	0.4	0.4	0.5
Infiltration (ach)	0.1	0.2	0.2	0.2	0.2	0.2	0.2	0.2	0.2	0.1
Heat recovery (%)	50%	0%	0%	0%	0%	0%	0%	0%	0%	55%
U-value of wall (W/m^2K)	0.18	0.22	0.28	0.50	0.40	0.20	0.70	0.75	0.34	0.23
U-value of window (W/m^2K)	1.74	1.650	1.98	4.50	2.35	1.81	4.50	3.49	2.60	2.40
U-value of roof (W/m^2K)	0.25	0.15	0.22	0.45	0.26	0.19	0.64	0.37	0.23	0.19
U-value of floor (W/m^2K)	0.28	0.15	0.36	0.57	0.39	0.20	0.74	0.53	0.63	0.63
U-value of building (W/m^2K)	0.33	0.26	0.38	0.74	0.47	0.29	0.92	0.72	0.52	0.45
Apartment building (APT)										
Ventilation	0.5	0.4	0.4	0.4	0.4	0.4	0.4	0.4	0.4	0.4
Infiltration	0.1	0.2	0.2	0.2	0.2	0.2	0.2	0.2	0.2	0.2
Heat recovery	50%	0%	0%	0%	0%	0%	0%	0%	0%	0%
U-value of wall (W/m^2K)	0.25	0.22	0.28	0.45	0.40	0.18	0.74	0.75	0.55	0.33
U-value of window (W/m^2K)	1.72	1.60	1.89	4.50	2.35	1.74	4.50	3.49	3.33	2.55
U-value of roof (W/m^2K)	0.21	0.15	0.22	0.40	0.26	0.10	0.70	0.37	0.47	0.29
U-value of floor (W/m^2K)	0.33	0.30	0.36	0.55	0.40	0.20	0.76	0.53	0.92	0.92
U-value of building (W/m^2K)	0.43	0.39	0.48	0.94	0.60	0.35	1.18	0.95	0.94	0.71
Row house mid unit (ROW-M)										
Ventilation	0.5	0.4	0.4	0.4	0.4	0.4	0.4	0.4	0.4	0.5
Infiltration	0.1	0.2	0.2	0.2	0.2	0.2	0.2	0.2	0.2	0.1
Heat recovery	50%	0%	0%	0%	0%	0%	0%	0%	0%	50%
U-value of wall (W/m^2K)	0.30	0.22	0.28	0.53	0.45	0.20	0.70	0.75	0.55	0.33
U-value of window (W/m^2K)	1.72	1.60	1.89	4.50	2.44	1.74	4.50	3.49	2.99	2.47
U-value of roof (W/m^2K)	0.28	0.15	0.22	0.44	0.28	0.16	0.60	0.37	0.20	0.14
U-value of floor (W/m^2K)	0.21	0.15	0.36	0.56	0.39	0.20	0.80	0.53	0.92	0.92
U-value of building (W/m^2K)	0.4	0.31	0.44	0.89	0.56	0.33	1.07	0.81	0.79	0.67
Row house end unit (ROW-E)										
Ventilation	0.5	0.4	0.4	0.4	0.4	0.4	0.4	0.4	0.4	0.5
Infiltration	0.1	0.2	0.2	0.2	0.2	0.2	0.2	0.2	0.2	0.1
Heat recovery	50%	0%	0%	0%	0%	0%	0%	0%	0%	50%
U-value of wall (W/m^2K)	0.23	0.22	0.28	0.44	0.40	0.20	0.64	0.75	0.55	0.33
U-value of window (W/m^2K)	1.72	1.60	1.89	4.50	2.44	1.74	4.50	3.49	3..05	2.52
U-value of roof (W/m^2K)	0.28	0.15	0.22	0.44	0.28	0.16	0.60	0.37	0.20	0.14
U-value of floor (W/m^2K)	0.21	0.15	0.36	0.56	0.39	0.20	0.70	0.53	0.92	0.92
U-value of building (W/m^2K)	0.36	0.30	0.42	0.81	0.54	0.32	0.97	0.83	0.77	0.62

Notes: * Natural gas

† Electricity

Table A1.5 *continued*

Climate Region	Temperate climate							
	England	Belgium	Austria	Netherlands1*	Netherlands2+	Germany	Switzerland	Scotland
Reference climate	Nottingham	Uccle	Vienna	Amsterdam	Amsterdam	Würzburg	Zürich	Glasgow
Degree Days	3086	3101	3169	3185	3185	3384	3480	3639
Detached single family house (SFH)								
Ventilation (ach)	0.4	0.4	0.4	0.4	0.5	0.4	0.4	0.4
Infiltration (ach)	0.2	0.2	0.2	0.2	0.1	0.2	0.2	0.2
Heat recovery (%)	0%	0%	0%	0%	45%	0%	0%	0%
U-value of wall (W/m^2K)	0.45	0.50	0.40	0.32	0.32	0.41	0.30	0.45
U-value of window (W/m^2K)	3.30	3.00	1.90	2.20	2.20	1.80	1.59	3.00
U-value of roof (W/m^2K)	0.25	0.35	0.2	0.27	0.32	0.25	0.3	0.25
U-value of floor (W/m^2K)	0.45	0.7	0.4	0.27	0.31	0.45	0.4	0.45
U-value of building (W/m^2K)	0.55	0.65	0.42	0.4	0.43	0.45	0.41	0.53
Apartment building (APT)								
Ventilation	0.4	0.4	0.4	0.4	0.5	0.4	0.4	0.4
Infiltration	0.2	0.2	0.2	0.2	0.1	0.2	0.2	0.2
Heat recovery	0%	0%	0%	0%	45%	0%	0%	0%
U-value of wall (W/m^2K)	0.45	0.50	0.40	0.27	0.32	0.59	0.28	0.45
U-value of window (W/m^2K)	3.30	3.00	1.90	1.90	2.80	1.80	1.34	3.00
U-value of roof (W/m^2K)	0.25	0.40	0.20	0.27	0.32	0.30	0.30	0.25
U-value of floor (W/m^2K)	0.45	0.90	0.40	0.27	0.31	0.60	0.37	0.45
U-value of building (W/m^2K)	0.75	0.86	0.54	0.46	0.61	0.67	0.43	0.71
Row house mid unit (ROW-M)								
Ventilation	0.4	0.4	0.4	0.4	0.5	0.4	0.4	0.4
Infiltration	0.2	0.2	0.2	0.2	0.1	0.2	0.2	0.2
Heat recovery	0%	0%	0%	0%	45%	0%	0%	0%
U-value of wall (W/m^2K)	0.45	0.60	0.40	0.27	0.32	0.62	0.30	0.45
U-value of window (W/m^2K)	3.30	3.00	1.90	2.70	2.80	1.80	1.59	3.00
U-value of roof (W/m^2K)	0.25	0.40	0.20	0.27	0.32	0.30	0.30	0.25
U-value of floor (W/m^2K)	0.45	0.90	0.40	0.27	0.31	0.60	0.40	0.45
U-value of building (W/m^2K)	0.66	0.87	0.48	0.5	0.56	0.62	0.46	0.63
Row house end unit (ROW-E)								
Ventilation	0.4	0.4	0.4	0.4	0.5	0.4	0.4	0.4
Infiltration	0.2	0.2	0.2	0.2	0.1	0.2	0.2	0.2
Heat recovery	0%	0%	0%	0%	45%	0%	0%	0%
U-value of wall (W/m^2K)	0.45	0.54	0.40	0.27	0.32	0.41	0.30	0.45
U-value of window (W/m^2K)	3.30	3.00	1.90	2.40	2.30	1.80	1.59	3.00
U-value of roof (W/m^2K)	0.25	0.40	0.20	0.27	0.32	0.30	0.30	0.25
U-value of floor (W/m^2K)	0.45	0.70	0.40	0.27	0.31	0.60	0.40	0.45
U-value of building (W/m^2K)	0.64	0.75	0.48	0.45	0.48	0.55	0.43	0.61

Notes: * Natural ventilation solar collector

+ Mechanical ventilation heat recovery

A1.5.2 U-values and resistances of the regional apartment buildings

The U-values shown in Tables A1.6 and A1.7 are average U-values of constructions according to the building codes of the year 2001 of countries in the three different climate regions. The U-values include the interior and exterior resistance (Rsi, Rse).

Table A1.6 *U-values of the regional apartment buildings*

Walls	Orientation	Area (m^2)	Country		Mild region	Temperate region	Cold region
	North	303.2	U-values (W/m^2K)		0.45	0.40	0.18
	South	239.2			0.45	0.40	0.18
	East	146.2			0.45	0.40	0.18
	West	146.2			0.45	0.40	0.18
	Average	834.8			0.45	0.40	0.18
Windows	North	80.0	U-values	Total	4.50	2.35	1.74
			(W/m^2K)	Glass 70%	5.70	2.63	1.75
				Frame 30%	1.70	1.70	1.70
			g-values		0.86	0.76	0.68
	South	144.0	U-values	Total	4.50	2.35	1.74
			(W/m^2K)	Glass 70%	5.70	2.63	1.75
				Frame 30%	1.70	1.70	1.70
			g-values		0.86	0.76	0.68
	Average	224.0	U-values	Total	4.50	2.35	1.74
			(W/m^2K)	Glass 70%	5.70	2.63	1.75
				Frame 30%	1.70	1.70	1.70
			g-values		0.86	0.76	0.68
Roofs		409.4	U-values	(W/m^2K)	0.40	0.26	0.10
Floors		409.4	U-values	(W/m^2K)	0.55	0.40	0.20
Total building	Average	1877.6	U-values	(W/m^2K)	0.94	0.60	0.35

The U- and g-values for windows are chosen to correspond to the following window types. The mild region has a single-pane window. The temperate region has a double-pane window with 4 mm glass, 30 mm air gap and 4 mm glass. The cold region has a triple-pane window consisting of one pane with 4 mm glass, 30 mm air gap and then a double pane with 4mm glass, 12 mm air gap and 4 mm glass.

The U-values include the interior and exterior resistances (Rsi, Rse) as described in the standard ISO13370. Therefore, the total resistance (R), which is the inverse of the U-value, has to be reduced by Rsi and Rse in order to obtain the resistance of the actual construction. The floor construction is reduced by Rsi = 0.17. The roof construction is reduced by both Rsi = 0.10 and Rse = 0.04. The wall construction is reduced by Rsi = 0.13 and Rse = 0.04.

Table A1.7 *Resistance of the regional apartment buildings*

Resistance R without Rsi or Rse for apartment building			
(1/U)-Rsi-Rse Mild region	(1/U)-Rsi-Rse Temperate region	(1/U)-Rsi-Rse Cold region	
2.05	2.33	5.39	Wall north
2.05	2.33	5.39	Wall south
2.05	2.33	5.39	Wall east
2.05	2.33	5.39	Wall west
2.36	3.71	9.86	Roof
1.65	2.33	4.83	Floor

A1.5.3 U-values of the regional row house mid units

The U-values shown in Table A1.8 are average U-values of constructions according to the building codes of the year 2001 of countries in the three different regions. The U-values include the interior and exterior resistance (Rsi, Rse).

Table A1.8 *U-values of the regional row house mid units*

Walls	Orient-ation	Area (m^2)	Country		Mild region	Temperate region	Cold region
	North	24.2	U-values		0.53	0.45	0.20
	South	15.2	(W/m^2K)		0.53	0.45	0.20
	Total Average	39.4			0.53	0.45	0.20
Windows	North	3.0	U-values	Total	4.50	2.44	1.74
			(W/m^2K)	Glass 70%	5.70	2.76	1.75
				Frame 30%	1.70	1.70	1.70
			g-values		0.86	0.76	0.68
	South	14.0	U-values	Total	4.50	2.44	1.74
			(W/m^2K)	Glass 70%	5.70	2.76	1.75
				Frame 30%	1.70	1.70	1.70
			g-values		0.86	0.76	0.68
	Average	17.0	U-values	Total	4.50	2.44	1.74
			(W/m^2K)	Glass 70%	5.70	2.76	1.75
				Frame 30%	1.70	1.70	1.70
			g-values		0.86	0.76	0.68
Roofs		60.0	U-values	(W/m^2K)	0.44	0.28	0.16
Floors		60.0	U-values	(W/m^2K)	0.56	0.39	0.20
Total Building	Total Average	176.4	U-values	(W/m^2K)	0.89	0.56	0.33

The U- and g-values for windows are chosen to correspond to the following window types. The mild region has a single-pane window. The temperate region has a double-pane window with 4 mm glass, 12 mm air gap and 4 mm glass. The cold region has a triple-pane window consisting of one pane with 4 mm glass, 30 mm air gap and then a double pane with 4mm glass, 12 mm air gap and 4 mm glass.

A1.5.4 U-values of the regional row house end units

The U-values in Table A1.9 are average U-values of constructions according to the building codes of the year 2001 of countries in the three different regions. The U-values include the interior and exterior resistance (Rsi, Rse).

The U- and g-values for windows are chosen to correspond to the following window types. The mild region has a single-pane window. The temperate region has a double-pane window with 4 mm glass, 12 mm air gap and 4 mm glass. The cold region has a triple-pane window consisting of one pane with 4 mm glass, 30 mm air gap and then a double pane with 4mm glass, 12 mm air gap and 4 mm glass.

Table A1.9 *U-values of the regional row house end units*

Walls	Orient-ation	Area (m^2)	Country		Mild region	Temperate region	Cold region
	North	24.2	U-values		0.53	0.45	0.20
	South	15.2	(W/m^2K)		0.53	0.45	0.20
	East or West	57.0			0.37	0.36	0.20
	Total Average	96.4			0.44	0.40	0.20
Windows	North	3.0	U-values	Total	4.50	2.44	1.74
			(W/m^2K)	Glass 70%	5.70	2.76	1.75
				Frame 30%	1.70	1.70	1.70
			g-values		0.86	0.76	0.68
	South	14.0	U-values	Total	4.50	2.44	1.74
			(W/m^2K)	Glass 70%	5.70	2.76	1.75
				Frame 30%	1.70	1.70	1.70
			g-values		0.86	0.76	0.68
	East or West	3.0	U-values	Total	4.50	2.44	1.74
			(W/m^2K)	Glass 70%	5.70	2.76	1.75
				Frame 30%	1.70	1.70	1.70
			g-values		0.86	0.76	0.68
	Average	20.0	U-values	Total	4.50	2.44	1.74
			(W/m^2K)	Glass 70%	5.70	2.76	1.75
				Frame 30%	1.70	1.70	1.70
			g-values		0.86	0.76	0.68
Roofs		60.0	U-values	(W/m^2K)	0.44	0.28	0.16
Floors		60.0	U-values	(W/m^2K)	0.56	0.39	0.20
Total Building	**Total Average**	**236.4**	**U-values**	**(W/m^2K)**	**0.81**	**0.54**	**0.32**

A1.5.5 Resistances of the regional row houses

The U-values in Table A1.10 include the interior and exterior resistances (Rsi, Rse). Therefore, the total resistance (R), which is the inverse of the U-value, has to be reduced by Rsi and Rse in order to obtain the resistance of the actual construction. The floor construction is reduced by Rsi = 0.17. The roof construction is reduced by both Rsi = 0.10 and Rse = 0.04. The wall construction is reduced by Rsi = 0.13 and Rse = 0.04.

Table A1.10 *Resistance of the row house*

Resistance R without Rsi or Rse for row house			
(1/U)-Rsi-Rse Mild region	(1/U)-Rsi-Rse Temperate region	(1/U)-Rsi-Rse Cold region	
1.72	2.05	4.83	Wall north
1.72	2.05	4.83	Wall south
2.53	2.61	4.83	Wall east or west
2.13	3.43	6.11	Roof
1.62	2.39	4.83	Floor

A1.5.6 U-values and resistances of the regional detached houses

The U-values in Tables A1.11 and A1.12 are average U-values of constructions according to the building codes of the year 2001 of countries in the three different regions. The U-values include the interior and exterior resistance (Rsi, Rse).

Table A1.11 *U-values of the regional detached houses*

Walls	Orient-ation	Area (m^2)	Country		Mild region	Temperate region	Cold region
	North	29.1	U-values		0.50	0.40	0.20
	South	23.1	(W/m^2K)		0.50	0.40	0.20
	East	33.7			0.50	0.40	0.20
	West	27.7			0.50	0.40	0.20
	Average	113.6			0.50	0.40	0.20
Windows	North	1.0	U-values	Total	4.50	2.35	1.81
			(W/m^2K)	Glass 70%	5.70	2.63	1.85
				Frame 30%	1.70	1.70	1.70
			g-values		0.86	0.76	0.71
	South	9.0	U-values	Total	4.50	2.35	1.81
			(W/m^2K)	Glass 70%	5.70	2.63	1.85
				Frame 30%	1.70	1.70	1.70
			g-values		0.86	0.76	0.71
	East	3.0	U-values	Total	4.50	2.35	1.81
			(W/m^2K)	Glass 70%	5.70	2.63	1.85
				Frame 30%	1.70	1.70	1.70
			g-values		0.86	0.76	0.71
	West	9	U-values	Total	4.50	2.35	1.81
			(W/m^2K)	Glass 70%	5.70	2.63	1.85
				Frame 30%	1.70	1.70	1.70
			g-values		0.86	0.76	0.71
	Average	22.0	U-values	Total	4.50	2.35	1.81
			(W/m^2K)	Glass 70%	5.70	2.63	1.85
				Frame 30%	1.70	1.70	1.70
			g-values		0.86	0.76	0.71
Roofs		129.7	U-values	(W/m^2K)	0.45	0.26	0.19
Floors		96.4	U-values	(W/m^2K)	0.57	0.39	0.20
Total Building	Average	361.7	U-values	(W/m^2K)	0.74	0.47	0.29

The U- and g-values for the windows are chosen to correspond to the following window types. The mild region has a single-pane window. The temperate region has a double-pane window with 4 mm glass, 30 mm air gap and 4 mm glass. The cold region has a double-pane window with 4 mm glass, 12 mm air gap, 4 mm glass and a low emissive layer.

The U-values include the interior and exterior resistances (Rsi, Rse). Therefore, the total resistance (R), which is the inverse of the U-value, has to be reduced by Rsi and Rse in order to obtain the resistance of the actual construction. The floor construction is reduced by Rsi = 0.17. The roof construction is reduced by both Rsi = 0.10 and Rse = 0.04. The wall construction is reduced by Rsi = 0.13 and Rse = 0.04.

Table A1.12 *Resistance of the regional detached house*

Resistance R without Rsi or Rse for detached house			
(1/U)-Rsi-Rse Mild region	(1/U)-Rsi-Rse Temperate region	(1/U)-Rsi-Rse Cold region	
1.83	2.30	4.83	Wall north
1.83	2.30	4.83	Wall south
1.83	2.33	4.83	Wall east
1.83	2.33	4.83	Wall west
2.08	3.71	5.12	Roof
1.58	2.39	4.83	Floor

A1.6 Results from calculations according to EN 832

As shown in Figures A1.5 to A1.8, the actual heat losses vary substantially within the different climate regions for all building types. The magnitude of the heat losses is dependent upon the U-values of the building envelope and the local climate conditions in each country. Relating the heat losses to the amount of degree days of each climate gives an indication of the energy performance due to the building code. Figures A1.9 to A1.12 show that the building codes of 2001 in Italy, England, Scotland, Belgium and Japan are less restrictive than the building codes in Austria, Switzerland, Germany, The Netherlands, Finland, Norway, Sweden and Canada.

A1.6.1 Heat gains and losses

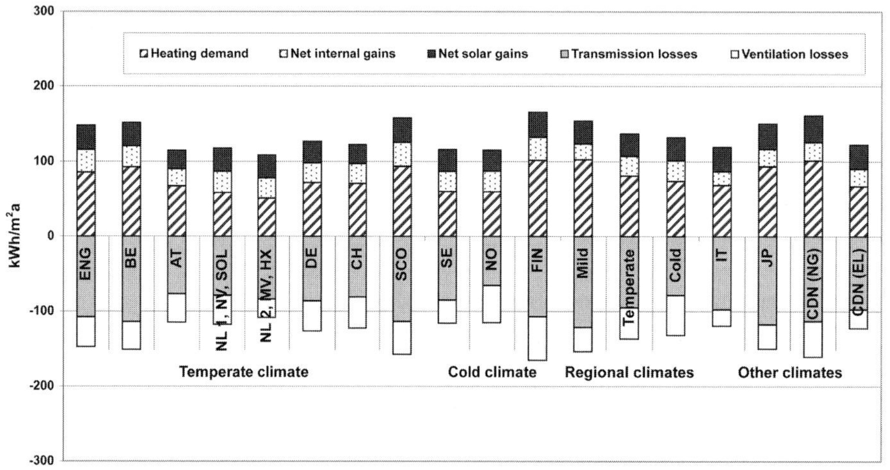

Source: Johan Smeds

Figure A1.5 *Heat gains and losses: Detached house*

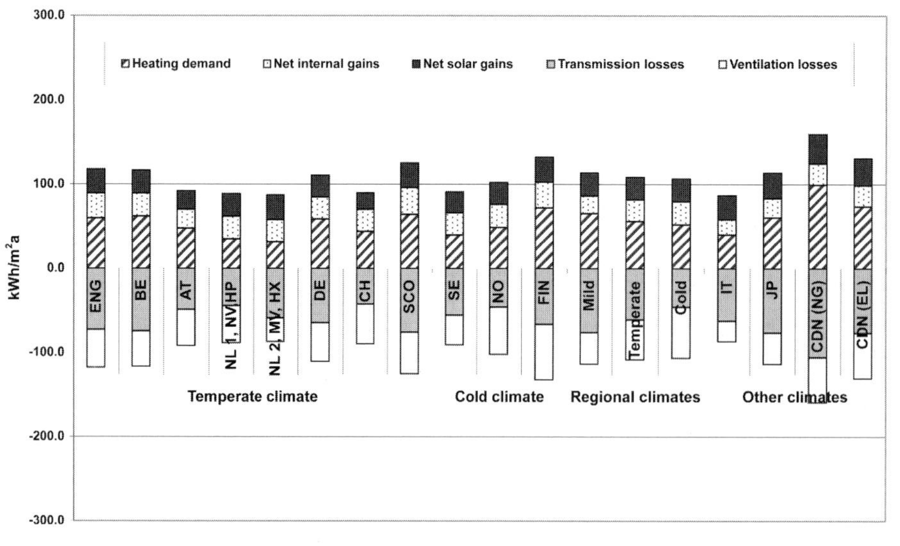

Source: Johan Smeds

Figure A1.6 *Heat gains and losses: Apartment building*

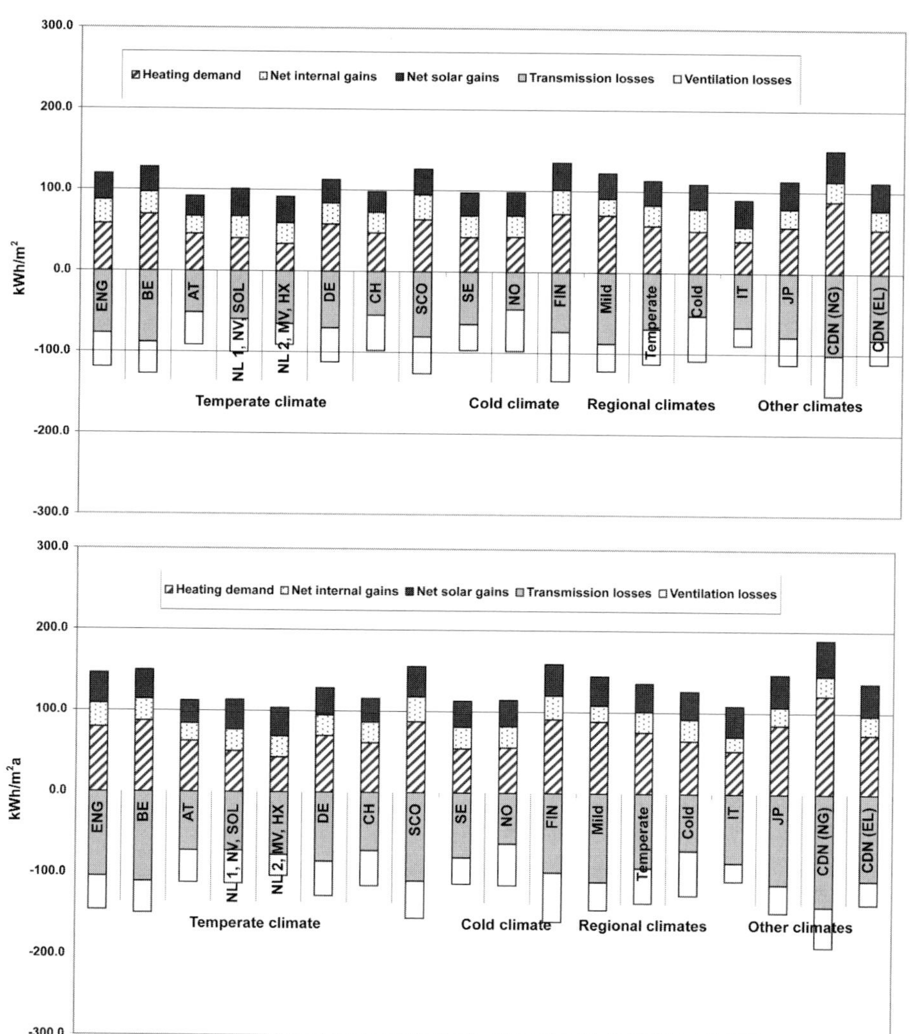

Source: Johan Smeds

Figure A1.7
Heat gains and losses: Row house mid unit

Source: Johan Smeds

Figure A1.8
Heat gains and losses: Row house end unit

A1.6.2 Heat gains and losses divided by degree days

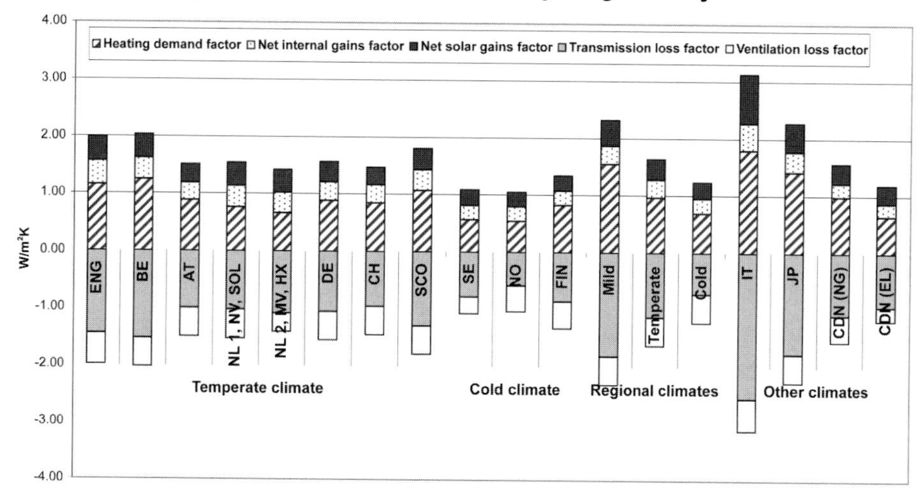

Source: Johan Smeds

Figure A1.9
Heat gains and losses divided by degree days: Detached house

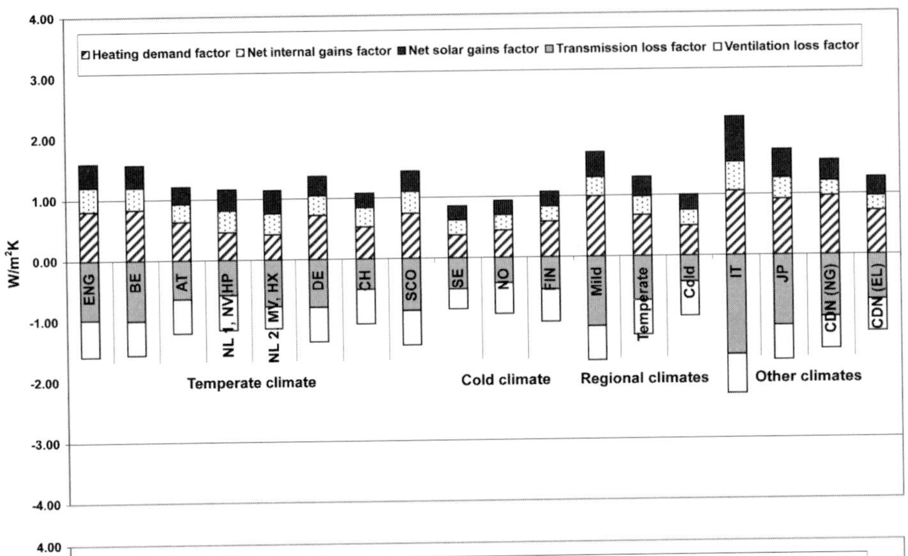

Source: Johan Smeds

Figure A1.10 *Heat gains and losses divided by degree days: Apartment building*

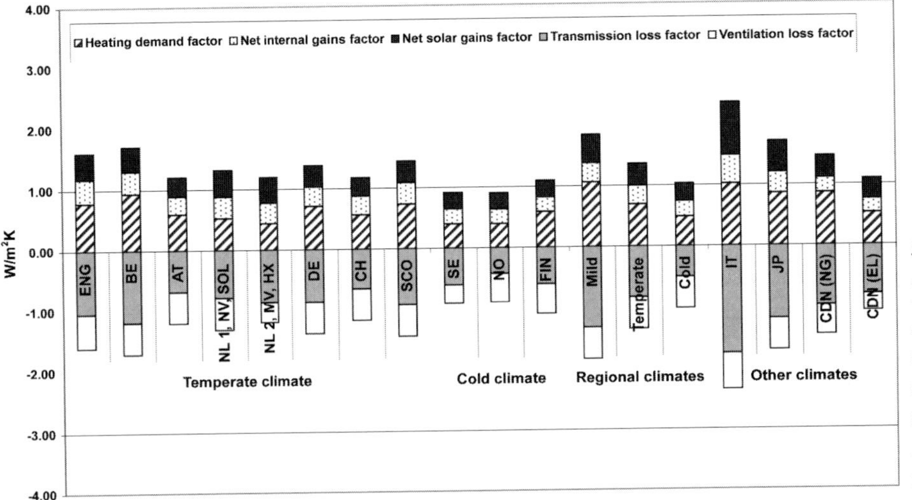

Source: Johan Smeds

Figure A1.11 *Heat gains and losses divided by degree days: Row house mid unit*

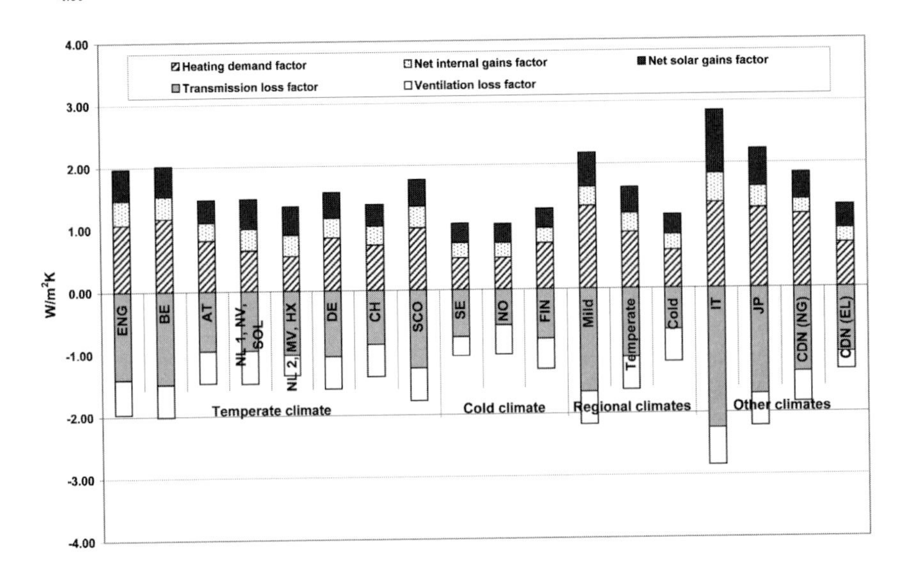

Source: Johan Smeds

Figure A1.12 *Heat gains and losses divided by degree days: Row house end unit*

References

Eichhammer, W. and Schlomann B. (1999) *Mure Database Case Study: A Comparison of Thermal Building Regulations in the European Union*, Fraunhofer Institute for Systems and Innovation Research, Karlsruhe, Germany, www.mure2.com/studies.shtml

Heidt, F. D. (1999) *Bilanz, Berechnungswerkzeug, NESA-Datenbank*, Fachgebiet Bauphysik und Solarenergie, Universität Siegen, Siegen, Germany

ISO 13370 (1998) *Thermal Performance of Buildings, Heat Transfer via the Ground, Calculation Methods*, International Organization for Standardization, Geneva, Switzerland

Appendix 2

Primary Energy and CO_2 Conversion Factors

Carsten Petersdorff and Alex Primas

The delivered and used energy in buildings for heating and DHW is conventionally fossil fuels (gas and oil), district heating, electricity or renewable resources that cause different CO_2 emissions when converted to heat. To judge the different environmental impacts of buildings during operation, two indicators are used in this book:

1. The primary energy: this is the amount of energy consumption on site, plus losses that occur in the transformation, distribution and extraction of energy.
2. CO_2 emissions: these are related to the heat energy consumption, including the whole chain from extraction to transformation of the energy carrier to heat. Using the CO_2 equivalent values (CO_2eq), not only CO_2 but all greenhouse gases are taken into account, weighted with their impact on global warming.

To determine the primary energy use or the related CO_2eq emissions, different methodologies are common. The purpose with this appendix is to describe the definitions and boundary conditions that are assumed for the simulations in this book:

- Only the non-renewable share of primary energy is taken into account.
- All factors are related to the lower heating value (LHV), not including condensation energy. This could mean that, theoretically, the efficiency of a heating system could exceed 100 per cent if a condensing gas furnace is used. However, we use 100 per cent efficiency for gas, 98 per cent for oil and 85 per cent for pellets. For DHW, the efficiency is 85 per cent.
- As a geographical boundary, the borderline of the building plot is chosen, which means that each energy carrier that is delivered to the house is weighted with factors for primary energy and CO_2eq emissions.
- For better comparison of the simulations, European average values are taken into account.

Table A2.1 presents the factors for primary energy and CO_2eq that are used in the simulations in this book, based on the GEMIS tool (GEMIS, 2004).

Table A2.1 *Primary energy factor (PEF) and CO_2 conversion factors*

Primary energy and CO_2 conversion factors	**PEF (kWh_{pe}/kWh_{end})**	**CO_2eq (g/kWh)**
Oil-lite	1.13	311
Natural gas	1.14	247
Hard coal	1.08	439
Lignite	1.21	452
Wood logs	0.01	6
Wood chips	0.06	35
Wood pellets	0.14	43
EU-17 electricity, grid	2.35	430
District heating combined heat and power (CHP) – coal condensation 70%, oil 30%	0.77	241
District heating CHP – coal condensation 35%, oil 65%	1.12	323
District heating, heating plant; oil 100%	1.48	406
Local district heating CHP – coal condensation 35%, oil 65%	1.10	127
Local district heating plant, oil 100%	1.47	323
Local solar	0.00	0
Solar heat (flat) central	0.16	51
Photovoltaic (multi)	0.40	130
Wind electricity	0.04	20

Note that primary energy and CO_2 conversion may differ for specific national circumstances. The different factors for electricity particularly influence the results on national levels (see Figures A1.1 and A1.2). On the other hand, the electricity market is international, which justifies average values for the EU-17 grid, for example.

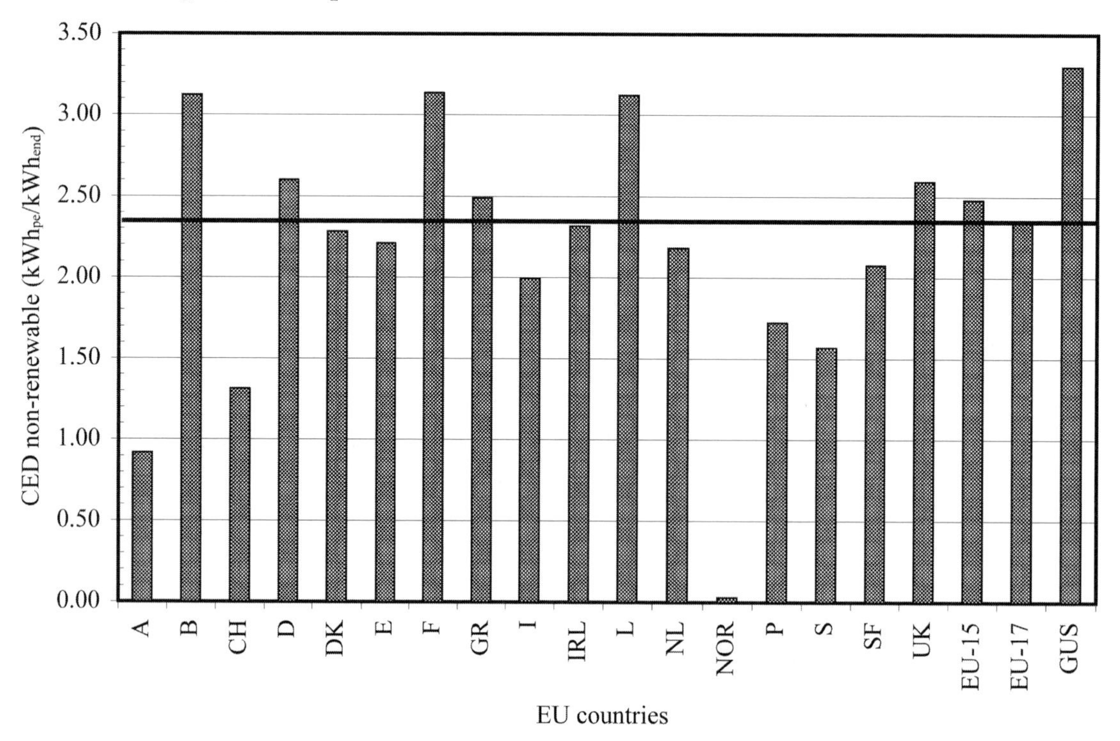

Source: Corsten Petersdorff and Alex Primas

Figure A2.1 *National primary energy factors for electricity; the line represents the EU-17 mix that is used in this book*

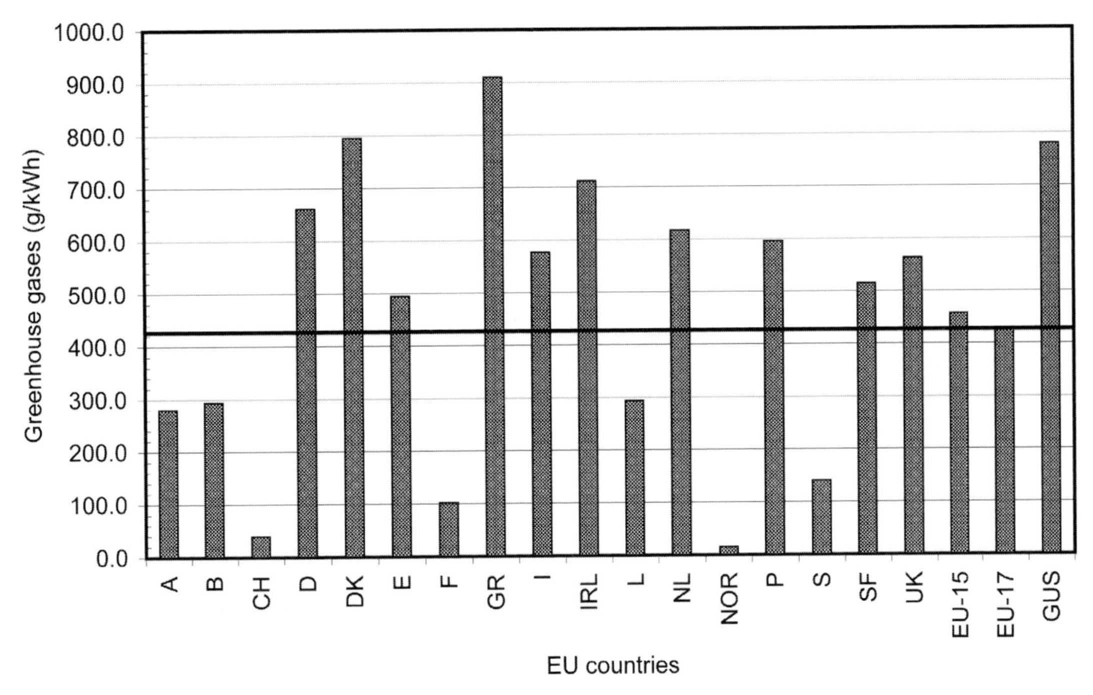

Source: Corsten Petersdorff and Alex Primas

Figure A2.2 *National CO_2 equivalent conversion factors for electricity; the line represents the EU-17 mix that is used in this book*

A2.1 Assumptions for the life-cycle analyses

In the life-cycle analyses (see Chapter 3 in this volume) the Union for the Coordination of Transmission of Electricity (UCTE) electricity mix was used. Table A2.2 shows the primary energy factors for electricity used for the life-cycle analyses (UCTE electricity mix) and the energy analyses of the typical solutions (EU 17 electricity mix). The difference between the two values is caused by the different production mix for electricity within the UCTE and the EU 17 countries. Further differences occur due to different definitions of the base (calorific value) and within the methodology of the two data sources (Frischknecht et al, 1996; GEMIS, 2004).

Table A2.2 *Primary energy factors for electricity (non-renewable)*

System	Base	Primary energy factor (PEF) (kWhpe/kWhend)	Data source
UCTE electricity mix	Gross calorific value	3.56	Frischknecht et al (1996)
EU 17 electricity mix	Net calorific value	2.35	GEMIS (2004)

References

Frischknecht, R., Bollens, U., Bosshart, S., Ciot, M., Ciseri, L., Doka, G., Hischier, R., Martin, A., Dones, R. and Gantner, U. (1996) *Ökoinventare von Energiesystemen, Grundlagen für den ökologischen Vergleich von Energiesystemen und den Einbezug von Energiesystemen in Ökobilanzen für die Schweiz*, Bundesamt für Energie, (BfE), Bern, Switzerland

GEMIS (2004) *GEMIS: Global Emission Model for Integrated Systems*, Öko-Institut, Darmstadt, Germany

APPENDIX 3

Definition of Solar Fraction

Tobias Boström

For the simulations of many of the solar systems, the program Polysun from the Solar Energy Laboratory SPF in Rapperswill, Switzerland, was used. The program offers the user the ability to vary nearly 100 different parameters. This flexibility enables a great variety of configurations of thermal solar active systems with diverse modes of operation to be produced.

For this publication, a link subroutine was created which allows the heat demand output file from the building simulation program Derob-LTH to be entered into Polysun. The hourly weather data used by Polysun is generated by Meteonorm. The whole building with solar system in a given climate can thus be simulated very accurately.

The most correct and fair definition of the solar fraction (SF) has to be in relation to the reference system. For example, the reference system can be a condensing gas or biomass heating system. To obtain the solar fraction, a simulation must first be made without collectors to determine the amount of auxiliary energy needed in kWh (Aux_0). Then the building with the solar system can be simulated to determine the amount of auxiliary energy needed in this case (Aux_1). The solar fraction can then be calculated with the following simple equation:

$$SF = 1 - \frac{Aux_1}{Aux_0} \qquad [A3.1]$$

Polysun defines the solar fraction with respect to the tank input and not the tank output. The solar fraction is then defined as the ratio of solar energy supplied to storage and the total energy supplied to storage, including auxiliary heat input:

$$SF_{alternative} = \frac{solar\ energy}{auxiliary\ energy\ +\ solar\ energy} = \frac{Q_{solar\ in\ store}}{Q_{aux\ in\ store} + Q_{solar\ in\ store}} \qquad [A3.2]$$

This results in the fact that Polysun *overestimates* the solar fraction of the solar active system because it does not reflect the system losses. However, all presented solar fractions are calculated according to the first definition and the Polysun definition is disregarded.

APPENDIX 4

The International Energy Agency

S. Robert Hastings

A4.1 Introduction

This book presents work completed within a framework of the International Energy Agency (IEA) under the auspices of two implementing agreements:

1 Solar Heating and Cooling (SHC); and
2 Energy Conservation in Buildings and Community Systems (ECBCS);

in a research project SHC Task 28/ECBCS Annex 38: Sustainable Solar Housing.

A4.2 International Energy Agency

The International Energy Agency (IEA) was established in 1974 as an autonomous agency within the framework of the Organisation for Economic Co-operation and Development (OECD), to carry out a comprehensive programme of energy cooperation among its 25 member countries and the commission of the European Communities.

An important part of the Agency's programme involves collaboration in the research, development and demonstration of new energy technologies to reduce excessive reliance on imported oil, to increase long-term energy security and to reduce greenhouse gas emissions. The IEA SHC's research and development activities are headed by the Committee on Energy Research and Technology (CERT) and supported by a small secretariat staff, headquartered in Paris. In addition, three working parties are charged with monitoring the various collaborative energy agreements, identifying new areas for cooperation and advising CERT on policy matters.

Collaborative programmes in the various energy technology areas are conducted under implementing agreements, which are signed by contracting parties (government agencies or entities designated by them). There are currently 42 implementing agreements covering fossil-fuel technologies, renewable energy technologies, efficient energy end-use technologies, nuclear fusion science and technology, and energy technology information centres.

IEA Headquarters
9, rue de la Federation
75739 Paris Cedex 15, France
Tel: +33 1 40 57 65 00/01
Fax: +33 1 40 57 65 59
info@iea.org

A4.3 Solar Heating and Cooling Programme

The Solar Heating and Cooling Programme was one of the first IEA implementing agreements to be established. Since 1977, members have been collaborating to advance active solar, passive solar and photovoltaic technologies and their application in buildings.

A total of 36 tasks have been initiated, 27 of which have been completed. Each task is managed by an operating agent from one of the participating countries. Overall control of the programme rests with an executive committee comprised of one representative from each contracting party to the implementing agreement. In addition, a number of special *ad hoc* activities – working groups, conferences and workshops – have been organized. The tasks of the IEA Solar Heating and Cooling Programme, both completed and current, are as follows.

Completed tasks:

1 Investigation of the Performance of Solar Heating and Cooling Systems
2 Coordination of Solar Heating and Cooling Research and Development
3 Performance Testing of Solar Collectors
4 Development of an Insolation Handbook and Instrument Package
5 Use of Existing Meteorological Information for Solar Energy Application
6 Performance of Solar Systems Using Evacuated Collectors
7 Central Solar Heating Plants with Seasonal Storage
8 Passive and Hybrid Solar Low Energy Buildings
9 Solar Radiation and Pyranometry Studies
10 Solar Materials Research and Development
11 Passive and Hybrid Solar Commercial Buildings
12 Building Energy Analysis and Design Tools for Solar Applications
13 Advance Solar Low Energy Buildings
14 Advance Active Solar Energy Systems
15 Photovoltaics in Buildings
16 Measuring and Modelling Spectral Radiation
17 Advanced Glazing Materials for Solar Applications
18 Solar Air Systems
19 Solar Energy in Building Renovation
20 Daylight in Buildings
21 Building Energy Analysis Tools
22 Optimization of Solar Energy Use in Large Buildings
23 Active Solar Procurement
24 Solar Assisted Air Conditioning of Buildings
25 Solar Combi-systems
26 Performance of Solar Façade Components
27 Solar Sustainable Housing.

Ongoing tasks:

28 Solar Crop Drying
29 Daylighting Buildings in the 21st Century
30 Advanced Storage Concepts for Solar Thermal
31 Systems in Low Energy Buildings
32 Solar Heat for Industrial Process
33 Testing and Validation of Building Energy Simulation Tools
34 Photovoltaics/Thermal Systems
35 Solar Resource Knowledge Management
36 Advanced Housing Renovation with Solar and Conservation.

To learn more about the IEA Solar Heating and Cooling Programme, visit the programme website: www.iea-shc.org, or contact the Executive Secretary, Pamela Murphy, pmurphy@MorseAssociatesInc.com.

A4.4 Energy Conservation in Buildings and Community Systems Programme

The IEA sponsors research and development in a number of areas related to energy. The mission of one of those areas, the Energy Conservation for Building and Community Systems Programme (ECBCS), is to facilitate and accelerate the introduction of energy conservation and environmentally sustainable technologies into healthy buildings and community systems through innovation and research in decision-making, building assemblies and systems, and commercialization. The objectives of collaborative work within the ECBCS research and development programme are directly derived from the ongoing energy and environmental challenges facing IEA countries in the area of construction, the energy market and research. ECBCS addresses major challenges and takes advantage of opportunities in the following areas:

- exploitation of innovation and information technology;
- impact of energy measures on indoor health and usability; and
- integration of building energy measures and tools into changes in lifestyles, work environment alternatives and business environments.

A4.4.1 The executive committee

Overall control of the programme is maintained by an executive committee, which not only monitors existing projects, but also identifies new areas where collaborative effort may be beneficial. To date, the following projects have been initiated by the executive committee on Energy Conservation in Buildings and Community Systems.

Completed annexes:

1 Load Energy Determination of Buildings
2 Ekistics and Advanced Community Energy Systems
3 Energy Conservation in Residential Buildings
4 Glasgow Commercial Building Monitoring
5 Energy Systems and Design of Communities
6 Local Government Energy Planning
7 Inhabitants Behaviour with Regard to Ventilation
8 Minimum Ventilation Rates
9 Building HVAC System Simulation
10 Energy Auditing
11 Windows and Fenestration
12 Energy Management in Hospitals
13 Condensation and Energy
14 Energy Efficiency in Schools
15 BEMS 1- User Interfaces and System Integration
16 BEMS 2- Evaluation and Emulation Techniques
17 Demand Controlled Ventilation Systems
18 Low Slope Roof Systems
19 Air Flow Patterns within Buildings
20 Thermal Modelling
21 Energy Efficient Communities

22 Multi Zone Air Flow Modelling (COMIS)
23 Heat, Air and Moisture Transfer in Envelopes
24 Real time HEVAC Simulation
25 Energy Efficient Ventilation of Large Enclosures
26 Evaluation and Demonstration of Domestic Ventilation Systems
27 Low Energy Cooling Systems
28 Daylight in Buildings
29 Bringing Simulation to Application
30 Energy-Related Environmental Impact of Buildings
31 Integral Building Envelope Performance Assessment
32 Advanced Local Energy Planning
33 Computer-Aided Evaluation of HVAC System Performance
34 Design of Energy Efficient Hybrid Ventilation (HYBVENT)
35 Retrofitting of Educational Buildings
36 Low Exergy Systems for Heating and Cooling of Buildings (LowEx)
37 Solar Sustainable Housing
38 High Performance Insulation Systems
39 Commissioning of Building HVAC Systems for Improved Energy Performance

Ongoing Annexes:

40 Air Infiltration and Ventilation Centre
41 Whole Building Heat, Air and Moisture Response (MOIST-ENG)
42 The Simulation of Building-Integrated Fuel Cell and Other Cogeneration Systems (COGEN-SIM)
43 Testing and Validation of Building Energy Simulation Tools
44 Integrating Environmentally Responsive Elements in Buildings
45 Energy-Efficient Future Electric Lighting for Buildings
46 Holistic Assessment Tool-kit on Energy Efficient Retrofit Measures for Government Buildings (EnERGo)
47 Cost Effective Commissioning of Existing and Low Energy Buildings
48 Heat Pumping and Reversible Air Conditioning
49 Low Exergy Systems for High Performance Built Environments and Communities
50 Prefabricated Systems for Low Energy / High Comfort Building Renewal

For more information about the ECBCS Programme, please visit the web site: www.ecbcs.org

A4.5 IEA SHC Task 28/ECBCS 38: Sustainable Solar Housing

Duration: April 2000–April 2005.
Objectives: the goal of this IEA research activity was to help participating countries achieve significant market penetration of sustainable solar housing by the year 2010, by researching and communicating marketing strategies, design and engineering concepts developed by detailed analyses, illustrations of demonstration housing projects and insights from monitoring projects. Results have been communicated in several forms through diverse channels, including:

- a booklet, *Business Opportunities in Sustainable Housing*, published on the IEA SHC website: www.iea-shc.org and also available in paper form from the Norwegian State Housing Bank: www.husbanken.no;
- brochures on 30 demonstration buildings published as PDF files on the IEA SHC website as a basis for articles in local languages (www.iea-shc.org);

Source: D. Enz, AEU GmbH, CH-8304 Wallisellen

Figure A4.1 *A very low energy house in Bruttisholz, CH by architect Norbert Aregger*

- a reference book, *Sustainable Solar Housing for Cooling Dominated Climates* (forthcoming);
- a book, *The Environmental Design Brief* (forthcoming)

A4.5.1 Active participants contributing to the IEA SHC Task 28/ECBCS Annex 38: Sustainable Solar Housing

PROGRAMME LEADER
S. Robert Hastings
(Sub-task B co-leader)
AEU Architecture, Energy and Environment Ltd
Wallisellen, Switzerland

AUSTRIA
Gerhard Faninger
University of Klagenfurt
Klagenfurt, Austria

Sture Larsen
Architekturbüro Larsen
A-6912 Hörbranz, Austria

Helmut Schöberl
Schöberl & Pöll OEG
Wien, Austria

AUSTRALIA
Richard Hyde
(Cooling Group Leader)
University of Queensland
Brisbane, Australia

BRAZIL
Marcia Agostini Ribeiro
Federal University of Minas Gerais
Belo Horizonte, Brazil

CANADA
Pat Cusack
Arise Technologies Corporation
Kitchener, Ontario
Canada

CZECH REPUBLIC
Miroslav Safarik
Czech Environmental Institute
Praha, Czech Republic

FINLAND
Jyri Nieminen
VTT Building and Transport
Finland

GERMANY
Christel Russ
Karsten Voss
(Sub-task D Co-leaders)
Andreas Buehring
Fraunhofer ISE
Freiburg, Germany

Hans Erhorn/
Johann Reiss
Fraunhofer Inst. für Bauphysik
Stuttgart, Germany

Frank D. Heidt/
Udo Giesler
Universität-GH Siegen, Germany

Berthold Kaufmann
Passivhaus Institut
Darmstadt, Germany

Joachim Morhenne
Ing.büro Morhenne GbR
Wuppertal, Germany

Carsten Petersdorff
Ecofys GmbH
Köln, Germany

IRAN
Vahid Ghobadian
(Guest expert)
Azad Islamic
Tehran, Iran

ITALY
Valerio Calderaro
University La Sapienza of
Rome, Italy

Luca Pietro Gattoni
Politecnico di Milano
Milan, Italy

Francesca Sartogo
PRAU Architects
Rome, Italy

JAPAN
Kenichi Hasegawa
Org. Akita Prefectural
University, Akita
Japan

Motoya Hayashi
Miyagigakuin Women's
College, Sendai
Japan

Nobuyuki Sunaga
Tokyo Metropolitan University
Tokyo, Japan

THE NETHERLANDS
Edward Prendergast/
Peter Erdtsieck
(Sub-task A Co-leaders)
MoBius consult bv.
Driebergen-Rijsenburg,
The Netherlands

NEW ZEALAND
Albrecht Stoecklein
Building Research Assoc.
Porirua, New Zealand

NORWAY
Tor Helge Dokka
SINTEF
Trondheim, Norway

Anne Gunnarshaug Lien
(Sub-task C Leader)
Enova SF
Trondheim, Norway

Trond Haavik
Segel AS
Nordfjordeid, Norway

Are Rodsjo
Norwegian State Housing
Bank
Trondheim, Norway

Harald N. Rostvik
Sunlab/ABB Building Systems
Stavanger, Norway

SWEDEN
Maria Wall
(Sub-task B Co-leader)
Lund University
Lund, Sweden

Hans Eek
Arkitekt Hans Eek AB,
Alingsås
Sweden

Tobias Boström
Uppsala University, Sweden

Johan Nilsson/Björn Karlsson
Lund University
Lund, Sweden

SWITZERLAND
Tom Andris
Renggli AG
Switzerland

Anne Haas
EMPA
Dübendorf, Switzerland

Annick Lalive d'Epinay
Fachstelle Nachhaltigkeit
Amt für Hochbauten
Postfach, CH-8021 Zürich
Switzerland

Daniel Pahud
SUPSI – DCT – LEEE
Canobbio, Switzerland

Alex Primas
Basler and Hofmann
CH 8029 Zurich, Switzerland

UK
Gökay Deveci
Robert Gordon, University of
Aberdeen, Scotland, UK

US
Guy Holt
Coldwell Banker
Kansas City MO, US